Lecture Notes in Mathematics

Edited by A. Dold and B. Eckmann

1297

Zhu You-lan Guo Ben-yu (Eds.)

Numerical Methods for Partial Differential Equations

Proceedings of a Conference held in Shanghai, P.R. China, March 25–29, 1987

Springer-Verlag

Berlin Heidelberg New York London Paris Tokyo

Editors

ZHU You-lan
The Computing Center, Academia Sinica
Beijing, 100080, People's Republic of China

GUO Ben-yu
Shanghai University of Science and Technology
Shanghai, People's Republic of China

Mathematics Subject Classification (1980): 65MXX, 65NXX

ISBN 3-540-18730-8 Springer-Verlag Berlin Heidelberg New York
ISBN 0-387-18730-8 Springer-Verlag New York Berlin Heidelberg

Printed in Germany

Printing and binding: Druckhaus Beltz, Hemsbach/Bergstr.
2146/3140-543210

Preface

This volume of Lecture Notes in Mathematics is the Proceedings of the First Chinese Conference on Numerical Methods for Partial Differential Equations, which was held at the Shanghai University of Science and Technology, Shanghai, China on March 25-29, 1987 and attracted about 100 participants from all parts of China. It includes 16 papers selected from 75 papers presented at the Conference. A complete list of the papers presented at the Conference is also given in this Proceedings. These papers are arranged in alphabetical order of the first author's name and every name is typewritten in "the Chinese way", i.e., the family name is typed first, followed by the given name.

We are indebted to our many colleagues and friends who helped us in preparing the Conference, during the meeting and in editing the Proceedings, but especially, to Yang Zhong-hua who supervised all of the local arrangements. We are also thankful to the Chinese Mathematical Society and the Chinese Society of Computational Mathematics for their support.

Finally, we would like to express our thanks to the series editors and the editorial staff of Springer-Verlag for valuable assistance in preparing the Proceedings.

July, 1987

Zhu You-lan and Guo Ben-yu
(Editors)

Organizing Committee

Co-chairmen

Guo Ben-yu and Zhu You-lan

Members of the Committee

Li De-yuan

Li Li-kang

Li Rong-hua

Lin Qun

Pan Zhong-xiong

Shi Zhong-ci

Wu Hua-mo

Ying Lung-an

Zhou Tian-xiao

List of Papers Presented at the First Chinese Conference on Numerical Methods for Partial Differential Equations

Cai Ti-qin [Dept. of Mechanics, Peking University, Beijing, China], Qin Su-di [Dept. of Mechanics, Peking University, Beijing, China], Fan Jia-hua [Institute of Water Conservancy and Hydroelectric Power Research, Beijing, China] and Wu Jiang-hang [Dept. of Mechanics, Peking University, Beijing, China]: Numerical modelling of flow field in a tidal river and the near field of intake for a nuclear power plant

Chang Qian-shun [Institute of Applied Mathematics, Academia Sinica, Beijing, China]: Applications of splitting schemes and multigrid methods to computation of separated flows

Chen Guang-nan [Institute of Applied Physics and Computational Mathematics, P.O. Box 8009, Beijing, China]: A tri-level difference Scheme for one dimensional parabolic systems

Chen Qi-kun [Dept. of Computer Science, Fuzhou University, Fuzhou, China]: A stable explicit difference scheme for solving second order parabolic partial equations with variable coefficients

Dai Jia-zun [Dept. of Mathematics, Physics and Mechanics, Nanjing Aeronautical Institute, Nanjing, China]: On the TVD feature of the discrete E-scheme

Dai Jia-zun and Hu Xiao-ping [Dept. of Mathematics, Physics and Mechanics, Nanjing Aeronautical Institute, Nanjing, China]: On the convergence of a class of high resolution finite difference schemes

Fang Bao-rong [Hohai University, Nanjing, China]: Finite analytic numerical solution for flexure problems of plates

Feng Guo-tai, Tang Sheng-li, Wang Zhong-qi and Wang Zun-zheng [Harbin University of Technologh, Harbin, China]: Some discussions on numerical methods for imcompressible viscous flows in nonorthogonal curvilinear coordinates

Gao Ying-cai and Feng Xiao-bing [Dept. of Mathematics, Xian Jiaotong University, Xian, China]: The Galerkin method for the problem of natural convection in an annular cavity

Ge Zhong and Feng Kang [Computing Center, Academia Sinica, Beijing, China]: On the approximation of linear Hamiltonian systems

Gu Li -zhen and Huang Bi-dan [Dept. of applied mathematics, Tsinghua University. Beijing, China]: Incomplete LU factorization for solving the steady-state two-dimensional Navier-Stokes equations

Guo Ben-yu [Shanghai University of Science and Technology, Shanghai, China]: The pseudo-spectral method for the M.K.D.V. equation

Guo Bo-ling [Institute of Applied Physics and Computational Mathematics, P.O. Box 8009, Beijing, China]: The spectral methods for Newton-Boussinesq equations in two dimensions

Han Hou-de [Dept. of Applied Mathematics, Tsinghua University, Beijing, China]: The

Boundary finite element methods for Signorini problems

He Guo-qiang [Dept. of Mathematics, Shanghai University of Science and Technology, Shanghai, China] and Chen Yung-ming [State University of New York at Stony Brook, Stony Brook, NY, 11794, USA]: A numerical study of inverse problems for KDV equations

Huang Ai-xiang and Zhang Bo [Institute of Computational and Applied Mathematics, Xian Jiaotong University, Xian, China]: Convergence of nodal expansion methods for neutron diffusion problem

Huang Ming-you [Dept. of Mathematics, Jilin University, Changchun, China]: A Hamiltonian approximation for nonlinear wave equation on n-dimensional spheres S^n

Huang Yu-ren and Wang Jian [Dept. of Mathematics, Shanghai University of Science and Technology, Shanghai, China]: Fully discrete finite element methods with third order accuracy in the time direction for M.K.D.V. equations

Jiang Jin-sheng [Dept. of Mathematics, Hangzhou University, Hangzhou, China]: A note for Lagrange multiplier finite element method of Bramble

Jiang Zhong-bing [Dept. of Mathematics, Tientsin University, Tianjin, China]: The infinite element analysis for wave load of offshore structures

Kang Li-shan [Dept. of Mathematics, Wuhan University, Wuhan, China]: Parallel algorithms and domain decomposition

Kang Li-shan and Chen Lu-juan [Dept. of Mathematics, Wuhan University, Wuhan, China]: The convergence rate of the Schwarz alternating procedure for multi-dimensional problems

Li Bo [Dept. of Applied Mathematics, Zhejiang University, Hangzhou, Zhejiang, China]: An analysis on the convergence of Wilson's nonconforming element

Li Chun-wang [Computing Center, Academia Sinica, Beijing, China]: Symplectic difference schemes for infinite dimensional Hamiltonian systems

Li De-yuan and Han Zhen [Institute of Applied Physics and Computational Mathematics, P.O. Box 8009, Beijing, China]: Difference method for degenerate hyperbolic equations of second order

Li Qian [Dept. of Mathematics, University of Shandong, Jinan, China]: The use of numerical integration in finite element methods for solving non-stable problems

Li Yi [Dept. of Mathematics, Sichuan University, Chengdu, China]: On three-level explicit difference schemes for dispersion equation $u_t = au_{xxx}$

Lin Peng-cheng [Dept. of Computer Science, Fuzhou University, Fuzhou, China]: Direct methods for solving mixed boundary value problem of three dimensional Poisson equation and solution of bending of rectangular plates under uniform load by two-direction trigonometric series

Lin Peng-cheng and Guo Wen [Dept. of Computer Science, Fuzhou University, Fuzhou, China]: Difference methods for solving a singular perturbation problem of a self-adjoint elliptic equation

Lin Peng-cheng and Jiang Ben-tian [Dept. of Computer Science, Fuzhou University, Fuzhou, China]: A singular perturbation problem for periodic boundary problems of elliptic-parabolic partial differential equations

Lin Qun and Xie-Rui-feng [Institute of Systems Science, Academia Sinica, Beijing, China]: Error expansions for finite element approximation and their applications

Lin Wen-xian [Dept. of Mathematics, Tianjin Normal University, Tianjin, China]: Finite difference methods of the boundary value problems for the systems of semilinear generalized Schrödinger type

Lu Bai-nian [Dept. of Mathematics, Shaanxi Normal University, Xian, Shaanxi, China]: A leap-frog finite difference scheme for a class of non-self-adjoint and nonlinear Schrödinger equation

Lu Jin-fu [Dept. of Applied mathematics, Tsinghua University, Beijing, China]: Some difference schemes for nonlinear convection diffusion equations

Ma Fu-ming [Institute of Mathematics, Jilin University, Changchun, China]: Limit point bifurcation with multiplicity two and its finite dimensional approximation

Ma He-ping and Guo Ben-yu [Shanghai University of Science and Technology, Shanghai, China]: The Fourier pseudospectral method with a restrain operator for the M.K.D.V. equation

Mao De-kang [Dept. of Mathematics, Shanghai University of Science and Technology, Shanghai, China]: A treatment to discontinuities in shock-capturing finite difference methods

Pan Zhong-xiong [Shanghai University of Science and Technology, Shanghai, China]: A numerical method for a class of nonlinear fourth order equations

Qin Meng-zhao [Computing Center, Academia Sinica, Beijing, China]: Calculation of chaotic behavior in Hamiltonian dynamical systems using symplectic schemes

Shen Ji-hong [Dept. of Mathematics, Jilin University, Changchun, China]: A high-accurate upwind scheme

Shen Long-jun [Institute of Applied Physics and Computational Mathematics, P.O.Box 8009, Beijing, China]: COnvergence conditions of the explicit and weak implicit finite difference schemes for parabolic systems

Shen Shu-min [Dept. of Mathematics, Suzhou University, Suzhou, China]: Finite element approximations for a variational inequality with a nonlinear monotone operator

Shi Jin-song [Hohai University, Nanjing, China]: An extended pressure method for a filtration problem

Sun Le-lin [Dept. of Mathematics, Wuhan University, Wuhan, China]: Domain decomposition-projection methods for solving some nonlinear PDEs

Sun Yu-ping and Wu Jiang-hang [Dept. of Mechanics, Peking University, Beijing, China]: Stability and convergence of the finite analytic method for convective diffusion equations

Wang Guo-ying [Dept. of Mathematics, Nanjing University, Nanjing, China]: A family of uniformly convergent difference schemes for solving the first boundary value problem of elliptic partial differential equation with a small parameter

Wang Ji-da and Lin Bing-chang [Anshan Institute of Steel Technology, Anshan, Liaoning, China]: The numerical analysis of the rate equation of chromatography

Wang Lie-heng [Computing Center, Academia Sinica, Beijing, China]: Some elements for the Stokes problem

Wang Ming-rui [Institute of Applied Physics and Computational Mathematics, P.O. Box 8009, Beijing, China]: The calculation and analysis for the projectile impacting on a target

Wang Shen-lin [Shandong University, Jinan, China]: Variational principles of the generalized difference methods and H^1-error estimates for parabolic and hyperbolic equations

Wang Shen-lin and Sun Shu-ying [Shandong University, Jinan, China]: Error estimates of Galerkin method and estimation of convergence rate of A.D.I. Galerkin method for some quasi-linear hyperbolic equations

Wu Hua-mo and Wu Yu-hua [Computing Center, Academia Sinica, Beijing, China]: Contour dynamics methods for discontinuous vortical flows

Wu Ji-ke and Li Hui [Dept. of Mechanics, Peking University, Beijing, China]: A numerical method of applying the meanvalue theorem

Wu Jiang-hang [Dept. of Mechanics, Peking University, Beijing, China]: A finite-difference solution of the two-dimensional unsteady convective diffusion and Navier-Stokes equations in a nonuniform triangle mesh

Wu Shi-xian [Dept. of Mathematics, Hebei Teachers University, Shijiazhuang, Hebei, China]: On the infinite element method for the parabolic equation

Xiang Xin-min [Dept. of Mathematics, Heilongjiang University, Harbin, China]: Spectral methods for a class of systems of multidimensional nonlinear wave equations and nonlinear Schrödinger equation

Xie De-xuan [Dept. of Applied Mathematics, Hunan University, Changsha, Hunan, China]: An new multigrid iteration for solving interface problems with small media ratio

Xu Hong-liang [Dept. of Applied Mathematics, Shanghai Jiaotong University, Shanghai, China]: An extrapolation algorithm for a class of stiff equations and its applications

Xu Jia-Mo [Wuhan University, Wuhan, China] and Lu Jun-an (Wuhan Institute of Hydraulic and Electric Engineering, Wuhan, China]: Optimum domain problems governed by a class of partial differential equations

Xu Meng-jie and Wang Chuan-shan [Dept. of Mathematics, Shanghai University of Science and Technology, Shanghai, China]: An algorithm for electron dose distribution in multiple-media

Yang Hai-liang and Wang Gang [Dept. of Mathematics, Inner Mongolia University, Huhehot, China]: Probability-numerical-methods for sovling differential equations

Yang Qing-min [Computing Center of Hunan Provence, Changsha, China]: A nonstandard finite element method and some results about stability theory of discrete schemes for operator equations

Yang Yi-du [Dept. of Mathematics, Guizhou University, Guiyang, China]: The effect of numerical integration on extrapolation for the finite element approximations of eigenvalues

Yang Zhong-hua [Dept. of Mathematics, Shanghai University of Science and Technology, Shanghai, China]: Folds of degree 4 and swallowtail catastrophe

Ying Lung-an [Dept. of Mathematics, Peking University, Beijing, China]: Convergence study for viscous splitting in bounded domains

Yong Wen-an and Zhu You-lan [Computing Center, Academia Sinica, Beijing, China]: Convergence of difference methods for nonlinear problems with moving boundaries

Yuan Ling-tao [Dept. of Mathematics, Shaanxi Normal University, Xian, Shaanxi,

China]: Difference methods for three-dimension parabolic-wave equations

Zhang Tie [Dept. of Mathematics, Northeast University of Technology, Shenyang, Ching]: L_p - error analysis for semidiscrete Galerkin approximations of parabolic equations

Zhang Xuan [Shenyang Architectural Engineering College, Shenyang, China]: Solving problems of subterranean water quality with a mixed finite element method

Zheng Jia-dong [Shanghai Institute of Computer Technologh, Shanghai, China]: The pseudospectral collocation methods for the generalized SRLW equations

Zhou Zheng-zhong [Hunan Computing Center, Changsha, China]: A spectral method for the Zakharov equation with periodic boundary conditions

Zhu Ben-ren and Jin Mao-yuan [Shandong University, Jinan, China]: An explicit scheme for an inverse scattering problem and its stability analysis

Zhu Jia-lin [Chongqing Institute of Architecture and Engineering, Chongqing, China]: Asymptotic error estimates for the 'BEM' to viscous flow problem in two dimensions

Zhu Jiang [Dept. of Mathematics and Mechanics, Nanjing Institute of Technology, Nanjing, China]: The characteristic numerical methods for nonlinear RLW equations

Zhu Qi-ding [Xiangtan University, Xiangtan, Hunan, China]: Local superconvergence estimates for the finite element method

CONTENTS

THE SYMPLECTIC METHODS FOR THE COMPUTATION OF HAMILTONIAN EQUATIONS*

Feng Kang and Qin Meng-zhao

(Computing Center, Academia Sinica, Beijing)

Abstract

The present paper gives a brief survey of results from a systematic study,undertaken by the authors and their colleagues, on the symplectic approach to the numerical computation of Hamiltonian dynamical systems in finite and infinite dimensions. Both theoretical and practical aspects of the symplectic methods are considered. Almost all the real conservative physical processes can be cast in suitable Hamiltonian formulation in phase spaces with symplectic structure, which has the advantages to make the intrinsic properties and symmetries of the underlying processes more explicit than in other mathematically equivalent formulations, so we choose the Hamiltonian formalism as the basis, together with the mathematical and physical motivations of our symplectic approach for the purpose of numerical simulation of dynamical evolutions. We give some symplectic difference schemes and related general concepts for linear and nonlinear canonical systems in finite dimensions. The analysis confirms the expectation for them to behave more satisfactorily, especially in the desirable conservation properties, than the conventional schemes. We outline a general and constructive theory of generating functions and a general method of construction of symplectic difference schemes based on all possible generating functions. This is crucial for the developments of the symplectic methods. A generalization of the above theory and method to the canonical Hamiltonian eqs. in infinite dimensions is also given. The multi-level schemes, including the leapfrog one,are studied from the symplectic point of view. We give an application of symplectic schemes, with some indications of their potential usefulness, to the computation of chaos.

CONTENTS

---

* Work supported by National Natural Science Foundation of China.

§ 1 Introduction

Recently it is evident that Hamiltonian formalism plays a fundamental role in the diverse areas of physics, mechanics, engineering, pure and applied mathmatics, e. g. geometrical optics, analytical dynamics, nonlinear PDE's of first order, group representations, WKB asymptotics, pseudodifferential and Fourier integral operators, classification of singularities, integrability of non-linear evolution equations, optimal control theory, etc. It is also under extension to infinite dimensions for various field theories, including electrodynamics, plasma physics,elasticity, hydrodynamics etc. It is generally accepted that all real physical processes with negligible dissipation could be expressed, in some way or other, by Hamiltonian formalism, so the latter is becoming one of the most useful tools in the mathematical physics and engineering sciences.

Hamiltonian formalism has the important property of being area-preserving (symplectic) i.e. the sum of the areas of canonical variable pairs, projected on any two-dimensional surface in phase space, is time invariant. In numerically solving these equations one hopes that the approximating equation will hold this property.

In DD-5 Beijing Conference the first author [1] propose an approach for computing Hamiltonian equation from the viewpoint of symplectic geometry. This paper is a brief survey of considerations and developments [1-11,15], obtained by the first author and his group, on the links between the Hamiltonian formalism and the numerical methods.

Now we will give a review of some facts from Hamiltonian mechanics which are fundamental to what follows. We consider the following canonical system of ordinary first order differential equations on R^{2n}

$$\frac{dp_i}{dt} = -\frac{\partial H}{\partial q_i}, \quad \frac{dq_i}{dt} = \frac{\partial H}{\partial p_i}, \quad i = 1,2,\cdots n, \tag{1.1}$$

where $H(p,q)$ is some real valued function. We call (1.1) a Hamiltonian system of differential equations (H - system). In the following, vectors are always represented by column matrices, matrix transpose is denoted by prime'. Let $z=(z_1 \cdots z_n, z_{n+1}, \cdots z_{2n})' = (p_1 \dots p_n, q_1 \dots q_n)'$, $H_z = [\frac{\partial H}{\partial p_1}, \dots, \frac{\partial H}{\partial p_n}, \frac{\partial H}{\partial q_1}, \dots, \frac{\partial H}{\partial q_n}]'$, $J_{2n} = J = \begin{bmatrix} 0 & I_n \\ -I_n & 0 \end{bmatrix}$, $J' = J^{-1} = -J$, where I_n is the $n \times n$ identity matrix. (1.1) can be written as

$$\frac{dz}{dt} = J^{-1} H_z, \tag{1.2}$$

defined in phase space R^{2n} with a standard symplectic structure given by the nonsingular anti-symmetric closed differential 2-form

$$\omega = \Sigma\, dz_i \wedge dz_{n+i} = \Sigma\, dp_i \wedge dq_i.$$

According to Darboux Theorem, the symplectic structure given by any non-singular antisymmetric closed differential 2-form can be brought to the above standard form, at least locally, by suitable change of coordinates.

The right side of equation (1.2) gives a vector field. At each point(p,q) of the phase space, there is a 2n-dimensional vector$(-H_q, H_p)$.

The fundamental theorem on Hamiltonian Formalism says that the solution z(t) of the canonical system (1.2) can be generated by a one-parameter group G(t), depending on given Hamiltonan H, of canonical transformations of R^{2n} (locally in t and z) such that

$$z(t)=G(t)z(0).$$

This group is also called the phase flow

$$G(t):\ (p(o),q(o)) \to\ (p(t),\ q(t))$$

where p(t), q(t) are the solution of Hamilton's system of equations(1.1). A transformation $z \to \hat{z}$ of R^{2n} is called canonical, or symplectic, if it is a local diffeomorphism whose Jacobian $\frac{\partial \hat{z}}{\partial z} = M$ is every-where symplectic i.e.

$$M'JM=J, \text{ ie. } M \in Sp(2n).$$

Linear canonical transformation is special symplectic transformation.

The canonicity of G(t) implies the preservation of 2-form ω, 4-form $\omega \wedge \omega$, $\cdots$, 2n-form $\omega \wedge \omega \wedge \cdots \wedge \omega$. They constitute the class of conservation laws of phase area of even dimensions for the Hamiltonian system (1.2).

Moreover, the Hamiltonian system possesses another class of conservation laws related to the energy H(z). A function $\psi(z)$ is said to be an invariant integral of (1.2) if it is invariant under (1.2)

$$\psi(z(t)) \equiv \psi(z(0))$$

which is equivalent to

$$\{\psi, H\} = 0,$$

where the Poisson **Brackets** for any pair of differentiable functions ϕ and ψ are defined as

$$\{\phi, \psi\} = \phi_z' J^{-1} \psi_z .$$

H itself is always an invariant integral, see, e.g. [12].

For the numerical study , we are less interested in (1.2) as a general system of ODE per se, but rather as a specific system with Hamiltonian structure. It is natural to look for those discretization systems which preserve as much as possible the characteristic properties and inner symmetries of the original continuous systems. To this end the transition $\hat{z} \to z$ from the k-th time step $z^k = \hat{z}$ to the next (k+1)-th time step $z^{k+1} = z$ should be canonical for all k and, moreover, the invariant integrals of the original system should remain invariant under these transitions.

Thus, a difference scheme may be regarded as a transformation from time t^k to time t^{k+1}. We have the following

Definition. A difference scheme may be called symplectic or canonical scheme if its transitional transformation is symplectic. We try to conceive, design, analyse and evaluate difference schemes and algorithms specifically within the framework of symplectic geometry. The approach proves to be quite successful as one might expect, we actually derive in this way numerous "unconventional" difference schemes.

An outline of the paper is as follows. In section 2 we review some symplectic difference schemes (S-scheme) for linear hamiltonian system (LH-system) and nonlinear hamiltonian system (NLH-system) and its related properties are given. In section 3 we systematically outline the general method of construction of S-scheme with any order accuracy via generating function. The constructive theory of generating function and the corresponding construction of S-schemes have been generalized to the case of phase space of infinite dimensions of the form $B^* \times B$, where B is a reflexive Banach space, B^* its dual [3] [8]. Section 4 contains the main idea. The multi-level difference S-schemes of hamiltonian type are described in §5. In §6 we show some computational results and comparison with R-K method. The last section is S-scheme and chaos. It is well known that canonical transformation is an area-preserving mapping. Therefore S-schemes are suitabe tools for studying chaotic behavior in hamiltonian mechanics.

§2. S-schemes for Linear and Nonlinear Hamiltonian Systems

Consider the case for which the Hamiltonian is a quadratic form

$$H(z) = \frac{1}{2} z'Sz, \qquad S' = S, \quad H_z = Sz \tag{2.1}$$

Then the canonical system

$$\frac{\partial z}{dt} = Lz, \qquad L = J^{-1}S \tag{2.2}$$

is linear, where L is infinitesimally symplectic, i.e. L satisfies $L'J + JL = 0$.

The solution of (2.2) is

$$z(t) = G(t)z(0)$$

where $G(t) = \exp tL$, as the exponential transform of infinitesimally symplectic tL, is symplectic.

It is easily seen that the weighted Euler scheme

$$\frac{1}{\tau}(z^{k+1} - z^k) = L(\alpha z^{k+1} + (1-\alpha)z^k)$$

for the linear system (2.2) is symplectic iff $\alpha = \frac{1}{2}$, i.e. it is the case of time-centered Euler scheme with the transition matrix F_τ,

$$z^{k+1} = F_\tau z^k, \qquad F_\tau = \psi(\tau L), \qquad \psi(\lambda) = \frac{1+\frac{\lambda}{2}}{1-\frac{\lambda}{2}}, \tag{2.3}$$

F_τ, as the Cayley transform of infinitesimally symplectic τL, is symplectic.

In order to generalize the time-centered Euler scheme, we need, apart from the exponential or Cayley transforms, other matrix transforms carrying infinitesimally symplectic matrices into symplectic ones.

Theorem 1. Let $\psi(\lambda)$ be a function of complex variable λ satisfying

(I) $\psi(\lambda)$ is analytic with real coefficients in a neighborhood D of $\lambda = 0$,

(II) $\psi(\lambda)\psi(-\lambda) \equiv 1$ in D,

(III) $\psi'(0) \neq 0$.

A is a matrix of order 2n, then $(\psi(\tau L))' A\, \psi(\tau L) = A$ for all τ with sufficiently small $|\tau|$, iff

$$L'A + AL = 0$$

If, more, $\exp \lambda - \psi(\lambda) = 0(|\lambda|^{m+1})$, then

$$z^{k+1} = \psi(\tau L)\, z^k \tag{2.4}$$

considered as an approximative scheme for the canonical system (2.2) is symplectic, of m-th order of accuracy and has the property that $z'A\,w$ is invariant under $\psi(\tau L)$ iff it is invariant under G_t of (2.2).

Remark 1. The last property is remarkable in the sense that, all the bilinear invariants of the system(2.2), no more and no less, are kept invariant under the scheme(2.4), in spite of the fact that the latter is only approximate.

Remark 2. The approximative scheme in theorem 1 becomes difference schemes only when $\psi(\lambda)$ is a rational function. As a concrete application to the construction of symplectic difference schemes, we take the diagonal Padé approximants to the exponential function

$$\exp \lambda - \frac{P_m(\lambda)}{P_m(-\lambda)} = 0(|\lambda|^{2m+1})$$

where $P_0(\lambda) = 1$, $P_1(\lambda) = 2 + \lambda$, $P_2(\lambda) = 12 + 16\lambda + \lambda^2$, $\cdots$ $P_m(\lambda) = 2(2m-1)P_{m-1}(\lambda) + \lambda^2 P_{m-2}(\lambda)$.

Theorem 1'. Difference schemes

$$z^{k+1} = \frac{P_m(\tau L)}{P_m(-\tau L)}\, z^k, \quad m = 1,2,\cdots \tag{2.5}$$

for the eq. (2.2) are symplectic, A-stable, of 2m-th order of accuracy, and having the same set of bilinear invariants as that of eq. (2.2), and the case m=1 is the centered Euler scheme [5].

For the general non-linear canonical system (1.2), the time-centered Euler scheme is

$$\frac{1}{\tau}(z^{k+1} - z^k) = J^{-1} H_z\left(\frac{1}{2}(z^{k+1} + z^k)\right). \tag{2.6}$$

The transition $z^{k+1} \rightarrow z^k$ is canonical with Jacobian

$$F_\tau = \left[I - \frac{\tau}{2} J^{-1} H_{zz}\left(\tfrac{1}{2}(z^{k+1}+z^k)\right)\right]^{-1}\left[1 + \frac{\tau}{2} J^{-1} H_{zz}\left(\tfrac{1}{2}(z^{k+1}+z^k)\right)\right]$$

symplectic everywhere. However, unlike the linear case, the invariant integrals ψ, not quadratic in z, including H(z), are conserved only approximately:

$$\psi(z^{k+1}) - \psi(z^k) = 0(\tau^3).$$

For the general non-linear separable system for which $H(p,q) = U(p) + V(q)$, U, V being kinetic and potential energies respectively, we define the staggered scheme

$$\frac{1}{\tau}(p^{k+1} - p^k) = -V_q(q^{k+\frac{1}{2}}),$$
$$\frac{1}{\tau}(q^{k+1+\frac{1}{2}} - q^{k+\frac{1}{2}}) = U_p(p^{k+1}). \qquad (2.7)$$

The p's are set at integer times $t = k\tau$, q's at half-integer times $t=(k+\frac{1}{2})\tau$. The transition

$$w^k = \begin{bmatrix} p^k \\ q^{k+\frac{1}{2}} \end{bmatrix} \to \begin{bmatrix} p^{k+1} \\ q^{k+1+\frac{1}{2}} \end{bmatrix} = w^{k+1} = F_\tau w^k,$$

$$F_\tau = \begin{bmatrix} I & 0 \\ -\tau M & I \end{bmatrix}^{-1} \cdot \begin{bmatrix} I & -\tau L \\ 0 & I \end{bmatrix},$$

$$M = U_{pp}(p^{k+1}), \qquad L = V_{qq}(q^{k+\frac{1}{2}}),$$

is symplectic of 2nd order accuracy and practically explicit, since p, q are computed at different times, we have to use certain synchronization, e.g. using

$$q^k = \frac{1}{2}(q^{k-\frac{1}{2}} + q^{k+\frac{1}{2}})$$

to compute the invariant integrals $\psi(p,q)$

$$\psi(p^{k+1}, q^{k+1}) - \psi(p^k, q^k) = O(\tau^3).$$

In [1] [9], a class of energy conservative schemes was constructed using the differencing of the Hamiltonian function, however, the symplectic property was not satisfactory. The problem of compatibility of energy conservation with phase area conservative in difference schemes is solved successfully for linear canonical system (Theorem 1). It seems difficult, however, for the general nonlinear system.

Suppose that the Hamiltonian function $H(p,q)$ does not depend on time.

$$H + h = 0, \qquad (2.8)$$

where h is a constant, is the integral of energy of the system.
We assume that (in some region) the equation (2.8) can be solved for some variable (no loss of generality) for example p_1, $H_{p_1} = \frac{dq_1}{dt} \neq 0$, so (2.8) is solved by

$$K(p_2, p_3 \cdots p_n, q_1 \cdots q_n, h) + p_1 = 0 \qquad (2.9)$$

and $t \to q_1(t)$ has an inverse $q_1 \to t(q_1)$, so we can use q_1 instead of t as "time". Equations (1.1) can be written as

$$\frac{dq_r}{dq_1} = \frac{\partial K}{\partial p_r}, \quad \frac{dp_r}{dq_1} = -\frac{\partial K}{\partial q_r}, \ (r=2,3,\cdots,n), \qquad (2.10)$$

$$\frac{dt}{dq_1} = \frac{\partial K}{\partial h}, \quad \frac{dh}{dq_1} = 0 = -\frac{\partial K}{\partial t}. \qquad (2.11)$$

The last two equations can be separated from the rest of the system, since the first (2n-2) equations do not involve t, and h is a constant. The phase trajectories of (1.1) on the energy surface $H(z)+h=0$ satisfy (2.10). Moreover (2.11), (2.10) combine into a new H-system $dw/dq_1 = J^{-1}K_w$ in "time" q_1 in the modified phase space with new canonical variables $w=(h,p_2,\dots,p_n;t,q_2,\cdots,q_n)$ with "time"-dependent Hamiltonian $K(w;q_1)$. Since K is independent of $w_{n+1}=t$, its conjugate variable $w_1=h$ is constant. We use the Euler scheme

$$\frac{1}{\Delta q_1}(w^{m+1}-w^m)= J^{-1}K_w\left(\frac{1}{2}(w^{m+1}+w^m);\ \frac{1}{2}(q_1^{m+1}+q_1^m)\right),\quad q_1^{m+1}=q_1^m+\Delta q_1 .$$

The transition $w^m \longrightarrow w^{m+1}$ is symplectic and preserves $w_1=h$, i.e. $h^m=h^{m+1}$. We compute, in addition, $K^m=K(w^m;\ q_1^m)$, $p_1^m = -K^m$. Owing to the identity

$$H(-K(p_2,\cdots,p_n,q_1,q_2,\cdots,q_n,h),\ p_2,\cdots,p_n,q_1,\dots,q_n)+h \equiv 0,$$

we have $H^m = H(p_1^m,p_2^m,\cdots,q_n^m)= -h^m= -h^{m+1}=H^{m+1}$. So we get energy conservative scheme which is also symplectic in a modified sense. Here the computed time steps $t^{m+1} - t^m$ are in general variable under the fixed "time" step Δq_1. This agrees with an idea of T.D. Lee [16], where time steps are to be solved to make energy conservative. The above approach to make S-schemes energy conservative is due to Qin Meng-zhao.

The problem of preservation of first integrals of system (1.2) under a S-scheme

$$z^{m+1} = S_H^\tau (z^m),$$

where S_H^τ is a symplectic transformation depending on $H(z)$ and step τ, is closely related to the invariance properties of S_H^τ under groups of symplectic transformations [7]. The scheme S_H^τ is said to be invariant under a group G of symplectic transformations if

$$g^{-1} \circ S_H^\tau \circ g = S_{H\circ g},\quad \forall\ g \in G.$$

<u>Theorem 2</u>. Let F be a first integral of system $dz/dt = J^{-1}H_z$. Then F is preserved up to a constant by the scheme S_H^τ, i.e.

$$F(z) \equiv F(S_H^\tau(z)) + c, \qquad c = \text{const}.$$

iff S_H^τ is invariant under the 1-parameter group G_F^t of the phase flow of the system $dz/dt = J^{-1}F_z$. The constant $c = 0$ if S_H^τ has a fixed point.

It is known that all linear first integrals (e.g., linear momenta) of H-system are preserved by any compatible difference scheme, symplectic or not. However for quadratic first integrals (e.g., angular momenta), this is by far not the case. In this aspect the symplectic schemes are distinguished as shown above at least for linear H-systems.

For general non-linear H-systems and for symplectic Euler schemes of arbitrary 2m-th order (the case m=1is (2.6), for higer order schemes see Theorem 8, §3), Ge Zhong, Wang Dao-liu, and Wu Yu-hua have proved the physically significant property of preservation of all the quadratic first integrals. This is easily seen for the case m=1: Let $F(z) = \frac{1}{2} z'Bz$, $B' = B$, be a first integral of (1.2), then $\{f, H\} = (Bz)' J^{-1} H_z(z) = 0$ for all z. Multiply (2.6) by $(B \frac{1}{2} (z^{k+1} + z^k))'$, we get $\frac{1}{\tau} \frac{1}{2} (z^{k+1} + z^k)' B(z^{k+1} - z^k) = 0$, then $F(z^{k+1}) = F(z^k)$.

§3 Constructive Theory of Generating Functions and S-schemes

In this section we reproduce almost verbally the results from [4]. The generalization to Poisson maps was done in [11]. The generalization to infinite dimensions was done in [8] and will be outlined in §4.

In order to develop a general method of construction of canonical difference schemes we first give a constructive generalization of the classical theory of generating function and Hamilton-Jacobi equations. Our approach in this part was inspired by the early works of Siegel [13] and Hua [14]. Every matrix

$$A = \begin{bmatrix} A_1 \\ A_2 \end{bmatrix} \in M(4n,2n), \quad A_1, A_2 \in M(2n), \qquad \text{rank}\, A = 2n$$

defines in R^{4n} a 2n-dimensional subspace $\{A\}$ spanned by its column vectors. $\{A\} = \{B\}$ iff $A \sim B$, i.e.

$$AP = B, \quad \text{i.e.} \quad \begin{bmatrix} A_1 P \\ A_2 P \end{bmatrix} = \begin{bmatrix} B_1 \\ B_2 \end{bmatrix}, \text{ for some } P \in GL(2n).$$

The spaces of symmetric and symplectic matrices of order 2n will be denoted by Sm(2n), Sp(2n) respectively. Let

$$J_{4n} = \begin{bmatrix} 0 & I_{2n} \\ -I_{2n} & 0 \end{bmatrix}, \qquad \tilde{J}_{4n} = \begin{bmatrix} -J_{2n} & 0 \\ 0 & J_{2n} \end{bmatrix}$$

$$X = \begin{bmatrix} X_1 \\ X_2 \end{bmatrix}, \quad Y = \begin{bmatrix} Y_1 \\ Y_2 \end{bmatrix} \in M(4n, 2n), \text{ of rank } 2n.$$

Subspace $\{X\} \subset R^{4n}$ is called J_{4n}-Lagrangian (and $\begin{bmatrix} X_1 \\ X_2 \end{bmatrix}$ is called a symmetric pair) if

$$X' J_{4n} X = O_{2n}, \quad \text{i.e.} \quad X_1' X_2 - X_2' X_1 = O_{2n}.$$

If, moreover, $|X_2| \neq 0$, then $X_1 X_2^{-1} = N \in Sm(2n)$ and $\begin{bmatrix} X_1 \\ X_2 \end{bmatrix} \sim \begin{bmatrix} N \\ I \end{bmatrix}$, where N is determined uniquely by the subspace $\{X\}$. Similarly, subspace $\{Y\} \subset R^{4n}$ is called $\tilde{J}_{4n}$ - Lagrangian (and $\begin{bmatrix} Y_1 \\ Y_2 \end{bmatrix}$ is called a symplectic pair) if

$$Y' \tilde{J}_{4n} Y = O_{2n}, \quad \text{i.e.} \quad Y_1' J_{2n} Y_1 - Y_2' J_{2n} Y_2 = O_{2n}.$$

If, moreover, $|Y_2| \neq 0$, then $Y_1 Y_2^{-1} = M \in Sp(2n)$ and $\begin{bmatrix} Y_1 \\ Y_2 \end{bmatrix} \sim \begin{bmatrix} M \\ I \end{bmatrix}$, where M is determined uniquely by the subspace $\{Y\}$.

A 2n-dimensional submanifold $U \subset R^{4n}$ is called J_{4n} - Lagrangian (respectively $\tilde{J}_{4n}$ - Lagrangian) if the tangent plane of U is a J_{4n}-Lagrangian (respectively $\tilde{J}_{4n}$-

Lagrangian) subspace of the tangent space at each point of U.

Let $z \to \hat{z} = g(z)$ be a canonical transformation in R^{2n}, with Jacobian $g_z = M(z) \in Sp(2n)$. The graph

$$V = \{ \begin{bmatrix} \hat{z} \\ z \end{bmatrix} \in R^{4n} \mid \hat{z} = g(z) \}$$

of g is a $\tilde{J}_{4n}$-Lagrangian submanifold, whose tangent plane is spanned by the symplectic pair $\begin{bmatrix} M(z) \\ I \end{bmatrix}$.

Similarly, let $w \to \hat{w} = f(w)$ be a gradient transformation in R^{2n}, the Jacobian $f_w = N(w) \in Sm(2n)$. This is equivalent to the (local) existence of a scalar function $\phi(w)$ such that $f(w) = \phi_w(w)$. The graph

$$U = \{ \begin{bmatrix} \hat{w} \\ w \end{bmatrix} \in R^{4n} \middle| \hat{w} = f'(w) \}$$

of f is a J_{4n}-Lagrangian submanifold with tangent planes spanned by the symmetric pair $\begin{bmatrix} N(w) \\ I \end{bmatrix}$.

<u>Theorem 3</u>. $T \in GL(4n)$ carries every $\tilde{J}_{4n}$-Lagrangian submanifold into J_{4n}-Lagrangian submanifold if and only if

$$T' J_{4n} T = \mu \tilde{J}_{4n}, \quad \text{for some} \quad \mu \neq 0,$$

i.e.

$$A_1 = -\mu^{-1} J_{2n} C', \qquad B_1 = \mu^{-1} J_{2n} A',$$

$$C_1 = \mu^{-1} J_{2n} D', \qquad D_1 = -\mu^{-1} J_{2n} B', \tag{3.1}$$

$$A = \begin{bmatrix} A & B \\ C & D \end{bmatrix}, \quad T^{-1} = \begin{bmatrix} A_1 & B_1 \\ C_1 & D_1 \end{bmatrix}.$$

The totality of T's in Theorem 3 will be denoted by $CSp(\tilde{J}_{4n}, J_{4n})$, the subset with $\mu = 1$ by $Sp(\tilde{J}_{4n}, J_{4n})$. The latter is not empty since $\tilde{J}_{4n}$ is congruent to J_{4n}. Fix $T_0 \in Sp(\tilde{J}_{4n}, J_{4n})$; then every $T \in CSp(\tilde{J}_{4n}, J_{4n})$ is a product

$$T = MT_0, \quad M \in CSp(4n) = \text{conformal symplectic group.}$$

T^{-1} for $T \in CSp(\tilde{J}_{4n}, J_{4n})$ carries J_{4n}-Lagrangian submanifolds into $\tilde{J}_{4n}$-Lagrangian submanifolds.

A major component of the transformation theory in symplectic geometry is the method of generating functions. Canonical transformations can in some way be expressed in implicit form, as gradient transformations with generating functions via suitable linear transformations. The graphs of canonical and gradient transformations in R^{4n} are $\tilde{J}_{4n}$-Lagrangian and J_{4n}-Lagrangian submanifolds respectively. Theorem 3 leads to the existence and construction of the generating functions, under non-exceptional conditions, for the canonical transformations.

<u>Theorem 4.</u> Let $T = \begin{bmatrix} A & B \\ C & D \end{bmatrix}$, $T^{-1} = \begin{bmatrix} A_1 & B_1 \\ C_1 & D_1 \end{bmatrix}$, $T \in CS_p(\tilde{J}_{4n}, J_{4n})$, which define linear transformations

$$\hat{w} = A\hat{z} + Bz, \qquad \hat{z} = A_1\hat{w} + B_1 w,$$
$$w = C\hat{z} + Dz, \qquad z = C_1\hat{w} + D_1 w.$$

Let $z \to \hat{z} = g(z)$ be a canonical transformation in (some neighborhood of) R^{2n}, with Jacobian $g_z = M(z) \in Sp(2n)$ and graph

$$V^{2n} = \{ \begin{bmatrix} \hat{z} \\ z \end{bmatrix} \in R^{4n} \mid \hat{z} - g(z) = 0 \}.$$

If (in some neighborhood of R^{4n})

$$|CM + D| \neq 0, \tag{3.2}$$

then there exists in (some neighborhood of) R^{2n} a gradient transformation $w \to \hat{w} = f(w)$ with Jacobian $f_w = N(w) \in Sm(2n)$ and graph

$$U^{2n} = \{ \begin{bmatrix} \hat{w} \\ w \end{bmatrix} \in R^{4n} \mid \hat{w} - f(w) = 0 \}$$

and a scalar function —— generating function —— $\phi(w)$ such that

(1) $f(w) = \phi_w(w)$;

(2) $N = (AM+B)(CM+D)^{-1}$, $M = (NC-A)^{-1}(B-ND)$;

(3) $T(V^{2n}) = U^{2n}$, $V^{2n} = T^{-1}(U^{2n})$.

This corresponds to the fact that, under the transversality condition (3.2),

$$[\hat{w} - \phi_w(w)]_{\hat{w}=A\hat{z}+Bz,\, w=C\hat{z}+Dz} = 0$$

gives the implicit representation of the canonical transformation $\hat{z}=g(z)$ via linear transformation T and generating function ϕ.

For the time-dependent canonical transformation, related to the time-evolution of the solutions of a canonical system (1.2) with Hamiltonian function H(z), we have the following general theorem on the existence and construction of the time-dependent generating function and Hamilton-Jacobi equation depending on T and H under some transversality condition.

Theorem 5. Let T be such as in Theorems 3 and 4. Let $z \to \hat{z}=g(z,t)$ be a time-dependent canonical transformation (in some neighborhood) of R^{2n} with Jacobian $g_z(z,t)=M(z,t) \in Sp(2n)$ such that

(a) $g(*,0)$ is a linear canonical transformation $M(z,0)=M_0$, independent of z,

(b) $g^{-1}(*,0)\ g(*,0)$ is the time-dependent canonical transformation carrying the solution z(t) at moment t to z(0) at moment t=0 for the canonical system. If

$$|CM_0+D| \neq 0, \tag{3.3}$$

then there exists, for sufficiently small $|t|$ and in (some neighborhood of) R^{2n}, a time-dependent gradient transformation $w \to \hat{w}=f(w,t)$ with Jacobian $f_w(w,t)= N(w,t) \in Sm(2n)$ and a time-dependent generating function $\phi(w,t)$ such that

(1) $[\hat{w}-f(w,t)]_{\hat{w}=A\hat{z}+Bz,\, w=C\hat{z}+Dz}=0$ is the implicit representation of the canonical transformation $\hat{z}=g(z,t)$;

(2) $N=(AM+B)(CM+D)^{-1}$, $M=(NC-A)^{-1}(B-ND)$;

(3) $\phi_w(w,t)=f(w,t)$;

$$(4) \quad \phi_t(w,t) = -\mu H(C_1\phi_w(w,t)+D_1 w), \quad w=C\hat{z} + Dz.$$

Equation (4) is the most general Hamilton-Jacobi equation abbreviated as H. J. equation for the Hamiltonian canonical system (1.2) and linear transformation $T \in CSP(\mathcal{J}_{4n}, J_{4n})$.

Special types of generating functions:

$$(I) \qquad T = \begin{pmatrix} -I_n & 0 & 0 & 0 \\ 0 & 0 & I_n & 0 \\ 0 & I_n & 0 & 0 \\ 0 & 0 & 0 & I_n \end{pmatrix}, \quad \mu = 1, \; M_0 = J_{2n}, \quad |CM_0+D| \neq 0;$$

$$w = \begin{pmatrix} \hat{q} \\ q \end{pmatrix}, \qquad \phi = \phi(\hat{q}, q, t);$$

$$\hat{w} = \begin{pmatrix} -\hat{p} \\ p \end{pmatrix} = \begin{pmatrix} \phi_{\hat{q}} \\ \phi_q \end{pmatrix}, \qquad \phi_t = -H(\phi_q, q)$$

are the generating function and H-J equation of the first kind [12].

$$(II) \qquad T = \begin{pmatrix} -I_n & 0 & 0 & 0 \\ 0 & 0 & 0 & -I_n \\ 0 & I_n & 0 & 0 \\ 0 & 0 & I_n & 0 \end{pmatrix}, \quad \mu = 1, \; M_0 = I_{2n}, \quad |CM_0+D| \neq 0;$$

$$w = \begin{pmatrix} \hat{q} \\ p \end{pmatrix}, \qquad \phi = \phi(\hat{q}, p, t);$$

$$\hat{w} = \begin{pmatrix} -\hat{p} \\ -q \end{pmatrix} = \begin{pmatrix} \phi_{\hat{q}} \\ \phi_p \end{pmatrix}, \qquad \phi_t = -H(p, -\phi_p)$$

are the generating function and H-J equation of the second kind [2].

$$(III) \qquad T = \begin{pmatrix} -J_{2n} & J_{2n} \\ \frac{1}{2} I_{2n} & \frac{1}{2} I_{2n} \end{pmatrix}, \quad \mu = -1, \quad M_0 = I_{2n}, \quad |CM_0+D| \neq 0;$$

$$w = \frac{1}{2}(z + \hat{z}), \quad \phi = \phi'(w,t);$$

$$\hat{w} = J(z - \hat{z}) = \phi_w, \quad \phi_t = H(w - \frac{1}{2} J \phi_w)$$

are a new type of generating functions and H-J equations, not encountered in the classical literature.

By recursions we can determine explicitly all possible time-dependent generating functions for analytic Hamiltonians [6].

<u>Theorem 6.</u> Let $H(z)$ depend analytically on z. Then $\phi(w,t)$ in Theorem 5 can be expressed as convergent power series in t for sufficiently small $|t|$:

$$\phi(w,t) = \sum_{k=0}^{\infty} \phi^{(k)}(w)t^k,$$

$$\phi^{(0)}(w) = \frac{1}{2} w' N_0 w, \quad N_0 = (AM_0 + B)(CM_0 + D)^{-1},$$

$$\phi^{(1)}(w) = -\mu H(E_0 w), \quad E_0 = (CM_0 + D)^{-1}, \tag{3.4}$$

$$k \geq 1: \quad \phi^{(k+1)}(w) = \frac{-1}{k+1} \sum_{m=1}^{k} \frac{\mu}{m!} \sum_{i_1,\dots,i_m=1}^{2n} H_{z_{i_1}\dots z_{i_m}}(E_0 w)$$

$$\times \sum_{\substack{k_1+\dots+k_m=k \\ k_j \geq 1}} (C_1 \phi_w^{(k_1)}(w))_{i_1} \cdots (C_1 \phi_w^{(k_m)}(w))_{i_m}.$$

Generating functions play the central role for the construction of canonical difference schemes for Hamiltonian systems. The general methodology for the latter is as follows: Choose some suitable type of generating function (Theorem 5) with its explicit expression (Theorem 6). Truncate or approximate it in some way and take gradient of this approximate generating function. Then we get automatically the implicit representation of some canonical transformation for the transition of the difference scheme. In this way one can get an abundance of canonical difference schemes. This methodology is unconventional in the ordinary sense, but natural from the point of view of symplectic geometry [6]. As an illustration we construct a family of canonical difference schemes of arbitrary order from the truncations of the Taylor series of the generating functions for each choice of $T \in CSp(\tilde{J}_{4n}, J_{4n})$ and $M_0 \in Sp(2n)$ satisfying (3.3).

Theorem 7. Using Theorems 5 and 6, for sufficiently small $\tau > 0$ as the time-step, we define

$$\psi^{(m)}(w, \tau) = \sum_{k=0}^{m} \phi^{(k)}(w)\tau^k, \quad m = 1,2,\cdots. \tag{3.5}$$

Then the gradient transformation

$$w \to \hat{w} = \psi_w^{(m)}(w, \tau) \tag{3.6}$$

with Jacobian $N^{(m)}(w, \tau) \in Sm(2m)$ satisfies

$$|N^{(m)} C - A| \neq 0 \tag{3.7}$$

and defines implicitly a canonical difference scheme $\hat{z} = z^k \to z^{k+1} = z$ of m-th order accuracy upon substitution

$$\hat{w} = Az^k + Bz^{k+1}, \quad w = Cz^k + Dz^{k+1}. \tag{3.8}$$

For the special case of type (III), the generating function $\phi(w,t)$ is odd in t. Then Theorem 7 leads to a family of canonical difference schemes of arbitrary even order accuracy, generalizing the centered Euler scheme, as follows.

Theorem 8. Using theorems 5 and 6, for sufficiently small $\tau > 0$ as the time-step, we define

$$\psi^{(2m)}(w, \tau) = \sum_{k=1}^{m} \phi^{(2k-1)}(w)\tau^{2k-1}, \quad m=1,2,\cdots. \tag{3.9}$$

Then the gradient transformation

$$w \to \hat{w} = \psi_w^{(2m)}(w, \tau) \tag{3.10}$$

with Jacobian $N^{(2m)}(w, \tau) \in Sm(2n)$ satisfies

$$|N^{(2m)} C - A| \neq 0$$

and defines implicitly a canonical difference scheme $\hat{z} = z^k \to z^{k+1} = z$ of 2m-th order accuracy upon substitution (3.8). The case m=1 is the centered Euler scheme (2.6).

For linear canonical system (2.1), (2.2) the type (III) generating function is the quadratic form

$$\phi(w,\tau) = \frac{1}{2}w' \; (2J \tanh(\frac{\tau}{2} L)) \qquad L = J^{-1}S, \quad S' = S, \tag{3.11}$$

where

$$\tanh \lambda = \lambda - \frac{1}{3}\lambda^3 + \frac{2}{15}\lambda^5 - \frac{17}{312}\lambda^7 + \cdots = \sum_{k=1}^{\infty} a_{2k-1}\lambda^{2k-1}.$$

$$a_{2k-1} = 2^{2k}(2^{2k}-1)B_{2n}/(2k)!, \quad B_{2k} \text{ — Bernoulli numbers},$$

$$2J \tanh(\frac{\tau}{2} L) \in Sm(2n).$$

(3.10) becomes symplectic difference schemes

$$z^{k+1} - z^k = (\sum_{k=1}^{m} a_{2k-1}(\frac{\tau}{2} L)^{2k-1}) \; (z^{k+1} + z^k). \tag{3.12}$$

The case m=1 is the centered Euler scheme (2.3).
Several specific S-schemes up to fourth order accuracy are given in Appendix 1.

§4 S-scheme for Infinite Dimensional Hamiltonian System

1. An Infinite Dimensional Hamiltonian Equation

Suppose B is a reflexive Banach space and B* its dual. E^n is an Euclidean space and n its dimension. The generalized coordinate in the Banach space is function $q(r,t)$: $E^n \times R \to R$, $\forall t \in R$. We have $q(r,t) \in B$. B corresponds to the configuration space. We introduce $p(r,t)$, generalized momentum, where $r \in E^n$, $t \in R$. For $\forall t \in R$, $p(r,t) \in B^*$. B* corresponds to momentum space. $B \times B^*$ is the phase space.

Let H be a functional in Hamiltonian mechanics. We have the Hamiltonian equation in $B \times B^*$:

$$\frac{dp}{dt} = -\frac{\delta H}{\delta q}(q,p,t),$$
$$\frac{dq}{dt} = \frac{\delta H}{\delta p}(q,p,t), \tag{4.1}$$

where $\frac{\delta}{\delta q}, \frac{\delta}{\delta p}$ are functional derivatives.

We denote $z=(q,p) \in B \times B^*$, then we get formally the form of (4.1)

$$\frac{dz}{dt} = \begin{bmatrix} 0 & I \\ -I & 0 \end{bmatrix} H_z .$$

We define the operator J as

$$J: \ B\times B^* \to (B\times B^*)^* = B^*\times B$$

$$J(q,p) = (-p,q)$$

then it is easy to prove that $J \in GL((B\times B^*), (B\times B^*)^*)$.

$$J = \begin{bmatrix} 0 & -I_{B^*} \\ I_B & 0 \end{bmatrix} \qquad J^{-1} = \begin{bmatrix} 0 & I_B \\ -I_{B^*} & 0 \end{bmatrix},$$

I_B: $B \to B$ identity operator,

I_{B^*}: $B^* \to B^*$ identity operator,

we write them breifly as I.

Then the H-system has the form

$$\frac{dz}{dt} = J^{-1} H_z, \tag{4.2}$$

where $H_z = (\frac{\delta H}{\delta q}, \frac{\delta H}{\delta p})$, with canonical 2-form:

$$\omega(\alpha_1, \alpha_2) = \langle J\alpha_1, \alpha_2 \rangle$$

where $\alpha_1, \alpha_2 \in B\times B^*$, and $\langle\,,\,\rangle$ is the dual product .

2. Generating Functional and Difference Schemes

Suppose $f: \ B\times B^* \to (B\times B^*)^*$

$$\hat{w} = f(w)$$

We call f a gradient transformation, or potential operator, if there is ϕ : $B\times B^* \to R$ such that:

$$\hat{w} = \phi_w(w)$$

Let g: $B\times B^* \to B\times B^*$ be a canonical transformation

$$z \to \hat{z} = g(z)$$

$T \in GL((B\times B^*)\times(B\times B^*),\ (B\times B^*)^*\times(B\times B^*))$

$$T = \begin{bmatrix} A & B \\ C & D \end{bmatrix}$$

in which $A, B \in GL\,(B\times B^*, (B\times B^*)^*)$, $C, D \in GL\,(B\times B^*, B\times B^*)$.

Lemma [17] If the following conditions are satisfied:

1. F is an operator: $E \to E^*$, E is a reflexive Banach space.

2. F has Gateaux differential $DF(x,h)$ in U which belongs to E and is defined by $|x - x_0| < r$.

3. Functional $\langle DF(x,h_1),h_2\rangle$ is continuous in U. Then F is a potential operator iff $\langle DF(x,h_1),h_2\rangle$ is symmetric for $\forall x \in U$:

$$\langle DF(x,h_1),h_2\rangle = \langle DF(x,h_2),h_1\rangle, \quad \forall\ h_1, h_2 \in E.$$

Theorem3': T carries every symplectic transformtion into gradient transformation, if:

$$T^{*}\begin{bmatrix}0 & I\\ -I & 0\end{bmatrix}T = \mu\begin{bmatrix}-J & 0\\ 0 & J\end{bmatrix}$$

$$\text{i.e.}\quad A_1 = -\mu^{-1} J C^{*},\qquad B_1 = \mu^{-1} JA^{*},$$
$$C_1 = \mu^{-1} J A^{*},\qquad D_1 = -\mu^{-1} J B^{*}, \tag{4.3}$$

$$T = \begin{bmatrix}A & B\\ C & D\end{bmatrix},\quad T^{-1} = \begin{bmatrix}A_1 & B_1\\ C_1 & D_1\end{bmatrix}.$$

With the aid of the above lemma, like the finite dimensional case, we have

Theorem 5': Let T be defined as above. $z \to \hat{z}=g(z,t)$ is a canonical transformation. $z(t)$ is the solution of (4.2).

$$M(z,t) = g_z(z,t).$$

1) $g(*, 0)$ is a linear canonical transformation,

$M(z, 0) = M_0$ is independent of z.

2) $g^{-1}(z, 0)\, g(*, t)$ is a canonical transformation dependent on t, and carries $z(t)$ into $z(0)$.

3) $CM_0 + D$ is nonsingular,

then there is a gradient transformation dependent on t: $w \to \hat{w} = f(w,t)$ and a generating functional $\phi(w, t)$ for sufficiently samll $|t|$, such that:

(1) $[\hat{w} - f(w,t)]_{\hat{w} = A\hat{z}+Bz,\ w=C\hat{z}+Dz} = 0$

(2) $\phi_t(w,t) = -\mu H(C_1 \phi_w(w,t) + D_1 w)\big|_{w=C\hat{z} + Dz}.$ (4.4)

Like the finite dimensional case, we can determine by recursion all possible time-dependent generating functionals for Hamiltonians $H(z)$ analytic in z.

$$\phi(w, t) = \sum_{k=0}^{\infty} \phi^{(k)}(w),$$

$$\phi^{(0)}(w,t) = \frac{1}{2}\langle w^{*}, N_0 \cdot w\rangle,\quad N_0 = (AM_0+B)(CM_0+D)^{-1},$$

$$\phi^{(1)}(w) = -\mu\, H(E_0 w),\quad E_0 = (CM_0+D)^{-1},$$

$$k \geq 1:\quad \phi^{(k+1)}(w) = -\frac{1}{k+1}\sum_{m=1} \frac{\mu}{m!} \sum_{k_1+k_2+\cdots+k_m=k} H^{(m)}(E_0 w) \tag{4.5}$$
$$\times (C_1 \phi_w^{(k_1)}) \cdots (C_1 \phi_w^{(k_m)}(w)).$$

Remark. All conclusion concerning construction of S-scheme via generating functional is true when the J-symplectic form is replaced by K-symplectic form in finite or infinite dimensional case.

Remark. When we take $B=R^n$, $q=(q_1,\cdots,q_n)\in R^n$, $p=(p_1,\ldots,p_n)\in (R^n)^{*}=R^n$, equations (4.1) become the canonical Hamiltonian equations (1.1). All the conclusion of the finite dimensional case can be regarded as a special case of infinite dimensional case.

Remark. Using above method, we get a semi-discrete S-scheme (involving time discretization only), in other words, we obtained an ∞-dim symplectic transformation in space $B \times B^*$. In order to obtain a fully discrete S-scheme in $R^n \times R^n$, we must approximate spatial derivative by central difference. As we have done in §6, a fully discrete S-scheme for wave equation is obtained.

Another way is a method of lines: discretizing spatial derivatives, then we get a system of Hamiltonian ODE's. For fully discrete S-scheme, we use methods of constructing S-scheme for finite dimensional H-system [1] [25].

§5 Multi-level S-schemes

We first consider an LH-system

$$\frac{du}{dt} = J^{-1}Au , \tag{5.1}$$

where A is symmetric matrix. For this system we consider a three-level difference scheme

$$u^{n+1} - \phi_1 u^n - \phi_2 u^{n-1} = 0. \tag{5.2}$$

Introducing a new variable $v^n = u^{n-1}$, we obtain equivalent two-level scheme

$$\begin{bmatrix} u^{n+1} \\ v^{n+1} \end{bmatrix} = \begin{bmatrix} \phi_1 & \phi_2 \\ I & 0 \end{bmatrix} \begin{bmatrix} u^n \\ v^n \end{bmatrix} . \tag{5.3}$$

In this case a equivalent H-system is given by

$$\frac{d}{dt}\begin{bmatrix} u \\ v \end{bmatrix} = \begin{bmatrix} 0 & J^{-1} \\ J^{-1} & 0 \end{bmatrix} \begin{bmatrix} 0 & A \\ A & 0 \end{bmatrix} \begin{bmatrix} u \\ v \end{bmatrix} . \tag{5.4}$$

Rewrite it in the form

$$\frac{dz}{dt} = K^{-1} Az \tag{5.5}$$

where

$$K^{-1} = \begin{bmatrix} 0 & J^{-1} \\ J^{-1} & 0 \end{bmatrix} .$$

We call scheme (5.2) a S-scheme, if scheme (5.3) is K-symplectic, i.e.

$$\begin{bmatrix} \phi_1^T & I_n \\ \phi_2^T & 0 \end{bmatrix} \begin{bmatrix} 0 & J \\ J & 0 \end{bmatrix} \begin{bmatrix} \phi_1 & \phi_2 \\ I_n & 0 \end{bmatrix} = \begin{bmatrix} 0 & J \\ J & 0 \end{bmatrix} . \tag{5.6}$$

After a short computation, we can prove

Lemma 5.1[7]. Scheme (5.3) is a S-scheme, iff ϕ_1 is an infinitesimal symplectic matrix, $\phi_2 = I$.

Theorem 9. Scheme

$$u^{n+1} = u^{n-1} + 2sh(\tau J^{-1}A)u^n \tag{5.7}$$

is a S-scheme with any order of accuracy, where

$$sh(\lambda) = \lambda + \frac{\lambda^3}{3!} + \frac{\lambda^5}{5!} + \cdots + \frac{\lambda^{2m-1}}{(2m-1)!} + \cdots .$$

m=1: It is the usual leap-frog scheme of second order

$$u^{n+1} = u^{n-1} + 2\tau J^{-1} A u^n. \tag{5.8}$$

m=2: It is the generalized leap-frog scheme of fourth order

$$u^{n+1} = u^{n-1} + 2\tau J^{-1} A u^n + \frac{\tau^3}{3!}(J^{-1}A)^3 u^n. \tag{5.9}$$

Leap-Frog scheme for LH-Systems can be generalized to the NLH-System(1.2). Consider a three-level scheme

$$z^{n+1} - \phi_1(z^n) - \phi_2(z^{n-1}) = 0 \tag{5.10}$$

Introducing a new variable $v^n = z^{n-1}$, then

$$T: \quad \begin{aligned} z^{n+1} &= \phi_1(z^n) + \phi_2(v^n) \\ v^{n+1} &= z^n \end{aligned} \tag{5.10'}$$

We say (5.10) is a S-scheme if the Jacobian

$$dT = \frac{\partial(z^{n+1}, v^{n+1})}{\partial(z^n, v^n)} \quad \text{satisfies}$$

$$(dT)' \begin{bmatrix} 0 & J^{-1} \\ J^{-1} & 0 \end{bmatrix} (dT) = \begin{bmatrix} 0 & J^{-1} \\ J^{-1} & 0 \end{bmatrix}.$$

Lemma 5.1'. Scheme (5.10) is a S-scheme, iff $\frac{\partial \phi_1}{\partial z^n}$ is an infinitesimal symplectic matrix, $\frac{\partial \phi_2}{\partial z^n} = I$.

For example, the leap-frog scheme is a S-scheme:

$$z^{n+1} = z^{n-1} - 2\tau J^{-1} H_z(z^n).$$

As application we consider two forms of H-System for wave equation $w_{tt} = w_{xx}$. The first form is to take its H-Functional as

$$H(v,u) = \frac{1}{2}\int (v^2 + u_x^2)dx, \text{ where } u=w,\ v=w_t.$$

We write the H-system

$$\frac{dz}{dt} = J^{-1}Az, \tag{5.11}$$

where $J^{-1}A = \begin{bmatrix} 0 & \frac{\partial^2}{\partial x^2} \\ 1 & 0 \end{bmatrix}$, $z = \begin{bmatrix} v \\ u \end{bmatrix}$, $A = \begin{bmatrix} 1 & 0 \\ 0 & -\frac{\partial^2}{\partial x^2} \end{bmatrix}$.

Another form is to put $v=w_t$, $u=w_x$,

$$\frac{dz}{dt} = \mathcal{D}\frac{\delta H}{\delta z}, \qquad H(v,u) = \frac{1}{2}\int (v^2 + u^2)dx, \tag{5.12}$$

where $\mathcal{D} = \begin{bmatrix} 0 & \frac{\partial}{\partial x} \\ \frac{\partial}{\partial x} & 0 \end{bmatrix}$, obviously it is a skew-adjoint operator.

Rewrite it in form

$$\frac{dz}{dt} = K^{-1}Az, \tag{5.13}$$

where

$$K^{-1}A = \begin{bmatrix} 0 & \frac{\partial}{\partial x} \\ \frac{\partial}{\partial x} & 0 \end{bmatrix}, \qquad A = \begin{pmatrix} I & 0 \\ 0 & I \end{pmatrix}, \qquad K^{-1} = \mathscr{D}.$$

Let

$$\Delta_2 u_m = \frac{u(m+1)-2u(m)+u(m-1)}{\Delta x^2} \tag{5.14}$$

$$\Delta_4 u_m = \frac{-u(m+2)+16u(m+1)-30u(m)+16u(m-1)-u(m-2)}{12\,\Delta x^2} \tag{5.15}$$

$$\nabla_2 u_m = \frac{u(m+1)-u(m-1)}{2\Delta x} \tag{5.16}$$

$$\nabla_4 u_m = \frac{-u(m+2)+8u(m+1)-8u(m-1)+u(m-2)}{12\,\Delta x} \tag{5.17}$$

For first kind of equation(5.11), the schemes (5.8),(5.9) have forms respectively.

$$\begin{bmatrix} v^{n+1} \\ u^{n+1} \end{bmatrix} = \begin{bmatrix} v^{n-1} \\ u^{n-1} \end{bmatrix} + 2\,\Delta t \begin{bmatrix} 0 & M \\ I & 0 \end{bmatrix} \begin{bmatrix} v^{n} \\ u^{n} \end{bmatrix}, \tag{5.18}$$

$$\begin{bmatrix} v^{n+1} \\ u^{n+1} \end{bmatrix} = \begin{bmatrix} v^{n-1} \\ u^{n-1} \end{bmatrix} + 2\Delta t \begin{bmatrix} 0 & M \\ I & 0 \end{bmatrix} \begin{bmatrix} v^{n} \\ u^{n} \end{bmatrix} + \frac{\Delta t^3}{3} \begin{bmatrix} 0 & M \\ I & 0 \end{bmatrix}^3 \begin{bmatrix} v^{n} \\ u^{n} \end{bmatrix}. \tag{5.19}$$

Here I is $n \times n$ identity matrix, M is $n \times n$ matrix. According to(5.14), (5.15) it may be put in two forms

$$M_1: \quad \frac{1}{\Delta x^2} \begin{bmatrix} -2 & 1 & 0 & \cdots & 1 \\ 1 & -2 & 1 & \cdots & 0 \\ & \ddots & \ddots & \ddots & \\ & & \ddots & \ddots & 1 \\ 1 & 0 & \cdots & 1 & -2 \end{bmatrix}$$

$$M_2: \quad \frac{1}{12\Delta x^2} \begin{bmatrix} -30 & 16 & -1 & \cdots\cdots & -1 & 16 \\ 16 & -30 & 16 & -1 & \cdots\cdots & -1 \\ -1 & 16 & -30 & 16 & -1 \cdots\cdots & 0 \\ & \ddots & \ddots & \ddots & \ddots & \ddots \\ & & \ddots & \ddots & \ddots & -1 \\ -1 & & & \ddots & \ddots & 16 \\ +16 & -1 & 0 & -1 & 16 & -30 \end{bmatrix}.$$

In scheme (5.18) if $M=M_1$ we express this scheme with accuracy $0(\Delta t^2+\Delta x^2)$ by SLFM (2,2) and if $M=M_2$ we express it by SLFM(2,4).

In scheme (5.19), if $M=M_1$ we express it by SLFM (4,2) and if $M=M_2$ we express it by SLFM (4.4).

For second kind of equation (5.13), the schemes (5.8), (5.9) become

$$\begin{bmatrix} v^{n+1} \\ u^{n+1} \end{bmatrix} = \begin{bmatrix} v^{n-1} \\ u^{n-1} \end{bmatrix} + 2\Delta\, t \begin{bmatrix} 0 & M \\ M & 0 \end{bmatrix} \begin{bmatrix} v^{n} \\ u^{n} \end{bmatrix},$$

$$\begin{bmatrix} v^{n+1} \\ u^{n+1} \end{bmatrix} = \begin{bmatrix} v^{n-1} \\ u^{n-1} \end{bmatrix} + 2\Delta t \begin{bmatrix} 0 & M \\ M & 0 \end{bmatrix} \begin{bmatrix} v^{n} \\ u^{n} \end{bmatrix} + \frac{\Delta t^3}{3} \begin{bmatrix} 0 & M \\ M & 0 \end{bmatrix}^3 \begin{bmatrix} v^{n} \\ u^{n} \end{bmatrix} .$$

Here M is n × n matrix, according to (5.16) (5.17) they are

$$M_3: \quad \frac{1}{2\Delta x} \begin{bmatrix} 0 & 1 & \cdots & \cdots & -1 \\ -1 & 0 & 1 & \cdots & 0 \\ & \ddots & \ddots & \ddots & 0 \\ 0 & & \ddots & \ddots & 1 \\ 1 & 0 & \cdots & -1 & 0 \end{bmatrix}$$

$$M_4: \quad \frac{1}{12\Delta x} \begin{bmatrix} 0 & 8 & -1 & \cdots & \cdots & 1 & -8 \\ -8 & 0 & 8 & -1 & & 0 & 1 \\ 1 & -8 & 0 & 8 & -1 & 0 & \\ & \ddots & \ddots & \ddots & \ddots & \ddots & -1 \\ & & \ddots & \ddots & \ddots & \ddots & 8 \\ -1 & 0 & & & & & \\ 8 & -1 & & 1 & -8 & & 0 \end{bmatrix}$$

In scheme (5.18), if $M=M_3$ we express it by KLFM (2,2): and if $M=M_4$ we express it by KLFM (2,2), and if $M=M_4$ we express it by KLFM (2,4).

In scheme (5.19), if $M=M_3$ we express it by KLFM (4,2) and if $M=M_4$ we express it by KLFM (4,4).

<u>Lemma</u>: Eigenvalues of matrices

$$\begin{bmatrix} 0 & M_1 \\ I & 0 \end{bmatrix}, \begin{bmatrix} 0 & M_2 \\ I & 0 \end{bmatrix}, \begin{bmatrix} 0 & M_3 \\ M_3 & 0 \end{bmatrix}, \begin{bmatrix} 0 & M_4 \\ M_4 & 0 \end{bmatrix} \quad \text{are}$$

$$\mu^{(1)} = \pm \frac{2i}{x} \sin\left(\frac{\pi k}{2N}\right), \quad \mu_k^{(2)} = \pm \frac{i}{\Delta x} \sqrt{4\sin^2 \frac{\pi k}{2N} + \frac{4}{3} \sin^4 \frac{\pi k}{2N}}$$

$$\mu_k^{(3)} = \pm \frac{i}{\Delta x} \sin\left(\frac{\pi k}{2N}\right), \quad \mu_k^{(4)} = \pm \frac{i}{\Delta x} \left(\frac{1}{6} \sin \frac{2\pi k}{2N} - \frac{8}{6} \sin \frac{\pi k}{2N}\right)$$

$k = 0,1,\cdots, 2N-1$, respectively.

Above eigenvalues and criterions in [20] are used to derive the stability conditions in the following Table 1.

Using above scheme, numerical solutions of problem (6.4) have been done, we will list the numerical results and test of stability in Appendix 2 Table (2-5).

<u>Remark</u>. Under the condition that the eigenvalues of infinitesimally symplectic matrices $J^{-1}A$ are purely imaginary the region of stability of the generalized leap-frog scheme (5.6) tends to the full real axis as the orders of accuracy increase to infinity [10].

Table 1

H-eq.	scheme	error of approximation	stability condition
$\frac{d}{dt}\begin{pmatrix} v \\ u \end{pmatrix}=\begin{pmatrix} 0 & -1 \\ 1 & 0 \end{pmatrix}\begin{pmatrix} v \\ -u_{xx} \end{pmatrix}$	SLFM(2,2)	$o(\Delta t^2+\Delta x^2)$	$\Delta t/\Delta x \leq \frac{1}{2}$
	SLFM(2,4)	$o(\Delta t^2+\Delta x^4)$	$\Delta t/\Delta x \leq 0.4330$
	SLFM(4,2)	$o(\Delta t^4+\Delta x^2)$	$\Delta t/\Delta x \leq 1.4237$
	SLFM(4,4)	$o(\Delta t^4+\Delta x^4)$	$\Delta t/\Delta x \leq 1.2330$
$\frac{d}{dt}\begin{pmatrix} v \\ u \end{pmatrix}=\begin{pmatrix} 0 & \frac{\partial}{\partial x} \\ \frac{\partial}{\partial x} & 0 \end{pmatrix}\begin{pmatrix} v \\ u \end{pmatrix}$	KLFM(2,2)	$o(\Delta x^2+\Delta x^2)$	$\Delta t/\Delta x \leq 1$
	KLFM(2,4)	$o(\Delta t^2+\Delta x^4)$	$\Delta t/\Delta x \leq 0.7287$
	KLFM(4,2)	$o(\Delta t^4+\Delta x^2)$	$\Delta t/\Delta x \leq 2.8473$
	KLFM(4,4)	$o(\Delta t^4+\Delta x^4)$	$\Delta t/\Delta x \leq 2.0750$

§6 Numerical Examples

1. Two-body problem.

In celestial mechanics, simplest and famous example is two-body problem. We outline its arguments here

(i) $M = R^2 \times \{R^2 - (0)\}$, phase space.

(ii) $m \in M$. initial data.

(iii) Hamiltonian, $H(p,q) = \|p\|^2/2 - 1/\|q\|$.

It has been proved that if angular momentum $G \neq 0$ the path is an ellipse, parabola, or hyperbola according to $E < 0$, $E=0$, $E > 0$. The case $G=0$ is degenerate, the path is a straight line.

We choose (A-III-4)

$$z^{n+1}=z^n + \tau J^{-1}H_z\left(\frac{z^{n+1}+z^n}{2}\right)+ \frac{\tau^3}{24}\left((J^{-1}H_z)'H_{zz}(J^{-1}H_z)\right)_z\left(\frac{z^{n+1}+z^n}{2}\right), \tag{6.1}$$

where $z^n = \begin{pmatrix} p \\ q \end{pmatrix}^n$, τ is step length. This scheme is fourth order. Because the scheme is symplectic , we keep $G= q_1p_2 - p_1q_2$ constant , i.e., keep the path's curvature unchanged. We have computed all the kinds of paths. The paths are the same as the theory predicts. In elliptic case, the R-K method cannot be carried through. As a matter of fact, it will overflow after short time. Because scheme (6.1) is A-stable, it gives good result in all cases. In Appendix-2 Table 6 we will list comparison data between R-K method and S-schemes (fourth order and second order).

2. Wave equation.

Let $B = H^1$, $P = H^1 \times H^{-1}$.

In $H^2 \times H^1$, a subspace of P, we have the Hamiltonian functional

$$H(u, v) = \int_{R^2} \frac{1}{2} (v^2 + (\nabla u)^2)\, dx .$$

The corresponding Hamiltonian vector field is

$$(v, \Delta u).$$

The associated H-equation is

$$\frac{dv}{dt} = -\frac{\delta H}{\delta u} = \Delta u, \qquad \frac{du}{dt} = \frac{\delta H}{\delta v} = v . \tag{6.2}$$

Choosing $T = \begin{bmatrix} -J & +J \\ \frac{1}{2}I & \frac{1}{2}I \end{bmatrix}$, we get a S-scheme

(A-III-4) of fourth order for wave equation [8]

$$z^{n+1} = z^n + \tau J^{-1} H_z - \frac{2\tau^3}{4!} J^{-1} H_{zz} J^{-1} H_{zz} J^{-1} H_z , \tag{6.3}$$

e.g.

$$\begin{bmatrix} v \\ u \end{bmatrix}^{n+1} = \begin{bmatrix} v \\ u \end{bmatrix}^{n} + \tau \begin{bmatrix} 0 & \Delta \\ I & 0 \end{bmatrix} \begin{bmatrix} \bar{v} \\ \bar{u} \end{bmatrix} - \frac{\tau^3}{12} \begin{bmatrix} 0 & \Delta \\ I & 0 \end{bmatrix}^3 \begin{bmatrix} \bar{v} \\ \bar{u} \end{bmatrix},$$

where $\bar{v} = \frac{1}{2}(v^{n+1} + v^n)$, $\bar{u} = \frac{1}{2}(u^{n+1} + u^n)$.

If we take Δ as the fourth-order centered difference approximation, using periodic boundary condition we get

$$z^{n+1} = z^n + \tau B\bar{z} - \frac{\tau^3}{12} B^3 \bar{z}.$$

where $B = \begin{bmatrix} 0 & M \\ I & 0 \end{bmatrix}$, $M = \frac{1}{12h^2} \begin{bmatrix} -30 & 16 & -1 & 0 & 0 & -1 & 16 \\ 16 & -30 & 16 & -1 & 0 & 0 & -1 \\ -1 & 16 & -30 & 16 & -1 & 0 & 0 \\ & \cdot & \cdot & \cdot & & & \\ & & \cdot & \cdot & \cdot & & \\ & & & \cdot & \cdot & \cdot & \\ -1 & 0 & & & \cdot & \cdot & \cdot \\ 16 & -1 & & & -1 & 16 & -30 \end{bmatrix}$

$$\bar{z} = \frac{z^{n+1} + z^n}{z} .$$

τ is the time step length, h is the space step length, M is an $N \times N$ matrix, I is the $N \times N$ unit matrix, $h = \frac{2\pi}{N}$.

Computations were done. The initial and boundary conditions are taken to be

$$u(0,x) = \sin x \qquad \text{for } 0 \le x \le 2\pi ,$$
$$u_t(0,x) = \cos x \qquad \text{for } 0 \le x \le 2\pi , \tag{6.4}$$
$$u(t,0) = u(t,2\pi).$$

The exact solution is $\sin(x + t) = u(t,x)$.

We list the numerical and corresponding exact results in Appendix 2 table 7.

§7 S-schemes and Chaos

Most systems with two or more degrees of freedom have some chaotic motions and some regular motions. The regular motions occur then the system is far away from resonances of low order. This is in accord with the conclusion of the KAM theorem. On the other hand, when the system is near such resonances, chaotic motion occurs.

Hénon and Heiles [21] considered a system with two degrees of freedom with coordinates $q_1(t)$ and $q_2(t)$ and conjugate momenta $p_1(t)$ and $p_2(t)$. Their Hamiltonian is

$$H(q_1,q_2,p_1,p_2) = \frac{1}{2}(p_1^2+p_2^2)+ \frac{1}{2}(q_1^2+q_2^2+2q_1^2q_2- \frac{2}{3} q_2^3). \tag{7.1}$$

They found that for total energy E less than 1/12, the solutions were regular for most initial conditions. As E increased the fraction of initial condition leading to chaotic solutions increased.

For (7.1), its canonical equations are

$$\frac{d}{dt}\begin{bmatrix} p_1 \\ p_2 \\ q_1 \\ q_2 \end{bmatrix} = \begin{bmatrix} 0 & 0 & 1 & 0 \\ 0 & 0 & 0 & 1 \\ -1 & 0 & 0 & 0 \\ 0 & -1 & 0 & 0 \end{bmatrix}^{-1} \begin{bmatrix} p_1 \\ p_2 \\ q_1+2q_1q_2 \\ q_2+q_1^2-q_2^2 \end{bmatrix}. \tag{7.2}$$

Using the explicit scheme (A-II-1), we obtain

$$\begin{aligned} p_1^{n+1} &= p_1^n - \tau(q_1^n + 2q_1^nq_2^n), \\ p_2^{n+1} &= p_2^n - \tau(q_2^n + (q_1^2)^n - (q_2^2)^n), \\ q_1^{n+1} &= q_1^n + \tau\, q_1^{n+1}, \\ q_2^{n+1} &= q_2^n + \tau\, p_2^{n+1}. \end{aligned} \tag{7.3}$$

Obviously this scheme is a symplectic transformation, approximating original system in second order precision, of course, it is regarded as an area-preserving mapping. As the time step we selected $\tau = 0.0005$ for 20000 cycles, the CPU time was 10 minutes (IBM-4341). The results obtained are in agreement with the results of Hénon-Heiles (1964).

In Appendix 2 Fig. 1, we list the graphics of surfaces of section computed with scheme (7.3) for E(or H) = 1/24, 1/12, 1/8, 1/6. Good results are also obtained by the implicit S-scheme(fourth order) (A-III-4).

Hénon [22] accepted Kruskal's suggestion [21] to study an area-preserving mapping directly.

$$T: \quad \begin{aligned} x_{n+1} &= 1 - ax_n^2 + y_n \\ y_{n+1} &= bx_n \end{aligned} \tag{7.4}$$

describes conservative system, dissipative system and logistic system when $|b| = 1$, $|b| < 1$ and $|b| = 0$ respectively .

In the case $b = \pm 1$, T is an area-preserving mapping and may be thought of as the Poincaré section for a H-system with two degrees of freedom.

Remark. The research of dynamical system with three degrees of freedom can be reduced to the study of a four dimensional mapping. Scheme (7.3) is an area-preserving mapping, therefore it is a good tool for studying dynamical system with three degrees of freedom.

Now we consider the equation

$$\frac{du}{dt} = u(1 - u) \tag{7.5}$$

with initial condition $u(0) = u_0$.

Using leap-frog scheme (5.8), we get a difference equation

$$\frac{u_{n+1} - u_{n-1}}{2h} = u_n(1 - u_n), \tag{7.6}$$

with initial condition $u(0) = u_0$ and $u_1 = u_0 + h\, u_0(1-u_0)$ computed by Euler's forward difference scheme. Using leap-frog method described in (5.10), we obtain

$$\begin{aligned} u_{n+1} &= 2h\, u_n(1 - u_n) + v_n \\ v_{n+1} &= u_n . \end{aligned} \tag{7.7}$$

Thus we have a mapping $\bar{T}: R^2 \to R^2$ defined by

$$\begin{aligned} X &= 2hx(1 - x) + y \\ Y &= x \end{aligned} \tag{7.8}$$

Scheme (7.6), in the sense of (5.6), is a S-scheme. The Jacobian matrix T is given by

$$d\bar{T} = \frac{\partial(X, Y)}{\partial(x, y)} = \begin{bmatrix} 2h(1-2x) & 1 \\ 1 & 0 \end{bmatrix}$$

Obviously $|d\bar{T}| = -1$, therefore it is an area-preserving scheme.

Let $u_{2m} = p_m$, $u_{2m-1} = q_m$. Rewrite the equation in the following form [9] [23]

$$S: \quad \begin{aligned} \frac{p_m - p_{m-1}}{2h} &= q_m(1 - q_m) \\ \frac{q_{m+1} - q_m}{2h} &= p_m(1 - p_m) \end{aligned} \tag{7.9}$$

The above system is H-system with $H(p,q) = \frac{1}{2}(p^2 - q^2) - \frac{1}{3}(p^3 - q^3)$. Symplectic transformation S(7.9) can be regarded as a composition transformation $\bar{T} \cdot \bar{T}$.

In general we must use the central difference for conservative systems. On the contrary, in order to describe a dissipative system we must employ mixed difference [24]

$$(1-\theta)\ \frac{u_{n+1}-u_{n-1}}{2h} + \theta\ \frac{u_{n+1}-u_n}{h} = u_n(1-u_n), \qquad 0\le\theta\le 1 , \tag{7.10}$$

or equivalently, a mapping

$$T_\theta : \quad \begin{aligned} u_{n+1} &= \frac{1-\theta}{1+\theta} v_n + \frac{2}{1+\theta} [(\theta+h)u_n - h\, u_n^2] \\ v_{n+1} &= u_n . \end{aligned} \tag{7.11}$$

Its Jacobian is

$$dT_\theta = \begin{pmatrix} 2(\theta+h-2hu_n)/(1+\theta) & (1-\theta)/(1+\theta) \\ 1 & 0 \end{pmatrix} .$$

Transform the mapping (7.11) by an affine transformation

$$u_n = \frac{\theta+h}{2h}\ [(\frac{h-\theta}{1+\theta})x_n + 1]$$

$$v_n = \frac{\theta+h}{2h}\ [(\frac{h-\theta}{1+\theta})\ y_n + 1]$$

into the form (7.4) with $a = (h^2 - \theta^2)/(1+\theta)^2$, $b=(1-\theta)/(1+\theta)$. In particular, we select $h=1.898$, $\theta = 0.538$ then the corresponding map (7.4) with $a=1.4$ and $b=1.3$ is a Hénon's strange attractor [24], see Fig.3(c) in Appendix 2.

The step length $h=1.898$ is too large as time step for numerical integration. But it is still significant in studying an area mapping.

Appendix 1

In this Appendix we collect S-schemes for H-system (1.2) with precision up to fourth order. According to method of construction of S-schemes described in §3, we have

$$\begin{aligned}
\phi^{(0)}(w) &= \frac{1}{2}\, w'\, N_0\, w, \\
\phi^{(1)}(w) &= -\mu\ H(E_0 w), \\
\phi^{(2)}(w) &= -\frac{\mu}{2}\ H_{z_i}\ (C_1\phi_w^{(1)})_i , \\
\phi^{(3)}(w) &= -\frac{\mu}{3}\ (\frac{1}{2!} H_{z_{i_1} z_{i_2}}\ (C_1\phi_w^{(1)})_{i_1} (C_1\phi_w^{(1)})_{i_2} + H_{z_i} (C_1\phi_w^{(2)})_i), \\
\phi^{(4)}(w) &= -\frac{\mu}{4}(\ \frac{1}{3!}\ H_{z_{i_1} z_{i_2} z_{i_3}} (C_1\phi_w^{(1)})_{i_1} (C_1\phi_w^{(1)})_{i_2} (C_1\phi_w^{(1)})_{i_3} \\
&\quad + H_{z_{i_1} z_{i_2}} (C_1\phi_w^{(2)})_{i_1} (C_1\phi_w^{(1)})_{i_2} + H_{z_i} (C_1\phi_w^{(3)})_i),
\end{aligned} \tag{A - 0}$$

In §3 case II, $M_0 = I_{2n}$, $\mu = 1$, by (3.1), we have

$$w = \begin{vmatrix} q \\ p \end{vmatrix}, \quad C_1 = \begin{bmatrix} 0 & 0 \\ 0 & -I \end{bmatrix}, \quad D_1 = \begin{bmatrix} 0 & I \\ 0 & 0 \end{bmatrix}, \quad \text{H-J equation is} \quad \phi_t = -H(p, -\phi_p) .$$

As matrices in theorem 6, N_0, E_0 are

$$N_0 = \begin{bmatrix} 0 & I \\ I & 0 \end{bmatrix}, \qquad E_0 = \begin{bmatrix} 0 & I \\ I & 0 \end{bmatrix}, \qquad E_0 w = \begin{bmatrix} p \\ \hat{q} \end{bmatrix}.$$

Substituting μ, N_0, E_0, C_1 into (A-O) yields

$$\phi^{(0)}(w) = \frac{1}{2} w' N_0 w = \frac{1}{2} \hat{q}_i p_i ,$$

where repeated subscripts i subject to Einstein summation convention.

$$\phi^{(1)}(w) = - H(p\hat{q}) ,$$

$$\phi^{(2)}(w) = - \frac{1}{2!} H_{q_i} H_{p_i} ,$$

$$\phi^{(3)}(w) = - \frac{1}{3!} (H_{q_i q_j} H_{p_i} H_{p_j} + H_{q_i p_j} H_{p_i} H_{q_j} + H_{p_i p_j} H_{q_i} H_{q_j}),$$

$$\begin{aligned} \phi^{(4)}(w) = - \frac{1}{4!} (&H_{q_i q_j q_k} H_{p_i} H_{p_j} H_{p_k} + H_{q_i q_j q_k} H_{p_i} H_{p_j} H_{q_k} \\ &+ H_{q_i p_j p_k} H_{p_i} H_{q_j} H_{q_k} + H_{p_i p_j p_k} H_{q_i} H_{q_j} H_{q_k} \\ &+ 5H_{q_i q_j} H_{p_j p_k} H_{p_i} H_{q_k} + H_{q_i p_j} H_{q_j p_k} H_{p_i} H_{q_k} \\ &+ 3H_{q_i p_j} H_{p_i p_k} H_{q_j} H_{q_k} + 3H_{q_i q_j} H_{q_k p_i} H_{p_j} H_{p_k}). \end{aligned}$$

Scheme of first order: Using (3.5), (3.6), we have

$$\psi^{(1)}(w, \tau) = \phi^{(0)}(w) + \phi^{(1)}(w)\tau$$

$$\hat{w} = \psi_w^{(1)}(w, \tau) = \phi_w^{(0)}(w) + \tau\phi_w^{(1)}(w)$$

$$\begin{bmatrix} -\hat{p} \\ -q \end{bmatrix} = - \begin{bmatrix} p \\ \hat{q} \end{bmatrix} + \tau \begin{bmatrix} -H_q(p,\hat{q}) \\ -H_p(p,\hat{q}) \end{bmatrix} ,$$

where $\hat{p} = p^0$, $\hat{q} = q^0$. Thus we obtain scheme of first order accuracy

$$\begin{aligned} p &= p^0 - \tau H_q(p, q^0), \\ q &= q^0 + \tau H_p(p, q^0). \end{aligned} \qquad \text{(A-II-1)}$$

Remark. This scheme will have 2-order accuracy when computing p in the n+1-th time step and q in the $n+1+\frac{1}{2}$-th step. Especially, for separable H-system it is the staggered explicit scheme (2.7).

By similar derivation, we have scheme of second order:

$$\psi^{(2)}(w, \tau) = \psi^{(1)}(w,\tau) + \tau^2 \phi^{(2)}(w),$$

$$\hat{w} = \psi_w^{(2)}(w, \tau) = \psi_w^{(1)}(w, \tau) + \tau^2 \phi_w^{(2)}(w).$$

$$p_i = p_i^0 - \tau H_{q_i}(p, q^0) - \frac{\tau^2}{2}\,[H_{q_j q_i} H_{p_j} + H_{q_j} H_{p_j q_i}],$$
$$q_i = q_i^0 + \tau H_{p_i}(p, q^0) + \frac{\tau^2}{2}\,[H_{q_j p_i} H_{p_j} + H_{q_j} H_{p_j p_i}]. \qquad \text{(A-II -2)}$$

Scheme of third order:

$$\psi^{(3)}(w, \tau) = \psi^{(2)}(w, \tau) + \tau^3 \phi^{(3)}(w),$$
$$\hat{w} = \psi^{(3)}_w(w, \tau) = \psi^{(2)}(w, \tau) + \tau^3 \phi^{(3)}_w(w).$$

$$\begin{aligned} p_i = p_i^0 &- \tau H_{q_i}(p, q^0) - \frac{\tau^2}{2!}(H_{q_j q_i} H_{p_j} + H_{p_j q_i} H_{p_j}) \\ &- \frac{\tau^3}{3!}[H_{q_k q_j q_i} H_{p_k} H_{p_j} + H_{q_k p_j q_i} H_{p_k} H_{q_j} + H_{p_k p_j q_i} H_{q_k} H_{q_j} \\ &+ H_{q_j p_k}(H_{p_j q_i} H_{q_k} + H_{p_j} H_{q_k q_i}) + 2H_{p_j p_k} H_{q_j q_i} H_{q_k} \\ &+ 2H_{q_j q_k} H_{p_j q_i} H_{p_k}], \end{aligned} \qquad \text{(A-II-3)}$$

$$\begin{aligned} q_i = q_i^0 &+ H_{p_i}(p, q^0) + \frac{\tau^2}{2!}(H_{q_j p_i} H_{p_j} + H_{q_j} H_{p_j p_i}) \\ &+ \frac{\tau^3}{3!}[H_{q_j p_k p_i} H_{p_j} H_{q_k} + H_{p_j p_k p_i} H_{q_j} H_{q_k} + H_{q_j q_k q_i} H_{p_j} H_{p_k} \\ &+ H_{q_j p_k}(H_{p_j p_i} H_{q_k} + H_{p_j} H_{q_k p_i}) + 2H_{p_j p_k} H_{q_j p_i} H_{q_k} \\ &+ 2H_{q_j q_k} H_{p_j p_i} H_{p_k}]. \end{aligned}$$

Scheme of fourth order:

$$\psi^{(4)}(w, \tau) = \psi^{(3)}(w, \tau) + \tau^4 \phi^{(4)}(w),$$
$$\hat{w} = \psi^{(4)}_w(w, \tau) = \psi^{(3)}_w(w, \tau) + \tau^4 \phi^{(4)}_w(w). \qquad \text{(A-II-4)}$$

In §3, case III, $\mu = -1$, $M_0 = I$, $w = \frac{1}{2}(z+\hat{z})$, $\hat{w} = J(z-\hat{z})$, by (3.1) we have $C_1 = \frac{1}{2} J^{-1} = \frac{1}{2}\begin{bmatrix} 0 & -I \\ I & 0 \end{bmatrix}$, $D_1 = \begin{bmatrix} I & 0 \\ 0 & I \end{bmatrix}$. H-J equation is $\phi_t = H(w - \frac{1}{2} J \phi_w)$,

$$N_0 = \begin{bmatrix} 0 & 0 \\ 0 & 0 \end{bmatrix}, \quad E_0 = \begin{bmatrix} I & 0 \\ 0 & I \end{bmatrix}, \quad E_0 w = \frac{1}{2}(z+\hat{z}) = \frac{1}{2}(z+z^0).$$

Substituting μ, N_0, E_0, C_1 into (A-O) yields

$$\phi^{(0)}(w) = 0,$$
$$\phi^{(1)}(w) = H(w),$$
$$\phi^{(2)}(w) = \frac{1}{2} H_z'\left(\frac{1}{2} J^{-1} H_z\right) = 0,$$

$$\phi^{(3)}(w) = \frac{1}{3} H_{z_i}(w)(C_1\phi_w^{(2)})_i + \frac{1}{6} H_{z_i z_j}(C_1\phi_w^{(1)})_i(C_1 \phi_w^{(1)})_j$$

$$= \frac{1}{6} H_{z_i z_j}(C_1 \phi_w^{(1)})_i (C_1\phi_w^{(1)})_j$$

$$= \frac{1}{4!} \{ H_{p_i p_j} H_{q_i} H_{q_j} - 2H_{p_i q_j} H_{p_i} H_{q_j} + H_{q_i q_j} H_{p_i} H_{p_j}\} , \quad \phi^{(4)}(w)=0.$$

By similar derivation as above, we have scheme of second order:

$$\psi^{(2)}(w, \tau) = \phi^{(1)}(w,\tau) + \tau^2\phi^{(2)}(w) = \psi^{(1)}(w,\tau) = \phi^{(0)}_{(w)} + \tau\phi^{(1)}(w),$$

$$\hat{w} = \psi_w^{(2)}(w,\tau) = \tau\phi_w^{(1)}(w) = \tau H_z(w).$$

i.e. $J(z - \hat{z}) = \tau H_z(\frac{z+\hat{z}}{2}),$

$$z = z^0 + \tau J^{-1} H_z(\frac{z+z^0}{2}). \tag{A-III-2'}$$

i.e.

$$p = p^0 - \tau H_q(\frac{p+p^0}{2}, \frac{q+q^0}{2}),$$

$$\text{(A-III-2)}$$

$$q = q^0 + \tau H_p(\frac{p+p^0}{2} , \frac{q+q^0}{2}).$$

Scheme of fourth order:

$$\hat{w} = \psi_w^{(4)}(w, \tau) = \tau\phi_w^{(1)}(w) + \tau^3\phi_w^{(3)}(w)$$

$$= \tau H_z(w) + \frac{\tau^3}{4!} [(J^{-1}H_z)' H_{zz}(J^{-1}H_z)]_z(w) , \tag{A-III-4'}$$

$$p_i = p_i^0 - \tau H_{q_i}(\frac{p+p^0}{2} , \frac{q+q^0}{2}) - \frac{\tau^3}{4!} \{ H_{p_j p_k q_i} H_{q_j} H_{q_k} + 2H_{p_j p_k} H_{q_j q_i} H_{q_k}$$

$$- 2H_{p_j q_k q_i} H_{p_j} H_{q_k} - 2H_{p_j q_k} H_{p_j q_i} H_{q_k} - 2H_{p_j q_k} H_{p_j} H_{q_k q_i}$$

$$+ 2H_{q_j q_k} H_{p_j q_i} H_{p_k} + H_{q_j q_k q_i} H_{p_j} H_{p_k} \} ,$$

$$q_i = q_i^0 + \tau H_{p_i} (\frac{p+p^0}{2}, \frac{q+q^0}{2}) + \frac{\tau^3}{4!} \{ H_{p_j p_k p_i} H_{q_j} H_{q_k} + 2H_{p_j p_k} H_{q_j p_i} H_{q_k}$$

$$-2H_{p_j q_k p_i} H_{p_j} H_{q_k} - 2H_{p_j q_k} H_{p_j p_i} H_{q_k} - 2H_{p_j q_k} H_{p_j} H_{q_k p_i}$$

$$+ H_{q_j q_k p_i} H_{p_j} H_{p_k} + 2H_{q_j q_k} H_{p_j p_i} H_{p_k} \} . \tag{A-III-4}$$

$$i = 1,2, \cdots, n$$

Appendix 2

This Appendix consists of Tables 2-7 and Figures 1-3, which give some computer results.

Table 2. Comparison of the exact solutions with the numerical solutions of eq. (5.11) with (6.4) and of (5.12) with initial data $u(0,x)=v(0,x)=\sin x$, using different leap-frog schemes, where $u(n,m)$ denotes value of $u(t,x)$ at $t=nDt$, $x=mDx$. Courant number $C=Dt/Dx=0.4$, $Dx=\frac{2}{40}$.

Table 3. Similar results as in Table 2, but only with schemes, SLFM(4,4) and KLFM (4,4) under C=1.1.

Table 4. Numerical tests of stability of scheme KLFM(4,4) with C=2.09 and 2.07, (critical C=2.075).

Table 5. Numerical tests of stability of scheme SLFM (4,4) with C=1.23 and 1.24, (critical C=1.233).

Table 6. Comparison of numerical results for 2-body problem by 5-6-th order R-K scheme and by 4-th and 2-nd order S-schemes. This shows that lower order S-schemes are comparable in precision with higher order R-K schemes.

Table 7. Comparison of the exact solution with the numerical results of eq. (5.11) with (6.4) using the scheme (6.3) adapted from A-III-4 scheme.

Fig. 1. Intersections of the orbits of Hénon-Heiles system with the Poincare section, computed by 2-nd order S-schemes. (A): $H=\frac{1}{24}$, all motions are regular. (B): $H=\frac{1}{12}$, all motions are regular. (C): $H=\frac{1}{8}$, most of the motions are regular but chaotic motions appear. (D): $H=\frac{1}{6}$, almost all motions are chaotic.

Fig. 2. Orbit of the conservative system (7.11) with $\theta=0$, (A): $h=0.4$ $u_0=0.58$. There are 2 rings, each ring contains 11 archipelagoes, each archipelago consists basically of 5 isles, (B): $h=0.1$, $u_0=0.54$. There are 2 rings with simple, regular structure. (C): $h=0.4$, $u_0=0.5$. (D): $h=0.4$, $u_0=0.6$. In (C), (D) the structure are grossly similar to that of (A), but more irregular, especially in (D).

Fig. 3. Orbit of dissipative system (7.11) with $\theta>0$, arising from mixed leap-frog schemes (7.10) for system (7.5). (A): $h=1.898$, $\theta=0.538$, $u_0=0.58$. This is the strange attractor of the Hénon map. (B): $h=0.1$, $\theta=0.01$, $u_0=0.6$. (C): $h=0.4$, $\theta=0.01$, $u_0=0.56$. (D): $h=0.5$, $\theta=0.01$, $u_0=0.56$. The last 3 graphs show interesting spiral structures. Here (B) is a dissipative counterpart of Fig. 2 (B), with 2 rings changed into 2 contracting spirals. In (C), (D), the spirals have more complicated structures than that of (B). Note that the number of branches of the contracting spiral is 8 in (C), 6 in (D).

Fig. 2 (A,C,D) and Fig. 3 (B,C,D) contain some interesting features, which seems, to the authors' knowledge, to be new.

Table 2

Dt=0.06283		Dx=0.15708	C=Dt/Dx=0.4	
SCHEME	u(99,0)	u(99,10)	u(99,20)	u(99,30)
KLFM(4,2)	0.06279	0.99777	-0.06678	-0.99777
KLFM(4,4)	0.06279	0.99815	-0.06279	-0.99815
KLFM(2,2)	0.04109	0.99149	-0.04109	-0.99149
KLFM(2,4)	0.03701	0.99930	-0.03701	-0.99930
SLFM(2,2)	0.06047	0.99817	-0.06048	-0.99817
SLFM(2,4)	0.05636	0.99841	-0.05635	-0.99841
SLFM(4,2)	0.06277	0.99803	-0.06277	-0.99803
SLFM(4,4)	0.06277	0.99803	-0.06279	-0.99803
ANALYTICAL SOLUTION	0.06279	0.99803	-0.06279	-0.99803

Table 3

Dt=0.17279		Dx=0.15708	C=Dt/Dx=1.1	
SCHEME	u(99,0)	u(99,10)	u(99,20)	u(99,30)
KLFM(4,4)	-0.98519	0.17145	0.98519	-0.17145
SLFM(4,4)	-0.98520	0.17111	0.98532	-0.17109
ANALYTICAL SOLUTION	-0.98511	0.17191	0.98512	-0.17192

Table 4

	Dx=0.15708	KLFM(4,4)		
C-number	u(30,0)	u(99,0)	u(30,10)	u(99,10)
2.09	-0.68332	overflow	-0.73430	overflow
2.07	-0.63119	0.99853	-0.77565	0.05482
ANALYTICAL SOLUTION	-0.63208	0.99870	-0.77490	0.05089

Table 5

Dx=0.15708 SLFM(4,4)				
C-number	u(30,0)	u(99,0)	u(30,10)	u(99,10)
1.23	-0.23288	0.75542	-0.97253	0.65509
1.24	0.09666	overflow	-0.55163	overflow
ANALYTICAL SOLUTION	-0.23344	0.75630	-0.97237	0.65423

Table 6

KUNGE-KUTTA-VERNER method(5 OR 6 ORDER)
initial data 1,1,1,1. step length: 0.05

	p1	p2	q1	q2
t=0.5	0.88041	0.88041	1.46630	1.46630
t=2.0	0.74699	0.74699	2.66716	2.66716
t=4.0	0.68254	0.68254	4.08757	4.08757
t=7.0	0.63992	0.63992	6.06290	6.06290
t=10.0	0.61794	0.61794	7.94673	7.94673

S-SCHEME 4 ORDER (A-III-4)
initial data 1,1,1,1. step length:0.05

	p1	p2	q1	q2
t=0.5	0.88041	0.88041	1.46629	1.46629
t=2.0	0.74699	0.74699	2.66713	2.66713
t=4.0	0.68254	0.68254	4.08750	4.08750
t=7.0	0.63991	0.63991	6.06281	6.06281
t=10.0	0.61793	0.61793	7.94660	7.94660

S-SCHEME 2 ORDER (A-III-2)
initial data 1,1,1,1. step length:0.05

	p1	p2	q1	q2
t=0.5	0.88045	0.88045	1.46635	1.46635
t=2.0	0.74706	0.74706	2.66732	2.66732
t=4.0	0.68262	0.68262	4.08789	4.08789
t=7.0	0.64001	0.64001	6.06348	6.06348
t=10.0	0.61803	0.61803	7.94756	7.94756

Table 7

h=2*3.14159/40   t=10.0

X=M*h	sin(x+t)	A-III-4 DT=0.15	A-III-4 DT=0.3
M=0	-0.54402	-0.54419	-0.54470
M=2	-0.77668	-0.77679	-0.77717
M=4	-0.93332	-0.93335	-0.93358
M=6	-0.99859	-0.99855	-0.99861
M=8	-0.96611	-0.96600	-0.96591
M=10	-0.83907	-0.83890	-0.83870
M=12	-0.62989	-0.62967	-0.62940
M=14	-0.35906	-0.35878	-0.35850
M=16	-0.05307	-0.05279	-0.05251
M=18	0.25811	0.25835	0.25657
M=20	0.54402	0.54422	0.54434
M=22	0.77668	0.77682	0.77680
M=24	0.93332	0.93337	0.93320
M=26	0.99886	0.99857	0.99823
M=28	0.96611	0.96601	0.96552
M=30	0.83907	0.83916	0.83829
M=32	0.62989	0.62968	0.62900
M=34	0.35906	0.35881	0.35809
M=36	0.05307	0.05282	0.05212
M=38	-0.25811	-0.25832	-0.25896
M=40	-0.54402	-0.54419	-0.54471

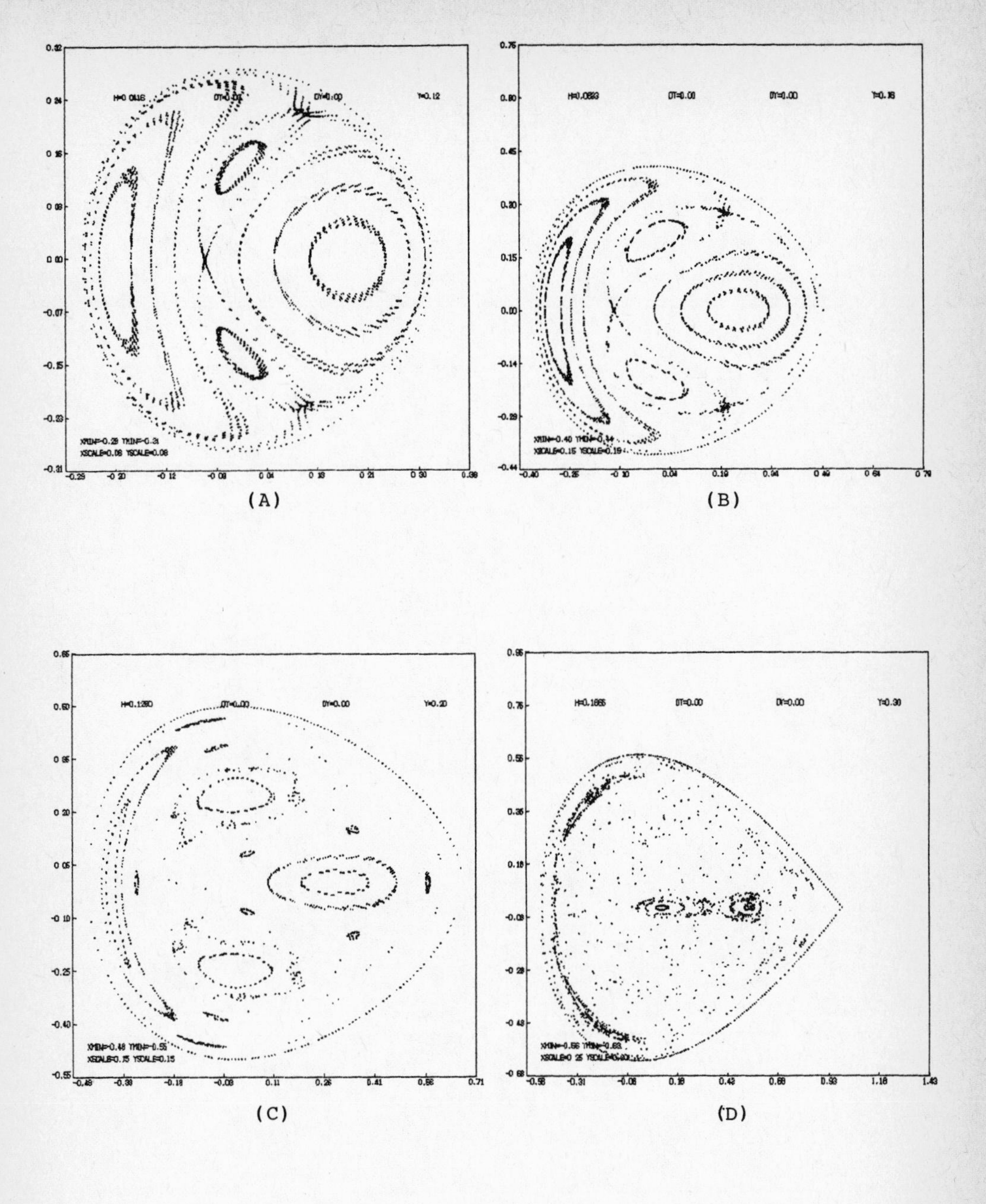

Fig.1.

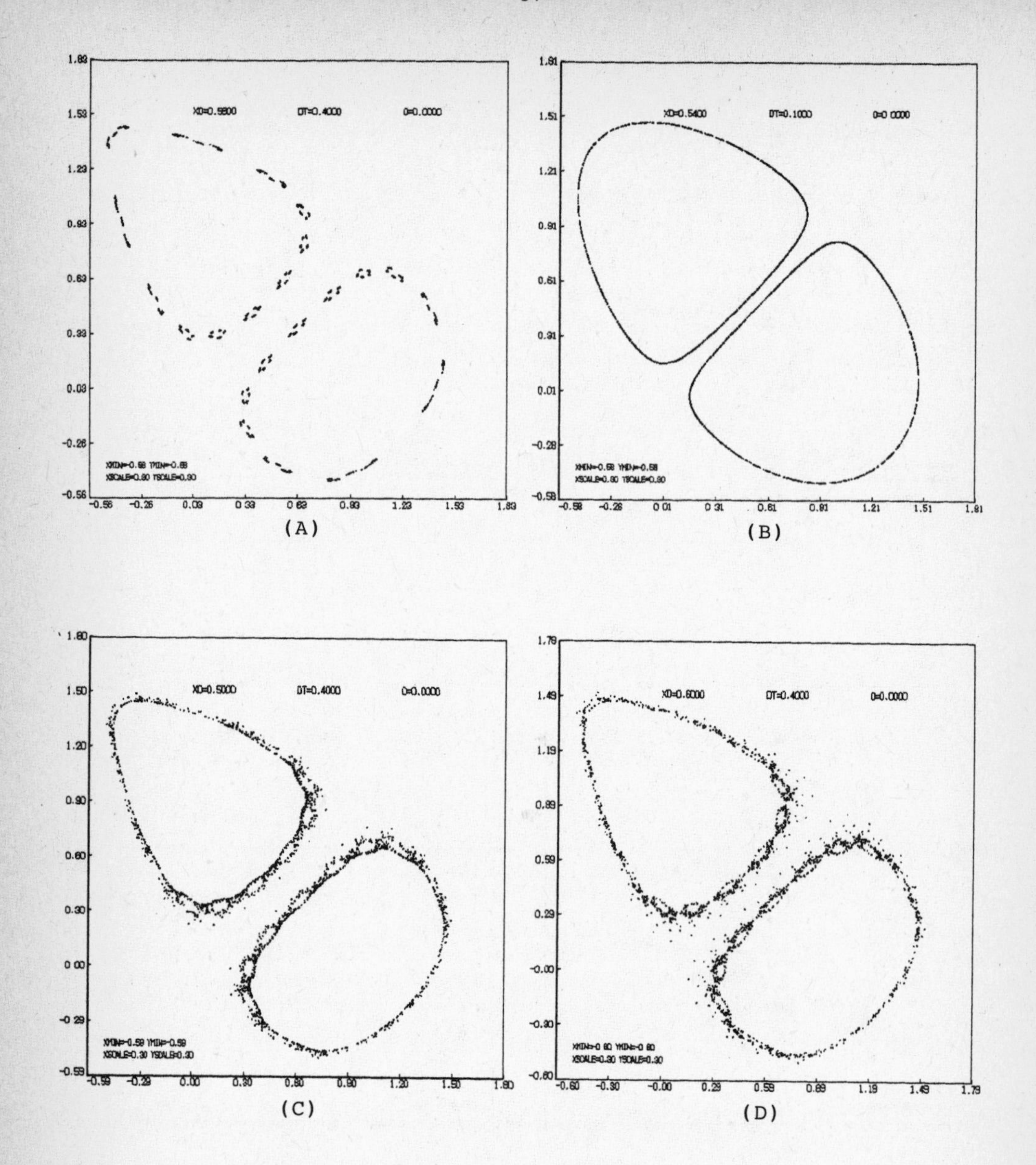
XD=0.5800
DT=0.4000
D=0.0000
XMIN=-0.56 YMIN=-0.56
XSCALE=0.30 YSCALE=0.30
(A)
XD=0.5400
DT=0.1000
D=0.0000
XMIN=-0.58 YMIN=-0.58
XSCALE=0.30 YSCALE=0.30
(B)
XD=0.5000
DT=0.4000
D=0.0000
XMIN=-0.59 YMIN=-0.59
XSCALE=0.30 YSCALE=0.30
(C)
XD=0.6000
DT=0.4000
D=0.0000
XMIN=-0.60 YMIN=-0.60
XSCALE=0.30 YSCALE=0.30
(D)

Fig.2.

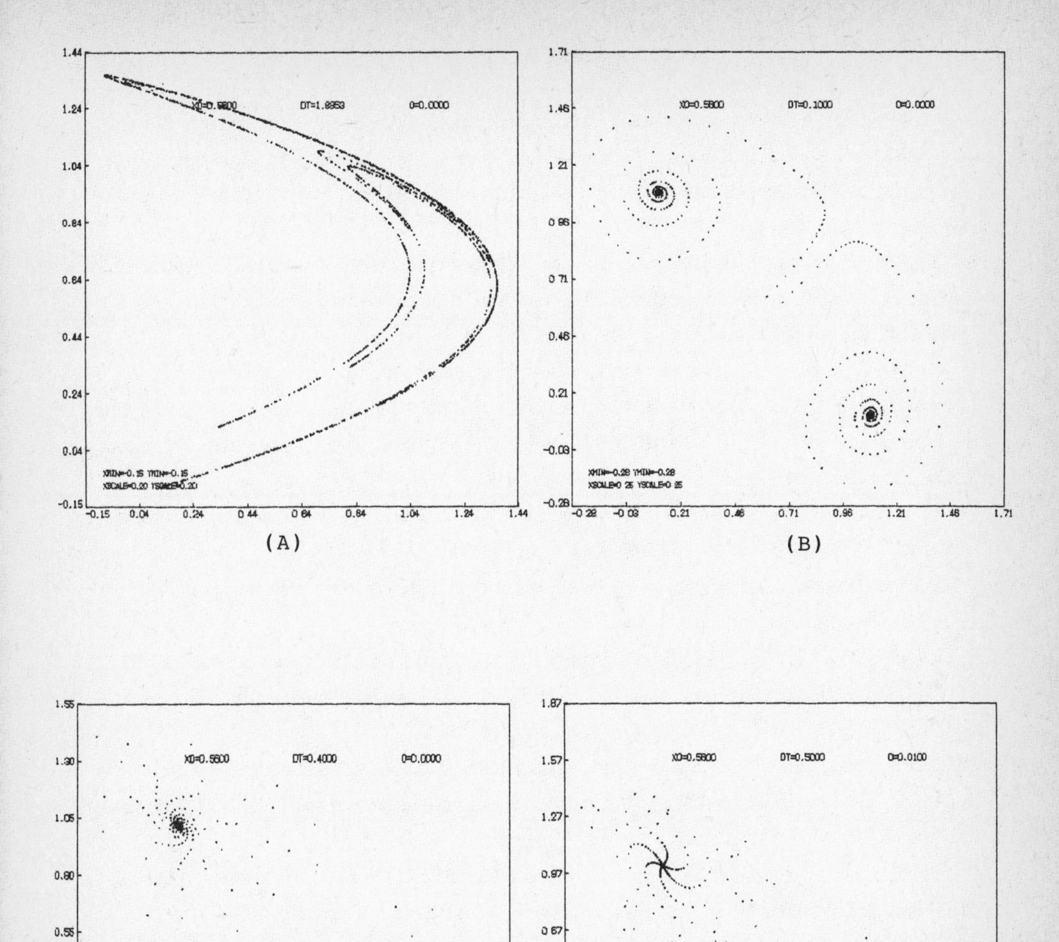
X0=0.5800
DT=1.8853
0=0.0000
XMIN=-0.15 YMIN=-0.15
XSCALE=0.20 YSCALE=0.20
X0=0.5800
DT=0.1000
0=0.0000
XMIN=-0.28 YMIN=-0.28
XSCALE=0.25 YSCALE=0.25

(A) (B)

X0=0.5600
DT=0.4000
0=0.0000
XMIN=-0.44 YMIN=-0.44
XSCALE=0.25 YSCALE=0.25
X0=0.5800
DT=0.5000
0=0.0100
XMIN=-0.52 YMIN=-0.52
XSCALE=0.30 YSCALE=0.30

(C) (D)

Fig.3.

References

[1] Feng Kang, On difference schemes and symplectic geometry, Proceedings of the 1984 Beijing Symposium on Differential Geometry and Differential Equations—— COMPUTATION OF PARTIAL DIFFERENTIAL EQUATIONS,Ed. Feng Kang, Science Press, Beijing, 1985, 42-58.

[2] Feng Kang, Canonical difference schemes for Hamiltonian canonical differential equation. International workshop on applied differential equation,Ed. Xiao Shutie, World Scientific 1985, 59-71.

[3] Feng Kang, Symplectic geometry and numerical methods in fluid dynamics. Lecture Notes in Physics V. 264. Proceedings of 10th International Conference on numerical methods in fluid dynamics, Beijing,1986, Ed. by F. G. Zhuang and Y. L. Zhu,Springer Verlag 1-7.

[4] Feng Kang, Difference schemes for Hamiltonian folmalism and symplectic geometry, Journal of Computational Mathematics, 4(1986) 279-289.

[5] Feng Kang, Wu Hua-mo,Qin Meng-zhao,Symplectic difference schemes for the linear Hamiltonian canonical systems,to appear.

[6] Feng Kang, Wu Hua-mo, Qin Meng- zhao,Wang Dao-liu,Construction of canonical difference schemes for Hamiltonian formalism via generating functions, to appear.

[7] Ge Zhong, Feng Kang, On the approximation of linear H-Systems,JCM, to appear.

[8] Li Chun- Wang, Qin Meng-Zhao, A symplectic difference scheme for the infinite demensional Hamilton system, JCM, to appear.

[9] Qin Meng-Zhao,A difference scheme for the Hamiltonian equation,JCM, 5(1987).

[10] Qin Meng-Zhao, Leap-frog schemes for wave equation, to appear in JCM.

[11] Ge Zhong,The generating function for the Poisson map, to appear.

[12] V. I. Arnold, Mathematical methods of classical mechanics, New york, 1978.

[13] C.L. Siegel, Symplectic geometry, Amer. Jour. Math. 65(1943),1-86.

[14] L. K. Hua, On the theory of automorphic functions of a matrix variable I, II, Amer. Jour. Math., 66(1944), 470-488, 531-563.

[15] Li Chun-Wang, Numerical calculation for the Hamiltonian System,Master's dissertation, Computing Center, Academia Sinica, (1986).

[16] Lee,T. D., Discrete Mechanics,CU-TP-267. A series of four Lectures given at the International School of Subnuclear Physics, Erice, August 1983.

[17] Vainberg, M., Variational method in nonlinear operator analysis, Holden Day, San Francisco, (1964).

[18] Abraham, R. and Marsden, J., Foundation of Mechanics, Addison-Wesley, Reading, Mass(1978).

[19] Chernoff, P. R, Marsden, J.E. Properties of infinite dimensional Hamiltonian, Lecture Notes in Mathematics 425.

[20] Miller, J.H.: On the Location of zeros of certain classes of polynomials with

application to numerical analysis, J.Inst.Math.Appl.Vol 8 (1971), 397-406.

[21] Hénon, M. and Heiles, C., The applicability of the third integral of motion, Astron. J., Vol 69 (1964), 73.

[22] Hénon, M., A two-dimensional mapping with a strange attractor, Comm. Math. Phys., 50(1976), 69-77.

[23] J. M. Sanz-Serna, Studies in numerical nonlinear instability I, SIAM. J. Sci. Stat. Comput., Vol 6, No 4 (1985), 923-938.

[24] M. Yamaguti and S. Ushiki, Chaos in numerical analysis of ordinary differential equations, Phys. 3D (1981), 618-626.

[25] Huang, M. Y. Hamiltonian approximation for the nonlinear wave equation. In these Proceedings.

THE BOUNDARY FINITE ELEMENT METHODS FOR SIGNORINI PROBLEMS*

Han Hou-de
Department of Applied Mathematics,
Tsinghua University, Beijing, China

In this paper, the boundary finite element method is used to solve Signorini problems, which arise in many practical problems. The equivalent boundary variational inequalities are given. Furthermore, the error estimates of the boundary finite element approximations for Signorini problems are obtained.

§1. Introduction

The Signorini problems are a class of very important variational inequalities, which arise in many practical problems, such as, the fluid mechanics problems in media with semi-permeable boundaries [1-2], the electropaint process[3] etc. The numerical solutions of Signorini problems have been discussed by the finite element method and finite difference method in literature[2,4]. In this paper, the boundary finite element will be used to solve the Signorini problems. We know that the solution of Signorini problem satisfies a linear partial differential equation in the domain though the Signorini problem is a nonlinear problem. Hence it is natural and advantageous to apply the boundary finite element method to Signorini problems.

As usual, let Ω be a bounded domain in $\mathbb{R}^2$ with a smooth boundary Γ. We define

$$V = H^1(\Omega) ,$$

$$a(u,v) = \int_\Omega \nabla u \cdot \nabla v dx + \int_\Omega uv dx ,$$

* This project is supported by National Natural Sciences Foundation of China.

$$L(v) = \int_\Omega fv\,dx + \int_\Gamma gv\,ds, \quad \forall f \in V', \quad g \in H^{-\frac{1}{2}}(\Gamma),$$

$$K = \{v \in H^1(\Omega), \quad v \geq 0 \quad \text{a.e. on } \Gamma\},$$

where $H^m(\Omega)$, $H^\alpha(\Gamma)$ denote the standard Sobolev spaces. We consider the following variational inequality:

$$\begin{cases} \text{Finding } u \in K, \text{ such that} \\ a(u, v-u) \geq L(v-u), \quad \forall \ v \in K. \end{cases} \tag{1.1}$$

Problem (1.1) is a model of the Signorini problems. We know that [4-6]

Lemma 1.1 Variational inequality (1.1) has a unique solution.

Lemma 1.2 Suppose the boundary of Ω is sufficiently smooth (or Ω is a convex polygonal domain) and $f \in L^2(\Omega)$, $g \in H^{\frac{1}{2}}(\Omega)$, then the solution of (1.1) belongs to $H^2(\Omega)$.

Let u_0 be the solution of the following boundary value problem

$$\begin{cases} -\Delta u_0 + u_0 = f, & \text{in } \Omega \\ u_0\big|_\Gamma = 0 . \end{cases} \tag{1.2}$$

Hence we have

$$-a(u_0, v+u_0-u) = -\int_\Omega f(v+u_0-u)\,dx - \int_\Gamma \frac{\partial u_0}{\partial n}(v+u_0-u)\,ds, \quad \forall \ v \in K. \tag{1.3}$$

From (1.1) and (1.3), we obtain

$$a(u-u_0, v-(u-u_0)) \geq \int_\Gamma \left(g - \frac{\partial u_0}{\partial n}\right)(v-(u-u_0))\,ds, \quad \forall \ v \in K.$$

Let $w = u-u_0$, $g^* = g - \frac{\partial u_0}{\partial n}$, then w satisfies the following problem:

$$\begin{cases} \text{Finding } w \in K, \text{ such that} \\ a(w, v-w) \geq \int_\Gamma g^*(v-w)\,ds, \quad \forall \ v \in K . \end{cases}$$

Therefore without loss of generality, we may suppose that $f \equiv 0$. In this case, the problem (1.1) is reduced to

$$\begin{cases} \text{Finding } u \in K, \text{ such that} \\ a(u, v-u) \geq \int_\Gamma g(v-u)ds , \quad \forall v \in K . \end{cases} \tag{1.1*}$$

For problem $(1.1)^*$, we have [4]

Lemma 1.3 The solution u of problem $(1.1)^*$ is characterised by

$$\begin{cases} -\Delta u+u = 0 , \quad \text{a.e. in } \Omega , \\ u\big|_\Gamma \geq 0 , \quad \frac{\partial u}{\partial n} \geq g , \qquad \text{a.e.} \quad \text{on } \Gamma , \\ u(\frac{\partial u}{\partial n} - g) = 0 \quad \text{a.e.} \quad \text{on } \Gamma . \end{cases} \tag{1.4}$$

§2. An equivalent boundary variational inequality of $(1.1)^*$.

We know that the fundamental solution of equation $-\Delta u+u=0$ is the modified Bessel function of order zero $K_0(|x-y|)$, where $r=|x-y| = \sqrt{(x_1-y_1)^2+(x_2-y_2)^2}$. It has the expansion near by $r=0$:

$$K_0(r) = \sum_{n=0}^{\infty} a_n r^{2n} \log \frac{1}{r} + \sum_{n=1}^{\infty} b_n r^{2n} , \quad a_0 = 1 .$$

At infinity, $\lim\limits_{r\to+\infty} K_0(r)=0$ and

$$K_0(r) = \sqrt{\frac{\pi}{2r}} \; e^{-r} + \cdots .$$

Let

$$u(y) = \int_\Gamma \rho(x) K_0(|x-y|)ds_x , \quad \forall y \in \Omega \tag{2.1}$$

be the solution of $(1.1)^*$, where $\rho(x)$ is a function on Γ to be determined. The problem $(1.1)^*$ can be reduced to a boundary variational inequality satisfied by $\rho(x)$. Consider the following operator

$$\mathcal{K}: \; H^\alpha(\Gamma) \to H^{\alpha+1}(\Gamma) , \quad \text{where} \quad -\frac{3}{2} \leq \alpha , \quad \text{and}$$

$$\mathcal{K}\phi = \int_\Gamma \phi(x) K_0(|x-y|)ds_x , \quad \forall \phi \in H^\alpha(\Gamma) .$$

We know that [7] $\mathcal{K} : H^\alpha(\Gamma) \to H^{\alpha+1}(\Gamma)$ is a bijective mapping and it is a continuous and bounded operator. Hence Banach Theorem implies the continuity and boundedness of the inverse $\mathcal{K}^{-1}$. For each $v \in H^1(\Omega)$, we have $v\big|_\Gamma \in H^{\frac{1}{2}}(\Gamma)$ by the Trace Theorem, then there exists a unique function $\phi \in H^{-\frac{1}{2}}(\Gamma)$ such that

$$v(y) = \mathcal{K}\phi = \int_\Gamma \phi(x) K_0(|x-y|) ds_x \,, \quad \forall\, y \in \Gamma \,.$$

For every function u given by (2.1) and each $v \in H^1(\Omega)$, a computation yields:

$$a(u,\, v-u) = \int_\Gamma \frac{\partial u}{\partial n_y} (v-u) ds_y$$

$$= \int_\Gamma \{ \pi\rho(y) + \int_\Gamma \rho(x) \frac{\partial k_0(|x-y|)}{\partial n_y} ds_x \} \{ \int_\Gamma (\phi(x) - \rho(x)) K_0(|x-y|) ds_x \} ds_y$$

$$= \pi \int_\Gamma \int_\Gamma K_0(|x-y|) \rho(y) (\phi(x) - \rho(x)) ds_x ds_y$$

$$+ \int_\Gamma \int_\Gamma J(x,y) \rho(y) (\phi(x) - \rho(x)) ds_x ds_y \equiv b(\rho,\, \phi - \rho) \,,$$

where

$$J(x,y) = \int_\Gamma K_0(|x-\xi|) \frac{\partial K_0(|y-\xi|)}{\partial n_\xi} ds_\xi \,.$$

For the bilinear form $b(\rho, \phi)$, we have

Lemma 2.1. (i) $b(\rho, \phi)$ is a bounded bilinear form on $H^{-\frac{1}{2}}(\Gamma) \times H^{-\frac{1}{2}}(\Gamma)$, i.e. there is a constant $M > 0$ such that

$$|b(\rho, \phi)| \leq M \|\rho\|_{-\frac{1}{2},\Gamma} \|\phi\|_{-\frac{1}{2},\Gamma} \,, \quad \forall \phi,\, \rho \in H^{-\frac{1}{2}}(\Gamma) \,. \tag{2.2}$$

(ii) There exists a constant $\nu > 0$, such that

$$b(\rho, \rho) \geq \nu \|\rho\|^2_{-\frac{1}{2},\Gamma} \,, \quad \forall \rho \in H^{-\frac{1}{2}}(\Gamma) \,. \tag{2.3}$$

(iii) $b(\rho, \phi)$ is symmetric

$$b(\rho, \phi) = b(\phi, \rho) \,, \qquad \forall,\, \rho,\, \phi \in H^{-\frac{1}{2}}(\Gamma) \,. \tag{2.4}$$

Proof If $\rho, \phi \in H^{-\frac{1}{2}}(\Gamma)$, then $\mathcal{K}\rho, \mathcal{K}\phi \in H^{\frac{1}{2}}(\Gamma)$.

We consider the following boundary value problems

$$\begin{cases} -\Delta u + u = 0 \ , & \text{in } \Omega \ , \\ u = \mathcal{K}\rho \ , & \text{on } \Gamma \ , \end{cases} \tag{2.5}$$

$$\begin{cases} -\Delta v + v = 0 \ , & \text{in } \Omega \ , \\ v = \mathcal{K}\phi \ , & \text{on } \Gamma \ . \end{cases} \tag{2.6}$$

The problem (2.5) (or (2.6)) has a unique generalized solution $u \in H^1(\Omega)$ (or $v \in H^1(\Omega)$) and

$$\|u\|_{1,\Omega} \leq C \|\mathcal{K}\rho\|_{\frac{1}{2},\Gamma} \leq C\|\rho\|_{-\frac{1}{2},\Gamma} \ ,$$

$$\|v\|_{1,\Omega} \leq C\|\mathcal{K}\phi\|_{\frac{1}{2},\Gamma} \leq C\|\phi\|_{-\frac{1}{2},\Gamma} \ ,$$

where C denotes a constant depending only on the domain Ω. In fact, the solution of problem (2.5) is given by (2.1) and

$$v(y) = \int_\Gamma \phi(x) K_0(|x-y|)\, ds_x \ , \qquad \forall\ y \in \Omega$$

is the solution of problem (2.6). Hence, we have

$$b(\rho, \phi) = a(u, v) \ , \qquad \forall \rho \ , \phi \in H^{-\frac{1}{2}}(\Gamma) \ ,$$

then we obtain the conclusion of this lemma immediately:

$$|b(\rho, \phi)| = |a(u, v)| \leq \|u\|_{1,\Omega} \|v\|_{1,\Omega} \leq C\|\rho\|_{-\frac{1}{2},\Gamma} \|\phi\|_{-\frac{1}{2},\Gamma} \ ,$$

$$b(\rho, \rho) = a(u, u) = \|u\|^2_{1,\Omega} \geq \gamma\|\mathcal{K}\rho\|^2_{-\frac{1}{2},\Gamma} \geq \nu\|\rho\|^2_{-\frac{1}{2},\Gamma} \ ,$$

$$b(\rho, \phi) = a(u, v) = a(v, u) = b(\phi, \rho) \ ,$$

where γ, ν are positive constants depending only on the domain Ω.

Let $B = \{\phi \in H^{-\frac{1}{2}}(\Gamma)$, and $\mathcal{K}\phi \geq 0$ a.e. on $\Gamma\}$.

It is easy to check that B is a closed convex set in $H^{-\frac{1}{2}}(\Gamma)$. We consider the following variational inequality on boundary Γ:

$$\begin{cases} \text{Finding } \rho \in B \ , \text{ such that} \\ b(\rho, \phi-\rho) \geq \int_\Gamma (\mathcal{K}g)(\phi-\rho)\, ds \ , \qquad \forall \phi \in B \ . \end{cases} \tag{2.7}$$

An application of Theorem 2.1 (in [2], Chapter I) yields the following result.

Theorem 2.1. The problem (2.7) has a unique solution ρ.

Suppose that $\rho \in B$ is the solution of problem (2.7), then by (2.1), we get the function $u \in K$. For $v \in H^1(\Omega)$, there is a unique $\phi \in H^{-\frac{1}{2}}(\Gamma)$ such that

$$v(y)\Big|_\Gamma = \int_\Gamma \phi(x) K_0(|x-y|) ds_x,$$

then we have

$$a(u, v-u) = b(\rho, \phi-\rho) \geq \int_\Gamma (\mathcal{K} g)(\phi-\rho) ds = \int_\Gamma g(v-u) ds, \quad \forall v \in K. \quad (2.8)$$

Namely u is the solution of $(1.1)^*$. On the other hand, suppose u is the solution of $(1.1)^*$, then there exists a unique $\rho \in B$ satisfying (2.1). It is easy to check that ρ is the solution of (2.7). Hence the boundary variational inequality problem (2.7) is equivalent to the problem $(1.1)^*$, and we have

Theorem 2.2. Suppose $g \in H^{\frac{1}{2}}(\Gamma)$, then the solution of problem (2.7) belongs to $H^{\frac{1}{2}}(\Gamma)$.

§3. The numerical approximation of the boundary variational inequality problem (2.7)

Suppose that boundary Γ is presented as

$$\begin{cases} x_1 = x_1(s), \\ \qquad\qquad\qquad 0 \leqq s \leqq L \\ x_2 = x_2(s) \end{cases}$$

and $x_j(0) = x_j(L)$, $j=1,2$.

Furthermore Γ is divided into some line segments $\{T\}$ by the points $x^i = (x_1(s_i), x_2(s_i))$, $i=1,2,\cdots,N$ and $s_1 = 0$, $s_{N+1} = L$, $h = \max_{1 \leq i \leq N} |s_{i+1} - s_i|$. Let $\mathcal{T}_h$ denote this partition. The space $H^{-\frac{1}{2}}(\Gamma)$ can be approximated by the space V_h, where

$$V_h = \{ \phi_h \mid \phi_h|_T \text{ is a constant}, \ \forall\, T \in \mathcal{T}_h \},$$

and we approximate B by

$$B_h = \{\phi_h \in V_h\ , \quad \mathcal{K}\phi_h \geq 0\ , \quad \text{a.e. on } \Gamma\}\ .$$

It is clear that V_h is a finite demensional subspace of $H^{-\frac{1}{2}}(\Gamma)$ and B_h is a closed convex nonempty subset of V_h, and $B_h \subset B$.

Now we consider the discrete problem:

$$\begin{cases} \text{Finding } \rho_h \in B_h\ , \text{ such that} \\ b(\rho_h\ , \phi_h - \rho_h) \geq \int_\Gamma (\mathcal{K}g)(\phi_h - \rho_h)ds, \quad \forall\ \phi_h \in B_h\ . \end{cases} \tag{3.1}$$

Then one can easily prove

<u>Theorem 3.1.</u> The problem (3.1) has a unique solution ρ_h.

Furthermore, we obtain the following error estimate.

<u>Theorem 3.2</u> Suppose $g \in H^{\frac{1}{2}}(\Gamma)$, then there is a constant C independent of h, such that

$$\|\rho-\rho_h\|_{-\frac{1}{2},\Gamma} \leq C \inf_{\phi_h \in B_h} \{\|\rho-\phi_h\|^2_{-\frac{1}{2},\Gamma} + \|\rho-\phi_h\|_{-\frac{3}{2},\Gamma}\}^{\frac{1}{2}}\ , \tag{3.2}$$

where ρ is the solution of problem (2.7) and ρ_h is the solution of problem (3.1).

<u>Proof</u> By (2.7), we obtain

$$b(\rho,\ \rho_h-\rho) \geq \int_\Gamma \mathcal{K}g(\rho_h-\rho)ds \tag{3.3}$$

and from (3.1), we have

$$b(\rho_h\ , \phi_h-\rho_h) \geq \int_\Gamma \mathcal{K}g(\phi_h-\rho_h)ds\ , \qquad \forall\ \phi_h \in B_h\ . \tag{3.4}$$

Now we consider

$$\begin{aligned} b(\rho-\rho_h\ , \rho-\rho_h) &= b(\rho-\rho_h\ , \rho-\phi_h) + b(\rho,\ \phi_h-\rho) \\ &\quad -b(\rho_h\ , \phi_h-\rho_h) - b(\rho,\ \rho_h-\rho)\ . \end{aligned}$$

By the inequalities (3.3) and (3.4), we get

$$b(\rho-\rho_h\ , \rho-\rho_h) \leq b(\rho-\rho_h\ , \rho-\phi_h) + b(\rho,\ \phi_h-\rho) - \int_\Gamma \mathcal{K}g(\phi_h-\rho)ds\ .$$

Furthermore, we know that

$$|b(\rho,\ \phi_h-\rho)| = |a(u,\ v_h-u)| \leq C\|u\|_{2,\Omega}\ \|v_h-u\|_{0,\Omega}$$

$$\leq C\|u\|_{\frac{3}{2},\Gamma}\|v_h-u\|_{-\frac{1}{2},\Gamma} = C\|\mathcal{K}\rho\|_{\frac{3}{2},\Gamma}\|\mathcal{K}(\phi_h-\rho)\|_{-\frac{1}{2},\Gamma}$$

$$\leq C\|\rho\|_{\frac{1}{2},\Gamma}\|\phi_h-\rho\|_{-\frac{3}{2},\Gamma} , \tag{3.5}$$

$$\left|\int_\Gamma \mathcal{K}g(\phi_h-\rho)ds\right| = \left|\int_\Gamma g\,\mathcal{K}(\phi_h-\rho)ds\right|$$

$$\leq C\|g\|_{\frac{1}{2},\Gamma}\|\mathcal{K}(\phi_h-\rho)\|_{-\frac{1}{2},\Gamma}$$

$$\leq C\|g\|_{\frac{1}{2},\Gamma}\|\phi_h-\rho\|_{-\frac{3}{2},\Gamma} , \tag{3.6}$$

$$|b(\rho-\rho_h , \rho-\phi_h)| \leq \frac{\nu}{2}\|\rho-\rho_h\|^2_{-\frac{1}{2},\Gamma} + C\|\rho-\phi_h\|^2_{-\frac{1}{2},\Gamma} . \tag{3.7}$$

Combining inequalities (3.5)-(3.7), we have

$$\frac{\nu}{2}\|\rho-\rho_h\|^2_{-\frac{1}{2},\Gamma} \leq C\{\|\rho-\phi_h\|^2_{-\frac{1}{2},\Gamma} + \|\rho-\phi_h\|_{-\frac{3}{2},\Gamma}\} , \quad \forall\ \phi_h \in B_h .$$

Hence, the conclusion (3.2) follows immediately.

§4. For the regular domains

Suppose Ω is a regular domain, such as a circle or a rectangle. We can get a simple boundary integral formulation for the Signorini problems. In this case, the canonical boundary finite element method proposed by Feng Kang [8] can be used to the Signorini problems.

Let $G(x,y)$ denote the Green function of the following boundary value problem:

$$\begin{cases} -\Delta\Phi + \Phi = 0 , & \text{in } \Omega , \\ \Phi = 0 , & \text{on } \Gamma . \end{cases} \tag{4.1}$$

Since Ω is a regular domain, such as a circle or a rectangle, it is not difficult to get the Green function $G(x,y)$ for the problem (4.1).

For any $u \in H^1(\Omega)$, if u satisfies

$$a(u,w) = 0, \quad \forall\, w \in H_0^1(\Omega) , \tag{4.2}$$

then $\frac{\partial u(x)}{\partial n_x}$ can be represented by the Green function and the boundary value of u through boundary canonical reduction [8]

$$\left.\frac{\partial u(x)}{\partial n_x}\right|_\Gamma = -\oint_\Gamma \frac{\partial G(x,y)}{\partial n_x\, \partial n_y} u(y)\, ds_y , \tag{4.3}$$

where n_x is the unit outward normal vector at point $X \in \Gamma$ and the integral in equation (4.3) possesses a non-integrable kernel, which is only defined as its finite part.

For any $u, v \in H^1(\Omega)$, if u satisfies the equation (4.2), we have

$$a(u,v) = \int_\Gamma \frac{\partial u}{\partial n} v\, ds = \int_\Gamma \oint_\Gamma - \frac{\partial^2 G(x,y)}{\partial n_x\, \partial n_y} u(y) v(x)\, ds_x ds_y \equiv b^*(u,v) .$$

$b^*(u,v)$ is a bilinear form on $H^{1/2}(\Gamma) \times H^{1/2}(\Gamma)$, and we can prove

<u>Lemma 4.1.</u> $b^*(u,v)$ is a bounded bilinear form on $H^{1/2}(\Gamma) \times H^{1/2}(\Gamma)$ and there exists two constants $M > 0$ and $\alpha > 0$, such that

$$|b^*(u,v)| \le M \|u\|_{1/2,\Gamma} \|v\|_{1/2,\Gamma} , \quad \forall\, u, v \in H^{1/2}(\Gamma) , \tag{4.4}$$

$$\alpha \|u\|^2_{1/2,\Gamma} \le b^*(u,u) , \quad \forall\, u \in H^{1/2}(\Gamma) . \tag{4.5}$$

<u>Proof:</u> For any $u, v \in H^{1/2}(\Gamma)$, u, v can be extended on Ω so that $u, v \in H^1(\Omega)$ and satisfy the equation (4.2). Hence we have

$$|b^*(u,v)| = |a(u,v)| = \left|\int_\Omega (\nabla u \cdot \nabla v + uv)\, dx\right|$$

$$\le \|u\|_{1,\Omega} \|v\|_{1,\Omega} \le M \|u\|_{1/2,\Gamma} \|v\|_{1/2,\Gamma} .$$

The last inequality is from the prior estimates of the solution of elliptic boundary value problems.

Similarly, we obtain

$$b^*(u,u) = a(u,u) = \|u\|^2_{1,\Omega} .$$

By the trace theorem in $H^1(\Omega)$, we know that there is a constant $\alpha > 0$, such that

$$\|u\|^2_{1,\Omega} \geq \alpha \|u\|^2_{\frac{1}{2},\Gamma} ,$$

then the inequality (4.5) follows immediately.

Let $B^* = \{u \mid u \in H^{\frac{1}{2}}(\Gamma), \text{ and } u \geq 0 \text{ a.e. on } \Gamma\}$. We consider the following variational inequality problem on the boundary Γ:

$$\begin{cases} \text{Finding } u \in B^*, \text{ such that} \\ b^*(u, v-u) \geq \int_\Gamma g(v-u)ds, \quad \forall v \in B^* . \end{cases} \tag{4.6}$$

It is not difficult to prove that the problem (4.6) has a unique solution and the boundary variational inequality (4.6) is equivalent to the problem (1.1)*.

For the partition $\mathcal{T}_h$ given in section 3, we construct a finite dimensional subspace of $H^{\frac{1}{2}}(\Gamma)$ as follows. Let

$$X_h = \{v_h \in C^{(0)}(\Gamma), \ v_h\big|_T \in P_1(T), \ \forall T \in \mathcal{T}_h\} ,$$

i.e., every $v_h \in X_h$ is a continuous function on Γ and is a linear function of the parameter s on each line segment. We know that X_h is a subspace of $H^{\frac{1}{2}}(\Gamma)$ with dimensionality N . Suppose

$$B^*_h = \{v_h \in X_h , \ v_h(x^i) \geq 0, \ i=1,2,\cdots, N\} , \tag{4.7}$$

then B^*_h is a nonempty convex set of X_h and $B^*_h \subset B$.

Consider the following approximation problem:

$$\begin{cases} \text{Finding } u_h \in B^*_h , \text{ such that} \\ b^*(u_h , v_h-u_h) \geq \int_\Gamma g(v_h-u_h)ds , \quad \forall v_h \in B^*_h . \end{cases} \tag{4.8}$$

By Lemma 4.1 , we know that problem (4.8) has a unique solution $u_h \in B^*_h$ and we can prove the following error estimate

Theorem 4.1. Suppose $g \in H^{\frac{1}{2}}(\Gamma)$, then the following error estimate holds

$$\| u-u_h \|^2_{\frac{1}{2},\Gamma} \leq C \inf_{v_h \in B_h^*} \{ \| u-v_h \|^2_{\frac{1}{2},\Gamma} + \| u-v_h \|_{-\frac{1}{2},\Gamma} \} , \tag{4.9}$$

where C is a constant depending only on the boundary Γ, $\| g \|_{\frac{1}{2},\Gamma}$ and constants M, α.

Proof Since B_h^* is a subset of B^*, which implies that the solution $u_h \in B^*$, we have

$$b^*(u, u_h-u) \geq \int_\Gamma g(u_h-u)ds .$$

On the other hand, we obtain

$$\begin{aligned} b^*(u_h-u, u_h-u) &= b^*(v_h-u, u_h-u) + b^*(u, v_h-u) \\ &\quad -b^*(u_h, v_h-u_h) - b^*(u, u_h-u) \\ &\leq b^*(v_h-u, u_h-u) + b^*(u, v_h-u) - \int_\Gamma g(v_h-u)ds , \quad \forall v_h \in B_h^* . \end{aligned} \tag{4.10}$$

Furthermore, we get

$$\begin{aligned} \left| b^*(u, v_h-u) \right| &= \left| \int_\Gamma \frac{\partial u}{\partial n}(v_h-u)ds \right| \\ &\leq \left\| \frac{\partial u}{\partial n}\Big|_\Gamma \right\|_{\frac{1}{2},\Gamma} \left\| v_h-u \right\|_{-\frac{1}{2},\Gamma} \\ &\leq C \left\| u \right\|_{2,\Omega} \left\| v_h-u \right\|_{-\frac{1}{2},\Gamma} \\ &\leq C \left\| g \right\|_{\frac{1}{2},\Gamma} \left\| v_h-u \right\|_{-\frac{1}{2},\Gamma} , \end{aligned} \tag{4.11}$$

and

$$\left| \int_\Gamma g(v_h-u)ds \right| \leq \| g \|_{\frac{1}{2},\Gamma} \| v_h-u \|_{-\frac{1}{2},\Gamma} . \tag{4.12}$$

Combining the inequalities (4.4), (4.5) and (4.10)-(4.12), the conclusion follows immediately.

The boundary finite element method (B.F.E.M) for the Signorini problems is a direct method. From the computational point of view, this method is much simpler than the indirect method in section 3. Unfortunately, we can only get the Green function $G(x,y)$ for some regular domains. Therefore this method is inapplicable to general domains. The direct B.F.E.M. without Green function for Signorini problems in general domains will be discussed in a separate paper.

REFERENCE

[1] Duvaut, G., Lions, J.L., Les inéquations en mécanique et en physique, Dunod, Paris, 1972.

[2] Glowinski, R., Lions, J.L. and Trémolières, R., Numerical Analysis of Variational Inequalities, North-Holland, Amsterdam, 1981.

[3] Aitchison, J.M., Lacey, A.A. and Shillor, M., A model for an electro-paint Process, IMA J. Appl. Math. (1984) 33, pp.17-31.

[4] Glowinski, R., Numerical Methods for Nonlinear variational problems.

[5] Brézis, H., Problémes unilateraux, J. de Math. Pures et Appliquées, 51(1972), pp.1-168.

[6] Caffarelli, L.A, Further regularity for the Signorini problem, Commun. P.D.E. 4(1979), pp.1067-1076.

[7] Hsiao, G.C., Wendland, W., A finite element method for some integral equations of the first kind, J. Math. Anal. Appl. 58(1977) pp.449--481.

[8] K. Feng, D.H. Yu, Canonical integral equations of elliptic boundary value problems and their numerical solutions, Proceedings of the China-France Symposium on Finite Element Methods, Beijing, China, 1982.

A HAMILTONIAN APPROXIMATION FOR NONLINEAR WAVE EQUATIONS ON N-DIMENSIONAL SPHERES S^n

Huang Ming-you
Department of Mathematics
Jilin University, Changchun

§1. Introduction

This paper discusses the nontrivial time periodic solution to the following semilinear wave equation

$$\begin{cases} u_{tt}-\Delta_n u+(\frac{n-1}{2})^2 u+f(t,x,u)=0 \\ u(t+2\pi,x)=(t,x), \end{cases} \tag{P}$$

where Δ_n is the Laplace-Beltrami operator on the n-dimensional sphere s^n. $f(t,x,u)\in C(R\times S^n\times R)$ is 2π-periodic with respect to t and satisfies some suitable conditions which we state later. Set

$$F(t,x,u)=\int_0^u f(t,x,v)dv,$$

then to seek the weak solution of (P) is reduced to solving a critical problem of the following indefinite nonlinear functional

$$I(u)=\int_{S^1\times S^n}\{\tfrac{1}{2}(|u_t|^2-|\nabla u|^2-(\frac{n-1}{2})^2|u|^2)-F(t,x,u)\}dtdx.$$

P.H.Rabinowitz ([1],1978) studied the one dimensional problem (n=1, i.e. the semilinear string equation), under condition that f is superlinear at $u=0,\infty$ and strictly increasing with respect to u. Recently, there are papers devoted to n-dimensional problem (P) for $n>1$ (see K. C.Chang and C.W.Hong,[2],1985, and Z.F.Zhou,[3],1986). To investigate this problem, some approximations are needed usually for the wave operator $A=\partial_{tt}-\Delta_n+(\frac{n-1}{2})^2$ may have one infinite eigenspace. In this paper, we propose a semidiscrete approximation of problem (P), and the stability and convergence properties of the approximate problem are analysed by means of the deformation method. As a consequence, the existence of nontrivial solution to problem (P) is demonstrated. The approximation suggested here is a Hamiltonian system of ordinary differential equations, which can be numerically integrated by symplectic difference schemes initiated by K. Feng (see [4]) or by Fourier method. Moreover, this approximation preserves the Hamiltonian structure and many properties, such as energy conservation of the original problem, so it would be interesting and useful to the numerical computations as well as to the theoretical study.

§2. Hamiltonian approximation

Set $\Omega=S^1\times S^n$ and consider the linear operator $A=\partial_{tt}-\Delta_n+(\frac{n-1}{2})^2$ with $D(A)=\{u;\ u\in C^2(\Omega)\}$. The eigenvalues of A are

$$\lambda(l,j)=(l+\frac{n-1}{2}+j)(l+\frac{n-1}{2}-j),\ j=0,\pm1,\dots,\ l=1,2,\dots$$

and the eigenfunctions corresponding to $\lambda(l,j)$ are

$$\phi_{l,m}\sin jt,\ \ \phi_{l,m}\cos jt,\ m=1,2,\dots,\ M(l,m)$$

where $\{\phi_{l,m}(x)\}$ are the spherical harmonics of degree l on S^n and

$$M(l,n)=\frac{(2l+n-1)\Gamma(l+n-1)}{\Gamma(l+1)\Gamma(n)}\ .$$

Here, functions $\{\phi_{l,m}e^{ijt}\}$ compose an orthogonal basis of $L(\Omega)$. For any $u\in L_2(\Omega)$, we have Fourier expansion

$$u(t,x)=\sum_{l,j,m}u_{l,j,m}\phi_{l,m}e^{ijt},\quad u_{l,-j,m}=\bar{u}_{l,j,m}.$$

Introduce the following Hilbert space

$$H=\{u\ L_2(\Omega);\quad \|u\|_H=\langle u,u\rangle_H^{\frac{1}{2}}<\infty\}$$

where the inner product of H is

$$\langle u,v\rangle_H=\sum_{l,j,m}|\lambda(l,j)|u_{l,j,m}\bar{v}_{l,j,m}+\sum_{(l,j)=0}u_{l,j,m}\bar{v}_{l,j,m}.$$

Also, we define subspaces H^0,H^- and H^+ of H by

$$H^0=\{u\in H;\ \ u_{l,j,m}=0 \text{ when } \lambda(l,j)\neq0\}$$

$$H^-=\{u\in H;\ \ u_{l,j,m}=0 \text{ when } \lambda(l,j)\geq0\}$$

$$H^+=\{u\in H;\ \ u_{l,j,m}=0 \text{ when } \lambda(l,j)\leq0\},$$

then $H^0\oplus H^-\oplus H^+$ is an orthogonal decomposition of H. Obviously, $\|u\|_H=\|u\|$ for $u\in H^0$.

<u>Lemma 1</u>. $H^-+H^+\subset W^{\frac{1}{2},2}(\Omega)$, the embedding $H^-+H^+\hookrightarrow L_p(\Omega)$ is compact if

$p \in (2, 2+2/(n-1))$.

Proof. By the inequality

$$|\lambda(l,j)| \geq \frac{1}{2}\left(1+\frac{n-1}{2}+|j|\right), \text{ for } \lambda(l,j) \neq 0$$

and the fact that $\{\sum_{l,j,m} (1+\frac{n-1}{2}+|j|)|u_{l,j,m}|^2\}^{\frac{1}{2}}$ defines an equivalent norm of Sobolev space $W^{\frac{1}{2},2}(\Omega)$ for $u \in H^- \oplus H^+$, we see $H^- \oplus H^+ \subset W^{\frac{1}{2},2}(\Omega)$. For the proof of the second part, see Lemma 2.1 of [3].

We are looking for the critical point of nonlinear functional $I(u)$ on Hilbert space H, which is the weak solution of (P). This problem is simply denoted by $I(u)|_H$. We use the expansion of $u \in H$ in form

$$u(t,x) = \sum_{l,m} q_{l,m}(t)\phi_{l,m}(x),$$

then $I(u)$ can be expressed as

$$I(q) = \int_0^{2\pi} \{\frac{1}{2}\sum_{l,m} \dot{q}_{l,m}^2 - \frac{1}{2}\sum_{l,m} \mu_l q_{l,m}^2 - b(t,q)\}dt,$$

where $\mu_l = (l+\frac{n-1}{2})^2$, $b(q,t) = \int F(t,x,u(t,x))dx$, $q = \{q_{l,m}\}$ and $\dot{q}_{l,m} = \frac{d}{dt} q_{l,m}$. Set $p_{l,m} = \dot{q}_{l,m}$ and define

$$h(p,q,t) = \frac{1}{2}\sum_{l,m} (p_{l,m}^2 + \mu_l q_{l,m}^2) + b(q,t),$$

then we have

$$I(u) = I(p,q) = \int_0^{2\pi} \{(p,\dot{q}) - h(p,q,t)\}dt,$$

and the problem $I(u)|_H$ is equivalent to system:

$$\frac{dp_{l,m}}{dt} = -\frac{\partial h}{\partial q_{l,m}}, \quad \frac{dq_{l,m}}{dt} = \frac{\partial h}{\partial p_{l,m}} \qquad l=1,2,\ldots,\ m=1,2,\ldots,M(l,n). \quad \text{(H)}$$

This shows that the problem (P) or $I(u)|_H$, in fact, is an infinite dimensional Hamiltonian system. Therefore, it is very natural to approximate this problem by a finite dimensional Hamiltonian system. For this purpose we choose the following subspace of H

$$H_N = \{u = \sum_{l,m} q_{l,m}\phi_{l,m}\ H;\ \ q_{l,m} = 0 \text{ when } l > N\},$$

and as the approximation of $I(u)|_H$, we consider problem $I(u)|_{H_N}$, i.e.

looking for the critical point of $I(u)$ on H_N. By the argument above, we see that $I(u)|_{H_N}$ is equivalent to system

$$\frac{dp_{l,m}}{dt}=-\frac{\partial h_N}{\partial q_{l,m}} \qquad \frac{dq_{l,m}}{dt}=\frac{\partial h_N}{\partial p_{l,m}} \tag{H_N}$$

$$l=1,2,\ldots N, \quad m=1,2,\ldots,M(l,n),$$

where

$$h_N(p,q,t)=\tfrac{1}{2}\sum_{l=1}^{N}\sum_{m=1}^{M(l,n)}(p_{l,m}^2+\mu_l q_{l,m}^2)+b_N, \quad b_N=b|_u H_N.$$

Evidently, approximation (H_N) preserves the Hamiltonian form of problem (P).

§3. Properties of the approximation

The analyses of this section for problems(H_N) and (H) are performed under the following assumptions on the nonlinear term $f(t,x,u)$ in (H):

(f_1) $|f(t,x,u)|\leqq a|u|^{p-1}+a_1$,

(f_2) $\frac{1}{2}f(t,x,u)-F(t,x,u)\geqq a_2F(t,x,u)-a_3$,

(f_3) $(f(t,x,u_1)-f(t,x,u_2))(u_1-u_2)\geqq a_4|u_1-u_2|^p$,

(f_4) $f(t,x,u)=o(|u|)$, when $u\to 0$,

where $p\in(2,2+2/(n-1))$ and a, a_i ($i=1,2,3,4$) are positive constants.

Since (f_1)-(f_3) imply: f is strictly increasing with respect to u and

$$F(t,x,u)\geqq a_5|u|^p-a_6, \text{ and } \tfrac{1}{2}f(t,x,u)-F(t,x,u)\geqq a_7|u|^p-a_6 \tag{1}$$

for some positive constants a_5, a_6, a_7, we see that these assumptions are similar to that made in [1] for Hamiltonian systems of ordinary differential equations. By Theorem 1 of [5] we know that for any fixed N, system (H_N) or problem $I(u)|_{H_N}$ has nontrival solution u_N. The problems are whether $\{u_N\}$ has convergent subsequences and whether their limits are nontrivial solution of (P). To answer these problems, some uniform estimations of $\{u_N\}$ are required. For showing these we need the following deformation lemma (Theorem (0.1) in [6]).

<u>Lemma 2.</u> Let E be a real Hilbert space, E_1 a closed subspace of E, and $E_2=E_1^{\perp}$. Suppose that $f\in C^1(E,R)$ and satisfies

(1) $f(u)=\frac{1}{2}(Lu,u)+b(u)$ where $u=u_1+u_2\in E_1+E_2$, $Lu=L_1u_1+L_2u_2$ and $L_i:E_i\to E_i$,

$i=1,2$ are (bounded) linear selfadjoint mappings.

(2) b is weakly continuous and is uniformly differentiable on bounded subsets of E.

(3) If for a sequence $\{u_m\}$, $f(u_m)$ is bounded from above and $f'(u_m) \to 0$ as $m\to\infty$, then $\{u_m\}$ is bounded,

(4) There are constants r_1, r_2, ρ, α, ω with $r_1>\rho$, $\alpha>\omega$ and r_1, r_2, $\rho>0$ and there is an $e\in\partial B_1\cap E_1$ such that

(i) $f\geqq\alpha$ on $S=\partial B_\rho\cap E_1$,

(ii) $f\leqq\omega$ on ∂Q where $Q=\{re|\ 0\leqq r\leqq r_1\}+(B_{r_2}\cap E_2)$.

Then f possesses a critical value $c\geqq\alpha$.

In the following, we denote the restrictions of $I(u)$ and

$$B(u)=\int_\Omega F(t,x,u(t,x))dtdx$$

on H_N by $I_N(u)$ and $B_N(u)$.

Lemma 3. If (f_1)-(f_4) hold, then

(1) $I_N(u)\in C^1(H_N,R)$;

(2) $B_N(u)$ is weakly continuous in H_N;

(3) On any bounded set of H_N, $B_N(u)$ is uniformly differentiable, i.e. for any R, $\varepsilon>0$, there exists a constant $\delta=\delta(R,\varepsilon)$ such that when u, $u+v\in B_R=\{u\in H_N;\ \|u\|_H\leqq R\}$ and $\|v\|\leqq\delta$,

$$|B_N(u+v)-B_N(u)-\langle B_N'(u),v\rangle|\leqq\varepsilon\ \|v\|_H.$$

The proof of this lemma is done by the similar arguments as in [6] for proving Lemma (3.10) and Proposition (3.12), here we only mention the following facts: subspace H_N has an orthogonal decomposition as H, i.e. $H_N=H_N^0\oplus H_N^-\oplus H_N^+$, where H_N^0 and H_N^+ are finite dimensional spaces (but, H^0 is infinite dimensional when n is even). By Lemma 1, $H^-+H^+\hookrightarrow L_p(\Omega)$ is compact. And since $\dim H_N^0<\infty$, then embedding $H_N\hookrightarrow L_p(\Omega)$ is also compact.

Lemma 4. When (f_1)-(f_4) hold, $I_N(u)$ satisfies the weak Palais-Smale condition, i.e. the condition (3) of Lemma 2.

Proof. Suppose $I_N(u_m)\leqq K$ and $I_N'(u_m)\to 0$ as $m\to\infty$, then for m large enough, by (1) we have

$$K+\|u_m\|_H\geqq I_N(u_m)-\tfrac{1}{2}\langle I_N'(u_m),u_m\rangle$$

$$=\tfrac{1}{2}\int_\Omega f(t,x,u_m)u_m dtdx-\int_\Omega F(t,x,u_m)dtdx\geqq a_7\|u_m\|_{L_p}^p-a_6 .$$

Therefore

$$\|u_m\|_{L_p}^p \leq C\{\|u_m\|_H+1\},$$

here and hereafter C stands for a positive constant independent of m.

Next, by the definition of I(u),

$$I_N(u)=\tfrac{1}{2}(\|u^-\|_H^2-\|u^+\|_H^2)-B_N(u)$$

$$\text{for } u=u^0+u^-+u^+\in H^0\oplus H^-\oplus H^+.$$

Hence from $I_N'(u_m)\to 0$ we know

$$\|u_m^+\|_H \geq -\langle I_N'(u_m),\ u_m^+\rangle = \|u_m^+\|_H^2+\int_\Omega f(t,x,u_m)u_m^+ dtdx$$

$$\geq \|u_m^+\|_H^2-\|u_i^+\|_{L_p}\cdot\|f(t,x,u_m\|_{L_{p'}} \quad (\frac{1}{p}+\frac{1}{p'}=1),$$

which yields

$$\|u_m^+\|_H^2 \leq \|u_m^+\|_H(1+\|f(t,x,u_m)\|_{L_{p'}}).$$

From (f_1) and the estimation of $\|u_m\|_{L_p}^p$ derived above, we have

$$\|f(t,x,u_m)\|_{L_{p'}} \leq C\{(\int_\Omega |u_m|^{p'(p-1)})^{\frac{1}{p'}}+1\}$$

$$\leq C\{\|u_m\|_{L_{p'}}^{p/p'}+1\} \leq C\{\|u_m\|_H^{1/p'}+1\}.$$

Therefore

$$\|u_m^+\|_H^2 \leq C\|u_m^+\|_H\{\|u_m\|_H^{1/p'}+1\},$$

and a similar inequality for u_m^-

$$\|u_m^-\|_H^2 \leq C\|u_m^-\|_H\{\|u_m\|_H^{1/p'}+1\}$$

can be proved in the same way. In addition,

$$\|u_m\|_{L_2}^2 \leq C\|u_m\|_{L_p}^2 \leq C\{\|u_m\|_H^{2/p}+1\}.$$

These inequalities give

$$\|u_m\|_H^2 = \|u_m^-\|_H^2+\|u_m^+\|_H^2+\|u_m^0\|_{L_2}^2 \leq C\{\|u_m\|_H^\gamma+1\},$$

where $\gamma=\max(1,\ \frac{2}{p},\ 1+\frac{1}{p'})<2$. This shows that $\{u_m\}$ is a bounded sequence in H_N.

Now we are ready to prove the main theorem of the paper.

Theorem 1. Under assumptions (f_1)-(f_4), the nonlinear functional $I_N(u)$ has at least one critical point $u_N \neq 0$, and the corresponding critical value $C_N = I_N(u_N)$ satisfies

$$0 < \alpha \leqq C_N \leqq \beta$$

where α and β are constants independent of N. Moreover, $\{u_N\}$ are uniformly bounded in H_N and in $L_p(\Omega)$ with respect to N.

Proof. The first part of the theorem will be proved by an application of Lemma 2 to $I_N(u)$. For this purpose, we introduce the continuous self-adjoint operator L in H associated with A by

$$(Lu,v) = \sum_{l,j,m} \lambda(l,j) u_{l,j,m} \bar{v}_{l,j,m} .$$

Then

$$I_N(u) = \tfrac{1}{2}(Lu,u) - B_N(u) = \tfrac{1}{2}(\|u^-\|_H^2 - \|u^+\|_H^2) - B_N(u). \tag{2}$$

Set $H_N = E$, $E_1 = H_N^-$, $E_2 = H_N^0 \oplus H_N^+$. The condition (1) of Lemma 2 is valid obviously, and the conditions (2) and (3) also hold by Lemma 3 and Lemma 4. We now verify the condition (4) of Lemma 2. Suppose $u \in S = \partial B_\rho \cap E_1$, then $u = u^-$ and by (2),

$$I_N(u) = \tfrac{1}{2} \|u\|_H^2 - \int_\Omega F(t,x,u(t,x))dtdx.$$

Since (f_1)-(f_4) imply that there exists a constant $C>0$ such that

$$|F(t,x,u)| \leqq \tfrac{1}{4}|u|^2 + C|u|^p,$$

then

$$\int_\Omega F(t,x,u(t,x))dtdx \leqq \tfrac{1}{4} \|u\|_{L_2}^2 + C \|u\|_{L_p}^p \leqq \tfrac{1}{4} \|u\|_H^2 + C \|u\|_H^p$$

for $H^- \hookrightarrow L_p(\Omega)$ being a compact embedding. Therefore, we have

$$I_N(u) \geqq \tfrac{1}{4}\|u\|_H^2 - C \|u\|_H^p, \quad \text{for } u \in S.$$

Here $p>2$, so there exist constants α and ρ such that

$$I_N(u) \geqq \tfrac{1}{4}\rho^2 - \rho^p \geqq \alpha, \quad \text{on } S = \partial B_\rho \cap E_1,$$

i.e. (4) (i) holds.

To verify (4) (ii), let $\omega = 0$ and e be an arbitrary element of set $\partial B_1 \cap E_1$. Suppose $u = re + v$, where $v \in E_2 = H_N^0 \oplus H_N^+$, then

$$I_N(u) = \tfrac{1}{2}r^2 - \|v^+\|_H^2 - \int_\Omega F(t,x,u(t,x))dtdx.$$

By (f_3), (f_4), inequality $F(t,x,u) \geqq C_0|u|^p$ holds for some constant $C_0 > 0$. Hence

$$I_N(u) \leq \tfrac{1}{2}r^2 - C_0 \|u\|_{L_p}^p \leq \tfrac{1}{2}r^2 - C\|u\|_{L_2}^p$$

$$= \tfrac{1}{2}r^2 - C(r^2\|e\|_{L_2}^2 + \|v\|_{L_2}^2)^{p/2}$$

$$\leq \tfrac{1}{2}r^2 - C(r^p\|e\|_{L_2}^p + \|v\|_{L_2}^p),$$

where C is a constant independent of N. Since $H_N^0 \oplus H_N^+$ is a finite dimensional space, there exists a constant C(N) such that

$$C\|v\|_{L_2}^p \geq C(N)\|v\|_H^p,$$

hence

$$I_N(u) \leq \tfrac{1}{2}r^2 - Cr^p\|e\|_{L_2}^p - C(N)\|v\|_H^p, \quad (p>2).$$

Now we choose $r_1>\rho$ such that

$$\tfrac{1}{2}r^2 - Cr^p\|e\|_{L_2}^p \leq 0, \quad \text{for } r \geq r_1,$$

and set

$$\beta = \max_{0\leq r\leq r_1} (\tfrac{1}{2}r^2 - Cr^p\|e\|_{L_2}^p), \quad r_2 = r_2(N) = (\beta/C(N))^p,$$

$$Q = \{re \mid 0 \leq r \leq r_1\} + (B_{r_2} \cap E_2),$$

then

$$I_N(u) \leq \omega = 0, \quad \text{on } \partial Q,$$

and the condition (4) of Lemma 2 is verified. By Lemma 2, $I_N(u)$ has a critical value $C_N \geq \alpha > 0$. From the proof of Theorem (0.1) in [6], we know

$$C_N = \inf_{h\in\Gamma} \max_{u\in Q} I_N(h(u))$$

where $\Gamma = \{h \in C(Q,H_N);\ h(u)|_{\partial Q} = u\}$. Since mapping h=Id belongs to Γ, we have

$$C_N \leq \max_{u\in Q} I_N(u) \leq \max_{r\in[0,r_1]} (\tfrac{1}{2}r^2 - Cr^p\|e\|_{L_2}^p - C(N)\|v\|_H^p) \leq \beta.$$

Let u_N be a critical point of $I_N(u)$ corresponding to C_N. Obviously, $u_N \neq 0$, since $I_N(u_N) = C_N \geq \alpha > 0$. Next, by the definition of u_N,

$$\langle I_N'(u_N), w\rangle = 0, \text{ for any } w \in H_N. \tag{3}$$

Taking $w=u_N$ and using inequality (1), we have

$$\beta \geq C_N = I_N(u_N) - \tfrac{1}{2}\langle I_N'(u_N), u_N\rangle$$

$$= \int_\Omega (\tfrac{1}{2}f(t,x,u_N)u_N - F(t,x,u_N))\,dt\,dx \geq a_7\|u\|_{L_p}^p - a_6,$$

which shows that $\{u_N\}$ is uniformly bounded in $L_p(\Omega)$ with respect to N. Next, by (3) we know that

$$0=\langle I_N'(u_N),u_N^-\rangle=\|u_N^-\|_H^2-\int_\Omega f(t,x,u_N)u_N^-dtdx$$

$$0=-\langle I_N'(u_N),u^+\rangle=\|u_N^+\|_H^2+\int_\Omega f(t,x,u_N)u_N^+dtdx,$$

so

$$\|u_N^-\|_H^2+\|u_N^+\|_H^2=\int_\Omega f(t,x,u_N)(u_N^- -u_N^+)dtdx$$

$$\leqq\|u_N^- -u_N^+\|_{L_p}\cdot\|f(t,x,u_N)\|_{L_{p'}}\leqq C\|u_N\|_H\cdot\|f(t,x,u_N)\|_{L_{p'}}$$

where $\frac{1}{p}+\frac{1}{p'}=1$. Notice that (f_1) yields

$$\|f(t,x,u_N)\|_{L_{p'}}\leqq C\{(\int_\Omega|u_N|^{p'(p-1)}dtdx)^{1/p'}+1\}$$

$$\leqq C\{\|u_N\|_{L_p}^{p/p'}+1\},$$

then by the uniform boundedness of $\|u_N\|_{L_p}$ we get

$$\|u_N^-\|_H^2+\|u_N^+\|_H^2\leqq C\|u_N\|_H.$$

Therefore,

$$\|u_N\|_H^2=\|u_N^0\|_{L_2}^2+\|u_N^-\|_H^2+\|u_N^+\|_H^2$$

$$\leqq C\{\|u_N\|_{L_p}^2+\|u_N\|_H\}\leqq C\{1+\|u_N\|_H\},$$

which shows that $\{u_N\}$ is also uniformly bounded in H with respect to N and the proof of the theorem is completed.

Lemma 5. Suppose that $\{u_N\}$ weakly converges to u in H and in $L_p(\Omega)$, then u is a critical point of I(u) on H.

Proof. It suffices to show

$$(Lu-f(v),u-v)\geqq 0 \quad \text{for any } v\in H\cap L_p(\Omega), \tag{4}$$

then the proof can be completed in a standard way. Here and hereafter, we write $f(u)=f(t,x,u)$ for notational simplicity.

By (f_3), we have for any $v\in H\cap L_p(\Omega)$

$$(f(u_N)-f(v),u_N-v)=(f(u_N),u_N)-(f(u_N),v)-(f(v),u_N-v). \tag{5}$$

Since $u_N\to u$ in $L_p(\Omega)$ weakly, It is easy to see

$$(f(v),u_N-v)\to(f(v),u-v) \quad \text{as } N\to\infty.$$

By the definition of u_N,

$$0=\langle I_N'(u_N),v_N\rangle=(Lu_N-f(u_N),v_N), \quad \text{for any } v_N\in H_N,$$

so $Lu_N=P_Nf(u_N)$, where $P_Nf(u_N)$ denotes the L_2 projection of $f(u_N)$ onto H_N.

Since $\|f(u_N)\|_{L_{p'}}$ is uniformly bounded and P_N is a bounded operator from $L_{p'}$ to $L_{p'}$, it follows that $\{Lu_N\}$ is uniformly bounded in $L_{p'}$. Without loss of qenerality, we may assume that $Lu_N \to Lu$ in $L_{p'}(\Omega)$ weakly. Moreover, $u_N \to u$ in H weakly implies that

$$u_N^+ \to u^+, \quad u_N^- \to u^- \quad \text{strongly in } L_{p'}(\Omega).$$

Therefore, we have

$$(f(u_N),u_N)=(Lu_N,u_N)=(Lu_N,u_N^+)+(Lu_N,u_N^-)$$
$$\to (Lu,u^+)+(Lu,u^-)=(Lu,u)$$

and

$$(f(u_N),v)=(P_N f(u_N),v)+((I-P_N)f(u_N),v)$$
$$=(Lu_N,v)+(f(u_N),(I-P_N)v)\to(Lu,v)$$

by facts that $(Lu_N,u_N^0)=0$ and $(I-P_N)v\to 0$ in $L_p(\Omega)$. Then (4) is derived from (5) by letting $N\to\infty$.

Finally, let us prove that $u\neq 0$. By (f_3) we have

$$a_4 \|u_N-u\|_{L_p}^p \leqq (f(u_N)-f(u),u_N-u)=(Lu_N,u_N)-(f(u_N),u)-(f(u),u_N-u)$$
$$\to (Lu,u)-(Lu,u)-(f(u),u-u)=0, \quad \text{as } N\to\infty,$$

so $u_N\to u$ in $L_p(\Omega)$ strongly and

$$0<\alpha\leqq I_N(u_N)-\tfrac{1}{2}\langle I'(u_N),u_N\rangle=\int_\Omega(\tfrac{1}{2}f(u_N)u_N-F(u_N))dtdx$$

$$\to\int_\Omega(\tfrac{1}{2}f(u)u-F(u))dtdx=I(u)-\langle I'(u),u\rangle=I(u),$$

which shows $u\neq 0$.

As a consequence of Theoem 1 and Lemma 5, we proved

Theorem 2. Suppose that (f_1)-(f_4) hold, then problem (P) has at least one nontrivial weak solution u in $H\cap L_p(\Omega)$. Moreover, if the nontrivial solution u is unique, then the approximate nontrivial solution $\{u_N\}$ obtained from (H_N) converges to u as $N\to\infty$.

References

[1] P.H.Rabinowitz, Free vibrations for a semilinear wave equation, Comm. Pure and App. Math., vol.31(1978), 31-68.

[2] K.C.Chang and C.W.Hong, Periodic solutions for the semilinear spherical wave equation, Acta Mathematica Sinica, New Series, 1(1) (1985), 87-96.

[3] Z.F.Zhou, The existence of periodic solution of nonlinear wave equations on S^n, Doctoral Thesis, Courant Institute of Mathematical Sciences, New York, 1986.

[4] K.Feng, On difference schemes and symplectic geometry, Proceedings of The 1984 Beijing Symposium on Differential Geometry and Differential Equations, Science Press, Beijing, 1985, 42-58.

[5] P.H.Rabinowitz, Periodic solutions of Hamiltonian systems, Comm. Pure and Appl. Math., 31(1978), 157-184.
[6] V.Benci and P.H.Rabinowitz, Critical point theorem for indefinite functionals, Ivent. Math., 52(1979), 242-273.

PARALLEL ALGORITHMS AND DOMAIN DECOMPOSITION*

Kang Li-shan
Department of Mathematics,
Wuhan University, China

1. INTRODUCTION

High-performance parallel computers are increasingly in demand in the areas of the large-scale scientific computations. Without super-power parallel computers, the challenge of large-scale scientific computation to advance human civilization can not be made within a reasonable time period. Most models in large-scale scientific computation lead to solve the multi-dimensional nonlinear partial differential equations, weather forecasting for example. For developing the solvers of these problems we have to turn our face to the super-computers. But now the architectures of the parallel computers are changing quickly. So the designers of the parallel algorithms can not always concentrate their attention to a concrete parallel computer. Therefore it is important to found a fundamental theory of the parallel algorithms for the MIMD machines. For this purpose, we focus our attention on the domain decomposition techniques, on which a class of new asynchronous parallel algorithms has been designed for solving the PDEs on parallel computers, especially, on MIMD machines [1].

There are several reasons why the domain decomposition techniques might be attractive: (1) Since the problem can be decomposed into relative independent subproblems mapping on different processors and the communication needed is limited to the pseudo-boundaries of the subdomains. (2) It gives us a flexible strategy, the "divide-and-conquer" strategy, for solving complicated problems on the MIMD machines, because special solution techniques might exist for solving the subproblems on the subdomains efficiently and the corresponding software could be stored in a distributed manner in the local memories of the processors. It is often the case that the finite difference technique is used for the regular subdomains and the finite element method for the irregular ones. (3) The ways for decomposing the domain

* This work was supported by the National Natural Science Foundation.

are flexible. It is very useful for forming the adaptive parallel algorithms for MIMD machines and for solving the equations whose solutions might have different natures in different subdomains. There are some papers [2] - [5] concerned with these problems.

In this paper, the Schwarz-type iterative methods are considered. In section 2, we consider the convergence rate of Schwarz algorithms for the model problems. The domain is decomposed into m strips. In section 3, the convergence rate of the Schwarz algorithm with relaxation factors are considered. In section 4, we discuss the convergence of the asynchronous parallel Schwarz algorithm S-COR for general problems. Numerical results can be found in [1] - [6].

2. CONVERGENCE RATE OF THE SCHWARZ ALGORITHM

Consider the Dirichlet problem:

$$\begin{cases} -\Delta u + q^2 u = f & \text{in } \Omega = \{(x,y) \mid 0 < x < 1,\ 0 < y < 1\} \\ u = \varphi & \text{on } \Gamma, \end{cases} \tag{1}$$

where q is constant and Γ is the boundary of Ω.

The domain Ω is decomposed into m subdomains:

$$\Omega_j = \{ (x,y) \mid x_j^{(1)} < x < x_j^{(2)},\ 0 < y < 1\}, \quad j = 1,2,\dots,m,$$

where $x_1^{(1)} = 0$, $x_m^{(2)} = 1$, and $x_j^{(1)} < x_j^{(2)}$, $x_k^{(1)} < x_{k+1}^{(1)} < x_k^{(2)}$ and $x_{k+1}^{(1)} < x_k^{(2)} < x_{k+1}^{(2)}$, $k = 1,2,\dots,m-1$.

$$\Gamma_j^{(s)} = \{ (x,y) \mid x = x_j^{(s)},\ 0 < y < 1\}, \quad s = 1,2.$$

For solving the problem (1), the Schwarz procedure proceeds as follows:

$$\begin{cases} -\Delta u_j^{(i+1)} + q^2 u_j^{(i+1)} = f & \text{in } \Omega_j \\ u_j^{(i+1)} = \varphi & \text{on } \Gamma \\ u_j^{(i+1)} = u_{j-1}^{(t)} & \text{on } \Gamma_j^{(1)} \\ u_j^{(i+1)} = u_{j+1}^{(i)} & \text{on } \Gamma_j^{(2)} \end{cases} \tag{2}$$

$$j = 1,2,\dots,m,$$
$$i = 0,1,2,\dots.$$

where $u_0^{(i)} = \varphi(0,y)$, $u_{m+1}^{(i)} = \varphi(1,y)$, and $u_j^{(o)}$, $j = 1,\dots,m$, are the initial guesses.

If $t = i$ in (2), then procedure (2) is called the Schwarz-Jacobi procedure that is a synchronous parallel algorithm. If $t = i+1$, then the procedure (2) is called the Schwarz-Gauss-Seidel procedure, especially, when $m = 2$, that is the Schwarz alternating procedure.

Now we consider the Schwarz-Jacobi method, that is the case for $t = i$ in (2).

Denote the solution of (1) by u^* and the errors of (2) by

$$e_j^{(i+1)} = u^* - u_j^{(i+1)} \quad \text{on } \bar{\Omega}_j, \quad j = 1,2,\ldots,m.$$

If the initial errors can be expressed as follows:

$$e_{j+1}^{(o)}(x_j^{(2)},y) = \sum_{k=1}^{\infty} a_{j+1,k}^{(o)} \sin k\pi y, \quad j = 1,2,\ldots,m-1 \tag{3}$$

and

$$e_{j-1}^{(o)}(x_j^{(1)},y) = \sum_{k=1}^{\infty} b_{j-1,k}^{(o)} \sin k\pi y, \quad j = 2,3,\ldots,m, \tag{4}$$

then the errors are

$$e_j^{(i+1)}(x,y) = \sum_{k=1}^{\infty} \left\{ b_{j-1,k}^{(i)} r_{j,k}^{(1)}(x) + a_{j+1,k}^{(i)} r_{j,k}^{(2)}(x) \right\} \sin k\pi y \tag{5}$$

$$j = 1,2,\ldots,m,$$
$$i = 0,1,2,\ldots.$$

where

$$\begin{aligned} r_{j,k}^{(1)}(x) &= \operatorname{sh}Q_k(x - x_j^{(2)}) / \operatorname{sh}Q_k(x_j^{(1)} - x_j^{(2)}), \\ r_{j,k}^{(2)}(x) &= \operatorname{sh}Q_k(x - x_j^{(1)}) / \operatorname{sh}Q_k(x_j^{(2)} - x_j^{(1)}), \\ Q_k &= (k^2\pi^2 + q^2)^{\frac{1}{2}}, \\ r_{1,k}^{(1)}(x) &= 0 \text{ and } r_{m,k}^{(2)}(x) = 0. \end{aligned} \tag{6}$$

Denote

$$\left\{ \begin{aligned} e_j^{(i)}(x_{j-1}^{(2)},y) &= \sum_{k=1}^{\infty} a_{j,k}^{(i)} \sin k\pi y, \\ e_j^{(i)}(x_{j+1}^{(1)},y) &= \sum_{k=1}^{\infty} b_{j,k}^{(i)} \sin k\pi y, \end{aligned} \right. \tag{7}$$

then from (5), we have

$$\left\{ \begin{aligned} a_{j,k}^{(i+1)} &= b_{j-1,k}^{(i)} r_{j,k}^{(1)}(x_{j-1}^{(2)}) + a_{j+1,k}^{(i)} r_{j,k}^{(2)}(x_{j-1}^{(2)}), \\ b_{j,k}^{(i+1)} &= b_{j-1,k}^{(i)} r_{j,k}^{(1)}(x_{j+1}^{(1)}) + a_{j+1,k}^{(i)} r_{j,k}^{(2)}(x_{j+1}^{(1)}). \end{aligned} \right. \tag{8}$$

$$k = 1,2,3,\ldots.$$
$$i = 0,1,2,\ldots.$$

Denote

$$E_k^{(i)} = (b_{1,k}^{(i)}, a_{2,k}^{(i)}, b_{2,k}^{(i)}, \ldots, a_{j,k}^{(i)}, b_{j,k}^{(i)}, \ldots, a_{m-1,k}^{(i)}, b_{m-1,k}^{(i)}, a_{m,k}^{(i)})^T,$$

then from (6) and (8), we have

$$E_k^{(i+1)} = T_k E_k^{(i)}, \qquad k = 1,2,3,\ldots. \quad i = 0,1,2,\ldots. \tag{9}$$

where $T_k = (\, t_{p,q}^{(k)})$, $\quad p,q = 1,2,\ldots,2(m-1)$,

and

$$t^{(k)}_{2j,2j-1} = r^{(1)}_{j+1,k}(x^{(2)}_j), \quad t^{(k)}_{2j,2j+2} = r^{(2)}_{j+1,k}(x^{(2)}_j),$$

$$t^{(k)}_{2j+1,2j-1} = r^{(2)}_{j+1,k}(x^{(1)}_{j+2}), \quad t^{(k)}_{2j+1,2j+2} = r^{(2)}_{j+1,k}(x^{(1)}_{j+2}).$$

From (6), we can prove that

$$\begin{cases} 0 \le r^{(s)}_{j,k}(x) \le 1, \quad s = 1,2, \quad x \in [x^{(1)}_j, x^{(2)}_j] \\ 0 < r^{(1)}_{j,k}(x) + r^{(2)}_{j,k}(x) < 1, \quad x \in (x^{(1)}_j, x^{(2)}_j) \end{cases} \tag{10}$$

and

$$r^{(s)}_{j,k+1}(x) < r^{(s)}_{j,k}(x), \quad s = 1,2; \quad k = 1,2,3,\ldots.$$

So we have

$$\rho(T_k) < 1 \tag{11}$$

and

$$\rho(T_{k+1}) < \rho(T_k), \quad k = 1,2,3,\ldots. \tag{12}$$

where $\rho(T_k)$ is the spectrum radius of the matrix T_k. It means that the convergence factor of the Schwarz-Jacobi procedure is the spectrum radius of T_1, and the procedure converges.

For the Schwarz-Gauss-Seidel procedure, the errors are

$$e^{(i+1)}_j(x,y) = \sum_{k=1}^{\infty} \{ b^{(i+1)}_{j-1,k}\, r^{(1)}_{j,k}(x) + a^{(i)}_{j+1,k}\, r^{(2)}_{j,k}(x) \} \sin k\pi y \quad \text{on } \bar{\Omega}_j \tag{13}$$

$$j = 1,2,\ldots,m,$$
$$i = 0,1,2,\ldots.$$

where the initial errors are

$$e^{(o)}_{j+1}(x^{(2)}_j, y) = \sum_{k=1}^{\infty} a^{(o)}_{j+1,k} \sin k\pi y, \quad j = 1,2,\ldots,m-1.$$

Denote $\overset{*}{E}{}^{(i)}_k = (a^{(i)}_{2,k}, a^{(i)}_{3,k}, \ldots, a^{(i)}_{j,k}, \ldots, a^{(i)}_{m,k})^T$,

then from (13), we have

$$\overset{*}{E}{}^{(i+1)}_k = \overset{*}{T}_k\, \overset{*}{E}{}^{(i)}_k \qquad i = 0,1,2,\ldots. \quad k = 1,2,3,\ldots. \tag{14}$$

where $\overset{*}{T}_k$ is a lower Hessenberg matrix with order (m-1).

$$\overset{*}{T}_k = (\overset{*}{t}{}^{(k)}_{p,q}),$$

where $p = 2,\ldots,m, \quad q = 1,2,\ldots,p,$

$$\begin{cases} \overset{*}{t}{}^{(k)}_{p,p} = r^{(2)}_{p,k}(x^{(2)}_{p-1}), \\ \overset{*}{t}{}^{(k)}_{p,q} = r^{(1)}_{p,k}(x^{(2)}_{p-1}) r^{(2)}_{q,k}(x^{(1)}_{q+1}) \cdot \prod_{j=q+1}^{p-1} r^{(1)}_{j,k}(x^{(1)}_{j+1}). \end{cases} \tag{15}$$

According to (10), we can prove that

$$\rho(\overset{*}{T}_k) < 1, \tag{16}$$

$$\rho(\overset{*}{T}_{k+1}) < \rho(\overset{*}{T}_k), \quad k = 1,2,3,\ldots. \tag{17}$$

and

$$\rho(\overset{*}{T}_1) < \rho(T_1), \tag{18}$$

where $\rho(\overset{*}{T}_k)$ is the spectrum radius of $\overset{*}{T}_k$.

From (16) and (17) we know that the convergence factor of the Schwarz-Gauss-Seidel procedure is $\rho(\overset{*}{T}_1)$ and the procedure is convergent.

The inequality (18) means that the convergence rate of the Schwarz-Gauss-Seidel procedure is higher than the rate of the Schwarz-Jacobi-procedure.

When m=2, that is the case of the Schwarz alternating procedure,it is easy to prove that the convergence factor is

$$\rho(\overset{*}{T}_1) = \frac{\operatorname{sh}Q_1 x_2^{(1)}}{\operatorname{sh}Q_1 x_1^{(2)}} \cdot \frac{\operatorname{sh}Q_1(1 - x_1^{(2)})}{\operatorname{sh}Q_1(1 - x_2^{(1)})}, \tag{19}$$

where $Q_1 = (\pi^2 + q^2)^{\frac{1}{2}}$.

Let $x_1^{(2)} = x_2^{(1)} + d$, the formula (19) can be rewritten as

$$\rho(\overset{*}{T}_1) = \frac{\operatorname{sh}Q_1 x_2^{(1)}}{\operatorname{sh}Q_1(x_2^{(1)} + d)} \cdot \frac{\operatorname{sh}Q_1(1 - x_2^{(1)} - d)}{\operatorname{sh}Q_1(1 - x_2^{(1)})}. \tag{20}$$

It describes all the convergence character of the Schwarz alternating procedure: the relation between the convergence factor and the positions of the pseudo-boundaries,the convergence factor and the overlapping d. One can see that the more the overlapping, the faster the convergence.

Now we consider the Neumann problem:

$$\begin{cases} -\Delta u + q^2 u = f & \text{in } \Omega = \{ (x,y) \mid 0 < x < 1,\ 0 < y < 1 \} \\ \dfrac{\partial u}{\partial n} = \varphi & \text{on } \Gamma, \end{cases} \tag{21}$$

where $q \neq 0$.

By using Schwarz alternating procedure (m=2) we solve it:

$$\begin{cases} -\Delta u_j^{(i+1)} + q^2 u_j^{(i+1)} = f & \text{in } \Omega_j \\ \dfrac{\partial u^{(i+1)}}{\partial n} = \varphi & \text{on } \Gamma \\ \dfrac{\partial u_j^{(i+1)}}{\partial x} = \dfrac{\partial u_{j-1}^{(i+1)}}{\partial x} & \text{on } \Gamma_j^{(1)} \\ \dfrac{\partial u_j^{(i+1)}}{\partial x} = \dfrac{\partial u_{j+1}^{(i)}}{\partial x} & \text{on } \Gamma_j^{(2)} \\ j = 1,2; \quad i = 0,1,2,\dots. \end{cases} \tag{22}$$

where

$\frac{\partial u_2^{(o)}}{\partial x}$ on $\Gamma_1^{(2)}$ is the initial guess, $\frac{\partial u_o^{(i+1)}}{\partial x} = \varphi$ on $\Gamma_1^{(1)}$,

and $\frac{\partial u_3^{(i)}}{\partial x} = \varphi$ on $\Gamma_2^{(2)}$.

Assume that the initial error can be expressed as follows

$$\frac{\partial e_2^{(o)}(x_1^{(2)},y)}{\partial x} = \sum_{k=1}^{\infty} \bar{b}_k^{(o)} \cos k\pi y, \tag{23}$$

then the errors are

$$\begin{cases} e_1^{(i+1)}(x,y) = \sum_{k=1}^{\infty} \left(\frac{\bar{a}_k^{(i+1)}}{Q_k} \cdot \frac{\mathrm{ch} Q_k x}{\mathrm{sh} Q_k x_1^{(2)}} \cos k\pi y \right) \text{ on } \bar{\Omega}_1 \\ e_2^{(i+1)}(x,y) = \sum_{k=1}^{\infty} \left(\frac{\bar{b}_k^{(i+1)}}{Q_k} \cdot \frac{\mathrm{ch} Q_k (x-1)}{\mathrm{sh} Q_k (x_2^{(1)}-1)} \cos k\pi y \right) \text{ on } \bar{\Omega}_2, \end{cases} \tag{24}$$

where $Q_k = (k^2\pi^2 + q^2)^{\frac{1}{2}}$.

Denote

$$a_k^{(i)} = \frac{\bar{a}_k^{(i)}}{Q_k} \cdot \frac{\mathrm{ch} Q_k x_2^{(1)}}{\mathrm{sh} Q_k x_1^{(2)}} \quad \text{and} \quad b_k^{(i)} = \frac{\bar{b}_k^{(i)}}{Q_k} \cdot \frac{\mathrm{ch} Q_k (x_1^{(2)}-1)}{\mathrm{sh} Q_k (x_2^{(1)}-1)} \tag{25}$$

From (24) and the pseudo-boundary conditions, we have

$$a_k^{(i+1)} = \rho_k a_k^{(i)} \quad \text{and} \quad b_k^{(i+1)} = \rho_k b_k^{(i)} \tag{26}$$

where

$$\rho_k = \frac{\mathrm{sh} Q_k x_2^{(1)}}{\mathrm{sh} Q_k x_1^{(2)}} \cdot \frac{\mathrm{sh} Q_k (x_1^{(2)} - 1)}{\mathrm{sh} Q_k (x_2^{(1)} - 1)} \tag{27}$$

and

$$\bar{a}_k^{(1)} = \bar{b}_k^{(o)} \quad \text{and} \quad \bar{b}_k^{(1)} = \frac{\mathrm{sh} Q_k x_2^{(1)}}{\mathrm{sh} Q_k x_1^{(2)}} \bar{b}_k^{(o)}. \tag{28}$$

So we have

$$\begin{cases} a_k^{(i+1)} = \rho_k^i a_k^{(1)} = \rho_k^i \frac{\mathrm{ch} Q_k x_2^{(1)}}{Q_k \mathrm{sh} Q_k x_1^{(2)}} \bar{b}_k^{(o)}, \\ b_k^{(i+1)} = \rho_k^i b_k^{(1)} = \rho_k^{i+1} \frac{\mathrm{ch} Q_k (x_1^{(2)}-1)}{Q_k \mathrm{sh} Q_k (x_1^{(2)}-1)} \bar{b}_k^{(o)}. \end{cases} \tag{29}$$

But

$$\rho_{k+1} < \rho_k, \quad k = 1,2,3,\ldots.$$

It means that the convergence factor of procedure (22) is

$$\rho_1 = \frac{\mathrm{sh} Q_1 x_2^{(1)}}{\mathrm{sh} Q_1 x_1^{(2)}} \cdot \frac{\mathrm{sh} Q_1 (x_1^{(2)}-1)}{\mathrm{sh} Q_1 (x_2^{(1)}-1)}, \tag{30}$$

where $Q_1 = (\pi^2 + q^2)^{\frac{1}{2}}$.

Comparing this with the convergence factor (19) for the Dirichlet problem, one can see that their convergence factors are the same.

3. SCHWARZ ALGORITHMS WITH RELAXATION FACTORS

There are three ways to introduce relaxation factors to accelerate the convergence rate. One is to introduce an overrelaxation factor as people usually do (1). Another is called pseudo-boundary relaxation [5]. Recently G.Rodrigue posed a new way that the relaxation parameters appear on the pseudo-boundaries as the parameters of the third boundary conditions [6].

Consider the problem (1) and the acceleration procedures used for the Schwarz alternating procedure. Firstly, the pseudo-boundary relaxation is discussed. In this case, m=2, the domain Ω is decomposed into two subdomains Ω_1 and Ω_2.

The Schwarz procedure with pseudo-boundary relaxation factor ω for solving problem (1) proceeds as follows:

$$\begin{cases} -\Delta u_j^{(i+1)} + q^2 u_j^{(i+1)} = f & \text{in } \Omega_j \\ u_j^{(i+1)} = \varphi & \text{on } \Gamma \\ u_j^{(i+1)} = u_{j-1}^{(i)} + \omega (u_{j-1}^{(i+1)} - u_{j-1}^{(i)}) & \text{on } \Gamma_j^{(1)} \\ u_j^{(i+1)} = u_{j+1}^{(i-1)} + \omega (u_{j+1}^{(i)} - u_{j+1}^{(i-1)}) & \text{on } \Gamma_j^{(2)} \\ \qquad j = 1,2; \quad i = 0,1,2,\ldots \end{cases} \tag{31}$$

where $u_0^{(i)} = \varphi$ on $\Gamma_1^{(1)}$ and $u_3^{(i)} = \varphi$ on $\Gamma_2^{(2)}$, and $u_1^{(o)}(x_2^{(1)},y)$, $u_2^{(-1)}(x_1^{(2)},y)$ and $u_2^{(o)}(x_1^{(2)},y)$ are the initial guesses.

Assume that the initial errors can be expressed as follows:

$$e_1^{(o)}(x_2^{(1)},y) = \sum_{k=1}^{\infty} a_k^{(o)} \sin k\pi y \tag{32}$$

$$e_2^{(i)}(x_1^{(2)},y) = \sum_{k=1}^{\infty} b_k^{(i)} \sin k\pi y \qquad i= -1,0, \tag{33}$$

then the errors of problems (31) are given by

$$\begin{cases} e_1^{(i+1)}(x,y) = \sum_{k=1}^{\infty} a_k^{(i+1)} r_{1,k}^{(2)}(x) \sin k\pi y, \\ e_2^{(i+1)}(x,y) = \sum_{k=1}^{\infty} b_k^{(i+1)} r_{2,k}^{(2)}(x) \sin k\pi y. \end{cases} \tag{34}$$

According to the pseudo-boundary conditions of (31), we have

$$e_1^{(i+1)}(x_1^{(2)},y) = \sum_{k=1}^{\infty} a_k^{(i+1)} \sin k\pi y = (1-\omega)e_2^{(i-1)}(x_1^{(2)},y) + \omega e_2^{(i)}(x_1^{(2)},y)$$
$$= \sum_{k=1}^{\infty} r_{2,k}^{(2)}(x_1^{(2)}) \{(1-\omega)b_k^{(i-1)} + \omega b_k^{(i)}\} \sin k\pi y,$$

i.e., $a_k^{(i+1)} = r_{2,k}^{(2)}(x_1^{(2)}) \{(1-\omega)b_k^{(i-1)} + \omega b_k^{(i)}\}$, $i=0,1,2,\ldots.$ (35)

and in the same way we have

$$b_k^{(i+1)} = r_{1,k}^{(2)}(x_2^{(1)}) \{(1-\omega)a_k^{(i)} + \omega a_k^{(i+1)}\}, \quad i=0,1,2,\ldots. \tag{36}$$

From (35) and (36) we have

$$\begin{cases} a_k^{(i+1)} = \rho_k \{\omega^2 a_k^{(i)} - 2\omega(\omega-1)a_k^{(i-1)} + (\omega-1)^2 a_k^{(i-2)}\}, & i=2,3,4,\ldots. \\ b_k^{(i+1)} = \rho_k \{\omega^2 b_k^{(i)} - 2\omega(\omega-1)b_k^{(i-1)} + (\omega-1)^2 b_k^{(i-2)}\}, & i=1,2,3,\ldots. \end{cases} \tag{37}$$

where $\rho_k = r_{1,k}^{(2)}(x_2^{(1)}).r_{2,k}^{(2)}(x_1^{(2)})$.

The characteristic equations of the difference equations (37) are

$$r^3 - \rho_k \omega^2 (r - s)^2 = 0, \quad k = 1,2,\ldots. \tag{38}$$

where $s = (\omega - 1)/\omega$.

From (34),(35) and (36), we can see that the convergence rate is determined by the low-frequency components. So we define the convergence factor of the procedure (31) as

$$r(\rho_1, \omega) = \max\{|r_1|, |r_2|, |r_3|\}, \tag{39}$$

where $r_j (j = 1,2,3)$ are the roots of the cubic equation

$$r^3 - \rho_1 \omega^2 (r - s)^2 = 0, \tag{40}$$

where $0 \le \rho_1 < 1$.

The optimum convergence factor is determined by

$$\rho_{opt} = \min_{\omega \in R} r(\rho_1, \omega) = 3(\omega_{opt} - 1)/\omega_{opt}, \tag{41}$$

where $\omega_{opt} = (3/\sqrt{\rho_1}) \cos\{[\arccos(-\sqrt{\rho_1}) + 4\pi]/3\}$, (42)

that is the optimum pseudo-boundary relaxation factor(see Evans and Kang [5]).

Secondly, we consider the relaxation factor as the parameters of the pseudo-boundary condition. The Schwarz procedure with the third pseudo-boundary conditions for solving problem (1) proceeds as follows (m=2):

$$\begin{cases} -\Delta u_j^{(i+1)} + q^2 u_j^{(i+1)} = f & \text{in } \Omega_j \\ u_j^{(i+1)} = \varphi & \text{on } \Gamma \\ u_j^{(i+1)} + \omega_j \dfrac{\partial u_j^{(i+1)}}{\partial x} = u_{j-1}^{(i+1)} + \omega_j \dfrac{\partial u_{j-1}^{(i+1)}}{\partial x} & \text{on } \Gamma_j^{(1)} \end{cases} \tag{43}$$

$$\left\{ u_j^{(i+1)} + \omega_j \frac{\partial u_j^{(i+1)}}{\partial x} = u_{j+1}^{(i)} + \omega_j \frac{\partial u_{j+1}^{(i+1)}}{\partial x} \quad \text{on} \quad \Gamma_j^{(2)} \right.$$

$$j = 1,2; \quad i = 0,1,2,\ldots.$$

where $u_0^{(i)} = \varphi(0,y)$, $u_3^{(i)} = \varphi(1,y)$, $\frac{\partial u_0^{(i)}}{\partial x} = \frac{\partial u_3^{(i)}}{\partial x} = 0$, and $u_2^{(o)}$ and $\frac{\partial u_2^{(o)}}{\partial x}$ on $\Gamma_1^{(2)}$ are the initial guesses.

Denote the errors $e_j^{(i)}(x,y) = u^*(x,y) - u_j^{(i)}(x,y)$ on $\bar{\Omega}_j$.

Assume the initial errors can be expressed as

$$e_2^{(o)}(x_1^{(2)},y) = \sum_{k=1}^{\infty} b_{k,2}^{(o)} \sin k\pi y, \qquad \frac{\partial e_2^{(o)}}{\partial x}(x_1^{(2)},y) = \sum_{k=1}^{\infty} \bar{b}_{k,2}^{(o)} \sin k\pi y, \tag{44}$$

then the errors have the form

$$e_j^{(i)}(x,y) = \sum_{k=1}^{\infty} \varphi_{k,j}^{(i)}(x) \sin k\pi y, \tag{45}$$

where $\varphi_{k,j}^{(i)}(x)$ satisfy the one-dimensional problems:

$$\begin{cases} -\varphi_{k,1}^{(i)''} + Q_k^2 \varphi_{k,1}^{(i)} = 0 \\ \varphi_{k,1}^{(i)}(0)=0, \ (\varphi_{k,1}^{(i)} + \omega_1 \varphi_{k,1}^{(i)'})_{x=x_1^{(2)}} = (\varphi_{k,2}^{(i-1)} + \omega_1 \varphi_{k,2}^{(i-1)'})_{x=x_1^{(2)}} \\ \qquad \equiv l_{k,2}^{(i-1)}(x_1^{(2)}) \end{cases} \tag{46}$$

and

$$\begin{cases} -\varphi_{k,2}^{(i)''} + Q_k^2 \varphi_{k,2}^{(i)} = 0 \\ \varphi_{k,2}^{(i)}(1)=0, \ (\varphi_{k,2}^{(i)} + \omega_2 \varphi_{k,2}^{(i)'})_{x=x_2^{(1)}} = (\varphi_{k,1}^{(i)} + \omega_2 \varphi_{k,1}^{(i)'})_{x=x_2^{(1)}} \\ \qquad \equiv l_{k,1}^{(i)}(x_2^{(1)}), \end{cases} \tag{47}$$

where $Q_k = (k^2\pi^2 + q^2)^{\frac{1}{2}}$ and $l_{k,2}^{(o)}(x_1^{(2)}) = b_{k,2}^{(o)} + \omega_1 \bar{b}_{k,2}^{(o)}$.

The solutions of (46) and (47) are

$$\begin{cases} \varphi_{k,1}^{(i)}(x) = \dfrac{\operatorname{sh} Q_k x}{\operatorname{sh} Q_k x_1^{(2)} + \omega_1 Q_k \operatorname{ch} Q_k x_1^{(2)}} \, l_{k,2}^{(i-1)}(x_1^{(2)}), \\ \varphi_{k,2}^{(i)}(x) = \dfrac{\operatorname{sh} Q_k (x-1)}{\operatorname{sh} Q_k (x_2^{(1)}-1) + \omega_2 Q_k \operatorname{ch} Q_k (x_2^{(1)}-1)} \, l_{k,1}^{(i)}(x_2^{(1)}), \end{cases} \tag{48}$$

$$k,i = 1,2,3,\ldots.$$

From (48), we have

$$\begin{cases} l_{k,1}^{(i)}(x_2^{(1)}) = \varphi_{k,1}^{(i)}(x_2^{(1)}) + \omega_2 \varphi_{k,1}^{(i)'}(x_2^{(1)}) = \rho_k(\omega_1, \omega_2) l_{k,1}^{(i-1)}(x_2^{(1)}) \\ l_{k,2}^{(i)}(x_1^{(2)}) = \varphi_{k,2}^{(i)}(x_1^{(2)}) + \omega_1 \varphi_{k,2}^{(i)'}(x_1^{(2)}) = \rho_k(\omega_1, \omega_2) l_{k,2}^{(i-1)}(x_1^{(2)}), \end{cases} \tag{49}$$

where

$$\rho_k(\omega_1,\omega_2)=\frac{shQ_kx_2^{(1)}+\omega_2Q_kchQ_kx_2^{(1)}}{shQ_kx_1^{(2)}+\omega_1Q_kchQ_kx_1^{(2)}}\cdot\frac{shQ_k(x_1^{(2)}-1)+\omega_1Q_kchQ_k(x_1^{(2)}-1)}{shQ_k(x_2^{(1)}-1)+\omega_2Q_kchQ_k(x_2^{(1)}-1)}. \tag{50}$$

From (48), (49) and (50) we get

$$\begin{cases} \varphi_{k,1}^{(i)}(x)=\dfrac{shQ_kx}{shQ_kx_1^{(2)}+\omega_1Q_kchQ_kx_1^{(2)}}[\rho_k(\omega_1,\omega_2)]^{i-1}(b_{k,2}^{(o)}+\omega_1\bar{b}_{k,2}^{(o)}) \\ \varphi_{k,2}^{(i)}(x)=\dfrac{shQ_k(x-1)}{shQ_k(x_1^{(2)}-1)+\omega_1Q_kchQ_k(x_1^{(2)}-1)}[\rho_k(\omega_1,\omega_2)]^{i}(b_{k,2}^{(o)}+ \\ \qquad +\omega_1\bar{b}_{k,2}^{(0)}). \end{cases} \tag{51}$$

The convergence factor of procedure (43) is defined by

$$\rho(\omega_1,\omega_2) = \max_{k\in N^+}|\rho_k(\omega_1,\omega_2)| . \tag{52}$$

Especially,

$$\rho(1,1) = \rho_1(1,1) = \rho(\overset{*}{T}_1),$$

where $\rho(\overset{*}{T}_1)$ is the convergence factor of Schwarz alternating procedure (see formula (19)).

For accelerating the convergence speed, we must choose the parameters ω_1 and ω_2 to make the convergence factor as small as possible, that is

$$\rho_{opt} = \min_{\omega_1,\omega_2\in R}\ \max_{k\in N^+}|\rho_k(\omega_1,\omega_2)| . \tag{53}$$

A useful and easy way to determine the parameters is to choose ω_1 and ω_2 such that the coefficients of lower-frequencies of the errors vanish.

By solving the system

$$\begin{cases} \rho_1(\omega_1,\omega_2) = 0 \\ \rho_2(\omega_1,\omega_2) = 0, \end{cases}$$

we have

$$\begin{cases} \omega_1 = \dfrac{shQ_1(1-x_1^{(2)})}{Q_1chQ_1(1-x_1^{(2)})} \\ \omega_2 = -shQ_2x_2^{(1)}/(Q_2chQ_2x_2^{(1)}) \end{cases} \tag{54}$$

or

$$\begin{cases} \omega_1 = shQ_2(1-x_1^{(2)})/(Q_2chQ_2(1-x_1^{(2)})) \\ \omega_2 = -shQ_1x_2^{(1)}/(Q_1chQ_1x_2^{(1)}) . \end{cases} \tag{55}$$

Up to date we wonder if the parameters determined by (54) or (55) are the solutions of problem (53). But Numerical results show that the choice is efficient in accelerating convergence (see [6]).

4. ASYNCHRONOUS PARALLEL SCHWARZ ALGORITHMS

Consider the Dirichlet problem of the second-order linear elliptic equation

$$\begin{cases} Au = f \quad \text{in} \quad \Omega \\ u|_\Gamma = \varphi, \ \Gamma \text{ is the boundary of } \Omega, \text{ where } \Omega \subset R^n \text{ is a bounded} \end{cases} \tag{56}$$

open set, and

$$Au \equiv - \sum_{i,k=1}^{n} \frac{\partial}{\partial x_i}(A_{i,k}\frac{\partial u}{\partial x_k}) + B(x)u.$$

The weak form of problem (56) is to find $u \in H^1_\varphi(\Omega)$ such that

$$a_\Omega(u,v) = f_\Omega(v) \quad \forall\ v \in H^1_0(\Omega), \tag{57}$$

where

$$H^1_\varphi(\Omega) = \{v \mid v \in H^1(\Omega);\ v = \varphi \text{ on } \Gamma\}.$$

It is equivalent to solving the variational problem

$$J_\Omega(u^*) = \inf_{v \in H^1_\varphi} J_\Omega(v), \tag{58}$$

where

$$a_\Omega(u,v) = \int_\Omega (\sum_{i,k=1}^{n} A_{i,k}\frac{\partial u}{\partial x_i}\frac{\partial v}{\partial x_k} + Buv)dx,$$

$$f_\Omega(v) = \int_\Omega fvdx,$$

$$J_\Omega(v) = \tfrac{1}{2}\, a_\Omega(v,v) - f_\Omega(v).$$

The domain Ω is decomposed into m overlapping subdomains Ω_1, Ω_2, ..., Ω_m, such that $\Omega = \bigcup_{j=1}^m \Omega_j$, and let the boundary of Ω_j be $\Gamma_j(\Gamma_j \subset \Gamma)$ and $\Gamma'_j(\Gamma'_j \subset \Omega)$, $\Gamma = \bigcup_{j=1}^m \Gamma_j$ and each Γ'_j divides Ω into at least two parts, $\Gamma'_j = \bigcup_i \Gamma_{ji}$, where $\Gamma_{ji} = \Gamma'_j \cap \Omega_i$.

Suppose

$$J_\Omega(v) = J_{\Omega_j}(v) + J_{\Omega-\Omega_j}(v). \tag{59}$$

Denote

$$K_j(u) = \{ v \mid v \in H^1_\varphi(\Omega);\ v = u \quad \text{on} \quad \Gamma'_j \}.$$

Obviously, $K_j(u) \subset H^1_\varphi(\Omega)$.

The asynchronous parallel Schwarz algorithm can be described as follows:

Given $u^0 \in H^1_\varphi(\Omega)$, set a sequence of functions $\{u^{i(j)}\}$ as follows:

$$\begin{cases} \hat{u}^{i(j)} = \{ v \mid \inf_{v \in K_j(u^{i-1})} J_{\Omega_j}(v) \ \text{ in } \Omega_j;\ v = u^{i-1} \text{ on } \Omega - \Omega_j \} \\ u^{i(j)} = u^{i-1} + \omega(\hat{u}^{i(j)} - u^{i-1}), \end{cases} \tag{60}$$

$$j \in M = \{1,2,\ldots,m\}; \quad i = 1,2,\ldots.$$

where ω is a relaxation factor.

The superscript i(j) for $\hat{u}^{i(j)}$ means that the approximate solution $\hat{u}^{i(j)}$ is obtained by relaxing the $j^{\underline{th}}$ subdomain Ω_j at the $i^{\underline{th}}$ step.

Divide the sequence $\{u^{i(j)}\}$ into m subsequences $\{u^{i_1(1)}\}$, $\{u^{i_2(2)}\}$, ..., $\{u^{i_m(m)}\}$.

Definition The algorithm (60) is called S-COR algorithm (Schwarz -chaotic overrelaxation), if as $i \to \infty$,

$$i_j(j) \to \infty \quad \forall j \in M. \tag{61}$$

Theorem Suppose that the bilinear form $a_\Omega(u,v)$ is symmetric continuous and uniformly elliptic over $H_0^1(\Omega)$ and the linear functional $f_\Omega(v)$ is continuous, and $0 < \omega < 2$, then the sequence of functions $\{u^{i(j)}\}$ given by algorithm (60) converges to the solution u^* of problem (56).

Proof Firstly, we want to prove that $\{J_\Omega(u^{i(j)})\}$ is a monotonically decreasing sequence and has the same limit with sequence $\{J_\Omega(\hat{u}^{i(j)})\}$.

From (60), we know $\forall \omega \in R$

$$u^{i(j)} = u^{i-1} + \omega(\hat{u}^{i(j)} - u^{i-1}) \in K_j(u^{i-1})$$

and

$$J_{\Omega_j}(u^{i(j)}) \geq J_{\Omega_j}(\hat{u}^{i(j)}).$$

$J_{\Omega_j}(u^{i(j)})$, as a function of ω, gets its minimum value at $\omega=1$. Therefore, we have

$$\left.\frac{\partial J_{\Omega_j}(u^{i(j)})}{\partial \omega}\right|_{\omega=1} = 0.$$

It leads to the related expression

$$a_{\Omega_j}(u^{i-1}, \hat{u}^{i(j)}) = a_{\Omega_j}(\hat{u}^{i(j)}, \hat{u}^{i(j)}) - f_{\Omega_j}(\hat{u}^{i(j)} - u^{i-1}) \tag{62}$$

If $\omega \in (0,2)$, then from (59),(60) and (62) we have

$$\begin{aligned} J_\Omega(u^{i(j)}) &= J_\Omega(u^{i-1}) - \frac{2-\omega}{2\omega} a_\Omega(u^{i(j)} - u^{i-1}, u^{i(j)} - u^{i-1}) \\ &\leq J_\Omega(u^{i-1}) \end{aligned} \tag{63}$$

It means that $\{J_\Omega(u^{i(j)})\}$ is a monotonically decreasing sequence.

From (60), we have

$$J_\Omega(u^{i(j)}) = J_\Omega(u^{i-1}) + \omega(2-\omega)[J_\Omega(\hat{u}^{i(j)}) - J_\Omega(u^{i-1})].$$

That is

$$J_\Omega(\hat{u}^{i(j)}) = J_\Omega(u^{i-1}) + \frac{1}{\omega(2-\omega)}[J_\Omega(u^{i(j)}) - J_\Omega(u^{i-1})]. \tag{64}$$

It shows that $\{J_\Omega(u^{i(j)})\}$ and $\{J_\Omega(\hat{u}^{i(j)})\}$ have the same limit .

Secondly, we want to prove that $\{u^{i(j)} - u^*\}$ and $\{\hat{u}^{i(j)} - u^*\}$ are bounded in $H_0^1(\Omega)$.

In fact,

$$\|u^{i(j)} - u^*\|_1^2 \le \frac{1}{r} a_\Omega(u^{i(j)} - u^*, u^{i(j)} - u^*)$$

$$= \frac{2}{r}\{J_\Omega(u^{i(j)}) - J_\Omega(u^*)\} \le \frac{2}{r}\{J_\Omega(u^0) - J_\Omega(u^*)\} = \text{const.}$$

and

$$\|\hat{u}^{i(j)} - u^*\|_1^2 \le \frac{1}{r} a_\Omega(\hat{u}^{i(j)} - u^*, \hat{u}^{i(j)} - u^*) = \frac{2}{r}\{J_\Omega(\hat{u}^{i(j)}) - J_\Omega(u^*)\}$$

$$= \frac{2}{r}\Big\{J_\Omega(u^{i-1}) - J_\Omega(u^*) + \frac{1}{\omega(2-\omega)}[J_\Omega(u^{i(j)}) - J_\Omega(u^{i-1})]\Big\}$$

$$\le \frac{2}{r}\{J_\Omega(u^{i-1}) - J_\Omega(u^*)\} \le \frac{2}{r}\{J_\Omega(u^0) - J_\Omega(u^*)\} = \text{const.}$$

Thus, a weakly convergent subsequence $\{\hat{u}^{i_k} - u^*\}$ can be chosen from $\{\hat{u}^{i(j)} - u^*\}$.

That is

$$\hat{u}^{i_k} - u^* \rightharpoonup w \in H_0^1(\Omega),$$

or $\hat{u}^{i_k} \rightharpoonup u^* + w = u \in H_\varphi^1(\Omega)$.

Thirdly, prove $u = u^*$.

If $\Omega_j \cap \Omega_i \ne 0$, then there exists a "chain" $j \leftrightarrow i$ between Ω_j and Ω_i. From the principle of the domain decomposition, we know that there exists at least a complete chain which links all the subdomains, e.g.,

$$1 \leftrightarrow 2 \leftrightarrow 3 \leftrightarrow \dots \leftrightarrow (m-1) \leftrightarrow m.$$

If that is not the case, we can renumber the subdomains.

We choose the weakly convergent subsequence as follows: $\{\hat{u}^{i_m(m)} - u^*\}$ is a bounded infinite sequence. From it, a weakly convergent subsequence can be chosen, still denote it by $\{\hat{u}^{i_m(m)} - u^*\}$. Because $\Omega_m \cap \Omega_{m-1} \ne 0$, from (60) and (61), $\{\hat{u}^{i_m(m)} - u^*\}$ restricted on $\Omega - \Omega_m$ must contain an infinite subsequence of $\{\hat{u}^{i_{m-1}(m-1)} - u^*\}$. From it, a weakly convergent subsequence can be chosen, still denote it by $\{\hat{u}^{i_{m-1}(m-1)} - u^*\}$. Along the chain $m \leftrightarrow (m-1) \leftrightarrow (m-2) \leftrightarrow \dots \leftrightarrow 2 \leftrightarrow 1$, at last, $\{\hat{u}^{i_2(2)} - u^*\}$ restricted on $\Omega - \Omega_2$ must contain a weakly convergent subsequence of $\{\hat{u}^{i_1(1)} - u^*\}$. We denote it by $\{\hat{u}^{i_1(1)} - u^*\}$. This sequence restricted on Ω_j contains an infinite subsequence of $\{\hat{u}^{i_j(j)} - u^*\}$, $\forall j \in M$. Denote the 'weak' limit of $\{\hat{u}^{i_j(j)} - u^*\}$ by $(\hat{u}_j - u^*)$, that is

$$\hat{u}^{i_j(j)} \rightharpoonup \hat{u}_j \quad \text{in } \Omega_j \quad \forall j \in M \tag{65}$$

and

$$\hat{u}_j = \hat{u}_i \quad \text{on } \Omega_j \cap \Omega_i.$$

Consider the set of problems

$$\begin{cases} Au_j = f & \text{in } \Omega_j \\ u_j = \varphi & \text{on } \Gamma_j \\ u_j = u_i & \text{on } \Gamma_{ji},\ \forall i \in \{i \in M,\ i \ne j;\ \Omega_j \cap \Omega_i \ne 0\}, \end{cases} \tag{66}$$

$j = 1,2,\ldots,m.$

Obviously, the solution of problem(56) satisfies the set of problems (66). The existence of the solution of problem (56) implies the existence of the solution of problem (66). But the solution of (66) is unique. So problem (56) is equivalent to the set of problems (66).

Analogously, the problem (57), weak form of problem (56), is equivalent to the weak form of (66):

$$\begin{cases} \text{To find } u_j \in K_j(u_i) \quad \text{such that} \\ a_{\Omega_j}(u_j,v) = f_{\Omega_j}(v) \qquad \forall v \in K_j(0) \end{cases} \tag{67}$$
$$j = 1,2,\ldots,m,$$

where

$$K_j(u_i) = \{ v \mid v \in H^1_\varphi(\Omega); \quad v = u_i \quad \text{on} \quad \Gamma_{ji}, \quad i \in M\},$$
$$K_j(0) = \{ v \mid v \in H^1_0(\Omega); \quad v = 0 \quad \text{on} \quad \Gamma_j' \},$$

or

$$J_{\Omega_j}(u_j) = \inf_{v \in K_j(u_i)} J_{\Omega_j}(v), \tag{68}$$
$$j = 1,2,\ldots,m.$$

From (60), we have

$$a_{\Omega_j}(\hat{u}^{i_j(j)},v) = f_{\Omega_j}(v) \quad \forall v \in K_j(0). \tag{69}$$

And set $H^1_\varphi(\Omega)$ is weakly closed and convex.

From (65) and (69), we have

$$\begin{cases} \hat{u}_j \in K_j(\hat{u}_i) \\ a_{\Omega_j}(\hat{u}_j,v) = f_{\Omega_j}(v) \quad \forall v \in K_j(0), \quad j \in M. \end{cases} \tag{70}$$

It means that $\hat{u}_j$ are the solutions of (67).

Denoting

$$u = \hat{u}_j \qquad \text{in} \quad \Omega_j \quad \forall j \in M,$$

we have

$$u = u^*. \tag{71}$$

Finally, we want to prove

$$u^{i(j)} \longrightarrow u^* \quad \text{strongly.}$$

If

$$\lim_{i\to\infty} J_\Omega(\hat{u}^{i(j)}) = \lim_{i\to\infty} J_\Omega(u^{i(j)}) = J,$$

then we have

$$\lim_{i_k\to\infty} J_\Omega(\hat{u}^{i_k}) = J.$$

That is

$$\lim_{i\to\infty} f_\Omega(u^{i(j)}) = \lim_{i\to\infty} f_\Omega(\hat{u}^{i(j)}) = f_\Omega(\bar{u}). \tag{72}$$

But

$$\lim_{i_k \to \infty} f_\Omega(\hat{u}^{i_k}) = f_\Omega(u^*). \qquad (73)$$

From (72) and (73), we have

$$\lim_{i \to \infty} f_\Omega(u^{i(j)} - u^*) = 0.$$

That is

$$u^{i(j)} \rightharpoonup u^* \quad \text{weakly}. \qquad (74)$$

It means that the convergence of $\{u^{i(j)}\}$ is not dependent on the initial guess u^0.

If we take $u^0 = u^*$, then we have $\bar{u} = u^*$.

That is

$$\lim_{i \to \infty} J_\Omega(u^{i(j)}) = J_\Omega(u^*). \qquad (75)$$

So we have

$$\lim_{i \to \infty} \|u^{i(j)} - u\|_1^2 \le \lim_{i \to \infty} \frac{1}{r} a_\Omega(u^{i(j)} - u^*, u^{i(j)} - u^*)$$

$$= \lim_{i \to \infty} \frac{2}{r} [J_\Omega(u^{i(j)}) - J_\Omega(u^*)] = 0.$$

It means

$$u^{i(j)} \rightarrow u \quad \text{strongly}.$$

The theorem is proved.

REFERENCES

(1) Kang Li-shan, Sun Le-lin and Chen Yu-ping: Asynchronous Parallel Algorithm for Mathematical Physics Problems, Science Press, China, 1985.

(2) D.J.Evans, Kang Li-shan, Shao Jian-ping and Chen Yu-ping: The convergence rate of the Schwarz alternating procedure (I)— For one-dimensional problems, Intern.J.Computer Math.,20,(1986),157-170.

(3) D.J.Evans, Shao Jian-ping, Kang Li-shan and Chen Yu-ping: The convergence rate of the Schwarz alternating procedure (II)— For two-Dimensional problems, Intern.J.Computer Math.,20,(1986),325-339.

(4) Kang Li-shan, D.J.Evans: The convergence rate of the Schwarz alternating procedure (III)— For Neumann problems, Intern.J.Computer Math., 21, (1987), 85-108.

(5) D.J.Evans, Kang Li-shan, Chen Yu-ping and Shao Jian-ping: The convergence rate of the Schwarz alternating procedure (IV)— With pseudo-boundary relaxation factor, Intern.J.Computer Math., 21, (1987), 185-203.

(6) Lin Guang-ming, Wu Zhijian and G.Rodrigue: Domain decomposition method with mixed pseudo-boundary conditions, to appear.

A DIFFERENCE METHOD FOR DEGENERATE HYPERBOLIC EQUATIONS OF SECOND ORDER

Li De-yuan and Han Zhen

Institute of Applied Physics and Computational Mathematics
Beijing, P.O.Box 8009

§1.

In recent years the finite difference method for solving the partial differential equations has been studied by Zhou in his series of works [1-13]. He has constructed some finite difference schemes for various evolution equations, appearing in the modern physics and other fields of applied science. He has proved the convergence of the solution of these finite difference schemes to the generalized solution of the corresponding differential equations and their uniqueness. Zhou proved not only the convergence of the solution, but also the convergence of the difference quotients of the solution of the finite difference schemes to the derivatives arising in the definition of the generalized solution.

As it is well known that many differential equations, describing the physical phenomena, are of degenerate type. One of the authors and his colleagues have studied the difference method for solving the degenerate parabolic equations, which are the model equations of the superthermal electronic equations in plasma physics [14-16]. In this paper, we study the finite difference method for solving the degenerate hyperbolic equations of second order, which are concerned in the ocean acoustics [17]. The existence and uniqueness of the generalized solutions of such equations have been proved by M.L.Krasnov by means of Galerkin method [18].

The authors thank Prof. Shen Long-jun for his helpful discussions.

§2.

Consider the degenerate hyperbolic equation of second order

$$\frac{\partial^2 u}{\partial t^2} - x^p a(x,t)\frac{\partial^2 u}{\partial x^2} = b(x,t)\,\frac{\partial u}{\partial x} + f(x,t,u,\frac{\partial u}{\partial t}) \qquad (1)$$

in the domain $Q=\{0< x < L, 0< t< T\}$, where $p \geqslant 1$. The initial and boundary conditions are given as follows:

$$u(x,0) = \phi(x) \quad , \qquad 0\leqslant x\leqslant L \qquad (2)$$

$$\left.\frac{\partial u}{\partial t}\right|_{t=0} = \psi(x) \quad , \qquad 0\leqslant x\leqslant L \qquad (3)$$

$$u(0,t) \text{ is finite on } 0\leqslant t\leqslant T \qquad (4)$$

$$u(L,t) = 0 \quad , \qquad 0\leqslant t\leqslant T \qquad (5)$$

Suppose that for the coefficients in the equation (1) and the initial functions, the following assumptions are valid

(I) $a(x,t)$ is differentiable with respect to x and t on $\overline{Q}$. Besides, $a(x,t)$ is positive on $\overline{Q}$. Then there exist constants A_o and C, such that

$$a(x,t)\geqslant A_o > 0 \quad , \qquad x^p a(x,t)\leqslant C \quad .$$

(II) $b(x,t)$ is a continuous function of $x\in[0,L]$, further, there are

$$\left|b(x,t)\right|\leqslant Bx^{\frac{p}{2}} \quad , \qquad \forall (x,t)\in \overline{Q}$$

$$\left|b(x,t_1)-b(x,t_2)\right|\leqslant C_b x^{\frac{p}{2}}\left|t_1-t_2\right| \quad , \ \forall x \in [0,L],\ t_1,t_2\in [0,T] \ .$$

(III) $f(x,t,u,r)$ is a continuous function in $(x,t,u,r)\in \overline{Q}\times \mathbb{R}^2$, and for any $(x,t)\in\overline{Q}$ and $u,r\in\mathbb{R}$, there are

$$\left|\frac{\partial}{\partial x}f(x,t,u,r)\right| \leqslant C_f(1+|u|+|r|)$$

$$\left|\frac{\partial}{\partial u}f(x,t,u,r)\right| \leqslant C_f \quad .$$

(IV) $f(x,t,u,r)$ is semi-bounded for r :

$$\operatorname{sgn}(r_1-r_2)\left[f(x,t,u,r_1)-f(x,t,u,r_2)\right] \leqslant C_r\left|r_1-r_2\right|,$$

$\forall(x,t)\in \overline{Q}$, $u\in \mathbb{R}$, $r_1,r_2\in \mathbb{R}$.

(V) For $x=L$, the function $f(x,t,u,r)$ is Lipschitz continuous with respect to $t\in [0,T]$.

(VI) The initial functions satisfy the consistent condition, i.e. , $\phi(L)=\psi(L) = 0$. In addition, the functions $\phi(x)$ and $\psi(x)$ are Lipschitz-continuously differentiable.

The coefficients B, C_b, C_f, C_r, arising in (II)–(IV), are all constants.

Solve the problem (1)—(5) by means of finite difference method.

Divide the interval $[0,L]$ and $[0,T]$ into J and N+1 parts respectively. The space step is $h=L/J$ and the time step is $k=T/(N+1)$. Let

$$\omega_k = \{t^n = nk \mid n=0,1, \ldots , N+1\} ,$$

and

$$\omega_h = \{x_j = jh \mid j=0,1, \ldots , J\} .$$

The set of all net points on the domain $\overline{Q}$ is denoted by $\omega_h \times \omega_k$.

Let $v(x,t)$ be a discrete function, defined on the set $\omega_h \times \omega_k$. Introduce the following notations

$$v_j^n = v(x_j, t^n) ,$$

$$v_{x,j}^n = \frac{1}{h}(v_{j+1}^n - v_j^n) , \quad v_{\overline{x},j}^n = \frac{1}{h}(v_j^n - v_{j-1}^n) , \quad v_{\mathring{x},j}^n = \frac{1}{2}(v_{x,j}^n + v_{\overline{x},j}^n) ,$$

$$v_{t,j}^n = \frac{1}{k}(v_j^{n+1} - v_j^n) , \quad v_{\overline{t},j}^n = \frac{1}{k}(v_j^n - v_j^{n-1}) , \quad v_{\mathring{t},j}^n = \frac{1}{2}(v_{t,j}^n + v_{\overline{t},j}^n) .$$

Define the norm of the function $v(x,t)$ and its difference quotients as follows

$$\|v^n\|_\infty = \max_{0\le j\le J} |v_j^n| , \qquad \|v^n\| = \left\{ \sum_{j=0}^{J} (v_j^n)^2 h \right\}^{\frac{1}{2}} ,$$

$$\|v_{\overline{x}}^n\| = \left\{ \sum_{j=1}^{J} (v_{\overline{x},j}^n)^2 h \right\}^{\frac{1}{2}} , \quad \|v_{\overline{x}x}^n\| = \left\{ \sum_{j=1}^{J-1} (v_{\overline{x}x,j}^n)^2 h \right\}^{\frac{1}{2}} ,$$

$$\|v_{\overline{t}}^n\| = \left\{ \sum_{j=0}^{J} (v_{\overline{t},j}^n)^2 h \right\}^{\frac{1}{2}} , \quad \|v_{\overline{t}t}^n\| = \left\{ \sum_{j=0}^{J} (v_{\overline{t}t,j}^n)^2 h \right\}^{\frac{1}{2}}$$

and so forth.

Let the discrete function $v(x,t)$,defined on the set $\omega_h \times \omega_k$, be the approximate solution by difference method. Construct the difference scheme

$$v_{\overline{t}t,j}^n - x_j^p a_j^n v_{\overline{x}x,j}^n = b_j^n v_{\mathring{x},j}^n + f_j^n , \quad j=1,2,\ldots,J-1; n=1,2,\ldots,N \tag{6}$$

$$v_{\overline{t}t,o}^n = f_o^n , \qquad n=1,2, \ldots , N \tag{7}$$

$$v_J^n = 0 , \qquad n=0,1, \ldots , N+1 \tag{8}$$

$$v_j^o = \phi(x_j) , \qquad j=0,1, \ldots , J \tag{9}$$

$$v_{\overline{t},j}^1 = \psi(x_j) , \qquad j=0,1, \ldots , J \tag{10}$$

where $f_j^n = f(x_j, t^n, v_j^n, v_{\overset{\circ}{t},j}^n)$.

The scheme (6),(7) is an explicit scheme when f(x,t,u,r) is a linear function of r.

In this paper, we shall prove that when the assumptions (I)–(VI) are fulfilled, the solution of the difference scheme (6)–(10) converges to the generalized solution of the equation (1) satisfying conditions (2)–(5) as the space and time steps h and k tend to zero simultaneously under certain stability condition.

We shall also discuss the case $p \in (0,1)$. In this case, the boundary condition can be given on x=0, i.e., the boundary condition (4) may be replaced by

$$u(0,t) = 0 \quad , \quad 0 \leq t \leq T \quad . \tag{4'}$$

Respectively, the difference equation (7) may also be replaced by

$$v_0^n = 0 \quad , \quad n=0,1, \ldots , N+1 \quad . \tag{7'}$$

It also may be proved that the solution of the difference scheme (6), (7′),(8),(9),(10) converges to the generalized solution of the equation (1), satisfying the conditions (2),(3),(4′) and (5) .

§ 3.

For the discrete functions, we have the following interpolation formulae (cf. [8]) .

<u>Lemma 1.</u> For any discrete function $\{y_j^n\}$, defined on the set $\omega_h \times \omega_k$, there is the following inequality

$$\|y^n\|_\infty^2 \leq 2\|y^n\| \left(\|y_{\bar{x}}^n\| + \frac{1}{L}\|y^n\| \right)$$

or in another form

$$\|y^n\|_\infty^2 \leq \left(\frac{2}{L} + c\right) \|y^n\|^2 + \frac{1}{c}\|y_{\bar{x}}^n\|^2 \quad ,$$

where c is an arbitrarily given constant. Besides, there is also

$$\|y^n\|^2 \leq 2nk \sum_{q=1}^{n} \|y_{\bar{t}}^q\|^2 k + 2\|y^0\|^2 \quad .$$

<u>Lemma 2.</u> Assume that the discrete function $\{y_j\}$, defined on ω_h , satisfies the boundary condition $y_o=0$ (or $y_J=0$), then we have

$$\|y\|_\infty \leqslant L^{\frac{1}{2}} \|y_x\|$$

and

$$\|y\| \leqslant L \|y_x\| \quad .$$

§4.

In this section,we are going to estimate the solution of the finite difference scheme (6)–(10) .

Multiplying $-v^n_{\bar{x}x\mathring{t},j}hk$ on both sides of the difference equation (6), and then summing up for j from 1 to J-1 and for n from 1 to m $(1\leqslant m\leqslant N)$, we get

$$-\sum_{n=1}^{m}\sum_{j=1}^{J-1} v^n_{\bar{t}t,j} v^n_{\bar{x}x\mathring{t},j}hk + \sum_{n=1}^{m}\sum_{j=1}^{J-1} x_j^p\, a_j^n\, v^n_{\bar{x}x,j} v^n_{\bar{x}x\mathring{t},j}hk =$$

$$= -\sum_{n=1}^{m}\sum_{j=1}^{J-1} b_j^n\, v^n_{\mathring{x},j} v^n_{\bar{x}x\mathring{t},j}hk - \sum_{n=1}^{m}\sum_{j=1}^{J-1} f_j^n\, v^n_{\bar{x}x\mathring{t},j}hk \quad . \tag{11}$$

The first term on the left side of the above equality (11) is

$$-\sum_{n=1}^{m}\sum_{j=1}^{J-1} v^n_{\bar{t}t,j} v^n_{\bar{x}x\mathring{t},j}hk = \frac{1}{2}\left(\left\|v^{m+1}_{\bar{x}\bar{t}}\right\|^2 - \left\|v^{1}_{\bar{x}\bar{t}}\right\|^2\right) + \sum_{n=1}^{m} v^n_{\bar{t}t,o} v^n_{x\mathring{t},o}k -$$

$$- \sum_{n=1}^{m} v^n_{\bar{t}t,J} v^n_{\bar{x}\mathring{t},J}k \quad , \tag{12}$$

where the last term turns to zero for the boundary condition (8). Now, for any discrete function $\{y^n\}$, defined on ω_k , we have

$$y^n y^n_{\mathring{t}} = \frac{1}{8}\left\{\left[(y^{n+1}+ y^n)^2\right]_{\bar{t}} - k^2\left[(y^{n+1}_{\bar{t}})^2\right]_{\bar{t}}\right\} \quad .$$

Then the second term on the left side of (11) is

$$\sum_{n=1}^{m}\sum_{j=1}^{J-1} x_j^p\, a_j^n\, v^n_{\bar{x}x,j} v^n_{\bar{x}x\mathring{t},j}hk =$$

$$= \frac{1}{8}\left(\left\|x^{\frac{p}{2}}\sqrt{a^m}(v^{m+1}_{\bar{x}x}+ v^m_{\bar{x}x})\right\|^2 - \left\|x^{\frac{p}{2}}\sqrt{a^o}(v^{1}_{\bar{x}x}+ v^o_{\bar{x}x})\right\|^2\right) +$$

$$-\frac{1}{8}\sum_{n=1}^{m}\sum_{j=1}^{J-1}x_j^p\, a^n_{\bar t,j}(v^n_{\bar x x,j}+v^{n-1}_{\bar x x,j})^2 hk-\frac{k^2}{8}\left(\left\|x^{\frac{p}{2}}\sqrt{a^m}\,v^{m+1}_{\bar x x\bar t}\right\|^2-\left\|x^{\frac{p}{2}}\sqrt{a^0}\,v^1_{\bar x x\bar t}\right\|^2\right)$$

$$+\frac{k^2}{8}\sum_{n=1}^{m}\sum_{j=1}^{J-1}x_j^p\, a^n_{\bar t,j}(v^n_{\bar x x\bar t})^2 hk \quad . \tag{13}$$

In order to estimate the first term on the right side of (11),we sum it by parts with respect to t, and obtain

$$-\sum_{n=1}^{m}\sum_{j=1}^{J-1}b_j^n\, v^n_{\mathring{x},j}v^n_{\bar x x\mathring{t},j}hk =$$

$$=\sum_{n=1}^{m}\sum_{j=1}^{J-1}b_j^n\, v^n_{\mathring{x}\mathring{t},j}v^n_{\bar x x,j}hk+\frac{1}{2}\sum_{n=1}^{m}\sum_{j=1}^{J-1}(b^{n+1}_{\bar t,j}v^{n+1}_{\mathring{x},j}+b^{n-1}_{t,j}v^{n-1}_{\mathring{x},j})v^n_{\bar x x,j}hk -$$

$$-\frac{1}{2}\sum_{j=1}^{J-1}(b_j^m\, v^m_{\mathring{x},j}v^{m+1}_{\bar x x,j}+b_j^{m+1}v^{m+1}_{\mathring{x},j}\, v^m_{\bar x x,j})h +$$

$$+\frac{1}{2}\sum_{j=1}^{J-1}(b_j^0\, v^0_{\mathring{x},j}v^1_{\bar x x,j}+b_j^1\, v^1_{\mathring{x},j}v^0_{\bar x x,j})h \quad .$$

Using Lemma 1 and the relations

$$v^{n+1}=\frac{1}{2}(v^{n+1}+v^n)+\frac{k}{2}\,v^{n+1}_{\bar t} \quad , \qquad v^n=\frac{1}{2}(v^{n+1}+v^n)-\frac{k}{2}\,v^{n+1}_{\bar t} \quad ,$$

we have

$$-\sum_{n=1}^{m}\sum_{j=1}^{J-1}b_j^n\, v^n_{\mathring{x},j}v^n_{\bar x x\mathring{t},j}hk \leqslant \frac{1}{2}(1+T^2+4c_1T)\sum_{n=1}^{m}\left\|v^n_{\bar x t}\right\|^2 k +$$

$$+\frac{B^2}{8c_1}\left(\left\|x^{\frac{p}{2}}(v^{m+1}_{\bar x x}+v^m_{\bar x x})\right\|^2+k^2\left\|x^{\frac{p}{2}}v^{m+1}_{\bar x x\bar t}\right\|^2\right)+$$

$$+\frac{1}{4}(B^2+C_b^2)\left(\sum_{n=1}^{m}\left\|x^{\frac{p}{2}}(v^{n+1}_{\bar x x}+v^n_{\bar x x})\right\|^2 k+k^2\sum_{n=1}^{m}\left\|x^{\frac{p}{2}}v^{n+1}_{\bar x x\bar t}\right\|^2 k\right)+R_1 \quad , \tag{14}$$

where c_1 is an arbitrarily chosen constant and R_1 is a constant independent of h and k .

Summing the last term of (11) by parts with respect to x, we get

$$-\sum_{n=1}^{m}\sum_{j=1}^{J-1}f_j^n\, v^n_{\bar x x\mathring{t},j}hk =$$

$$=\sum_{n=1}^{m}\sum_{j=1}^{J}f^n_{\bar x,j}v^n_{\bar x\mathring{t},j}hk-\sum_{n=1}^{m}f_J^n\, v^n_{\bar x\mathring{t},J}k+\sum_{n=1}^{m}f_0^n\, v^n_{x\mathring{t},0}k \quad , \tag{15}$$

where the last term is cancelled with the last term of (12) due to the difference equation (7). Since

$$f^n_{\bar{x},j}h = f(x_j,t^n,v^n_j,v^n_{\bar{t},j})-f(x_{j-1},t^n,v^n_j,v^n_{\bar{t},j})+f(x_{j-1},t^n,v^n_j,v^n_{\bar{t},j})-$$

$$-f(x_{j-1},t^n,v^n_{j-1},v^n_{\bar{t},j})+f(x_{j-1},t^n,v^n_{j-1},v^n_{\bar{t},j})-f(x_{j-1},t^n,v^n_{j-1},v^n_{\bar{t},j-1}) =$$

$$=\int_0^1 \frac{d}{d\lambda}f(\lambda x_j+(1-\lambda)x_{j-1},t^n,v^n_j,v^n_{\bar{t},j})d\lambda +$$

$$+\int_0^1 \frac{d}{d\lambda}f(x_{j-1},t^n,\lambda v^n_j+(1-\lambda)v^n_{j-1},v^n_{\bar{t},j})d\lambda +$$

$$+ f(x_{j-1},t^n,v^n_{j-1},v^n_{\bar{t},j}) - f(x_{j-1},t^n,v^n_{j-1},v^n_{\bar{t},j-1}) \quad ,$$

under the assumptions (III),(IV),(VI) , we obtain

$$\sum_{n=1}^{m}\sum_{j=1}^{J} f^n_{\bar{x},j}v^n_{\bar{x}\bar{t},j}hk \leqslant$$

$$\leqslant \frac{1}{2}(3C_f^2T^2L + C_f^2T^2 + 3C_f^2L + 2C_r + 4)\sum_{n=1}^{m+1}\left\|v^n_{\bar{x}\bar{t}}\right\|^2 k + R_2 \quad , \qquad (16)$$

where R_2 is a constant independent of h and k. Summation by parts gives

$$-\sum_{n=1}^{m} f^n_J\, v^n_{\bar{x}\bar{t},J}k = \frac{1}{2}\sum_{n=1}^{m-1} f^{n+1}_{\bar{t},J}(v^{n+1}_{\bar{x},J}+ v^n_{\bar{x},J})k -$$

$$- \frac{1}{2}f^m_J(v^{m+1}_{\bar{x},J}+ v^m_{\bar{x},J}) + \frac{1}{2}f^1_J(v^1_{\bar{x},J}+ v^o_{\bar{x},J}) \quad , \qquad (17)$$

where appear the terms $v^{n+1}_{\bar{x},J}+ v^n_{\bar{x},J}$ $(n=1,2, \dots , m)$. Put $J_o= \left[\frac{J+1}{2}\right]$. We use the Lemma 1 for the discrete function $\{y_{J_o},y_{J_o+1}, \dots ,y_J\}$, where

$$y = v^{n+1}_{\bar{x}}+ v^n_{\bar{x}}$$

and

$$(J-J_o)h \geqslant \frac{L}{2} - \frac{h}{2} > \frac{L}{4} \quad .$$

Thus , we have

$$(v^{n+1}_{\bar{x},J}+ v^n_{\bar{x},J})^2 \leqslant (\frac{2}{L/4} +c)\sum_{j=J_o}^{J}(v^{n+1}_{\bar{x},J}+ v^n_{\bar{x},J})^2h + \frac{1}{c}\sum_{j=J_o}^{J-1}(v^{n+1}_{\bar{x}x,j}+ v^n_{\bar{x}x,j})^2h$$

$$2(\frac{8}{L} + c)(\left\|v^{n+1}_{\bar{x}}\right\|^2 + \left\|v^n_{\bar{x}}\right\|^2) + \frac{1}{c}\sum_{j=J_o}^{J-1}(\frac{x_j}{L/2})^p(v^{n+1}_{\bar{x}x,j}+ v^n_{\bar{x}x,j})^2h$$

$$4(\frac{8}{L} +c)(2n+1)k\sum_{q=1}^{n+1}\left\|v^q_{\bar{x}\bar{t}}\right\|^2 k + \frac{1}{c}(\frac{2}{L})^p\left\|x^{\frac{p}{2}}(v^{n+1}_{\bar{x}x}+ v^n_{\bar{x}x})\right\|^2 +8(\frac{8}{L} +c)\left\|v^o_{\bar{x}}\right\|^2 \quad . \qquad (18)$$

Using this estimate and the assumptions (III),(V) , we obtain from (17) that

$$-\sum_{n=1}^{m} f_J^n v_{\bar{x}\hat{t},J}^n k \leqslant$$

$$\leqslant \frac{1}{4}\left(\frac{2}{L}\right)^p \sum_{n=1}^{m-1} \left\|x^{\frac{p}{2}}(v_{\bar{x}x}^{n+1}+v_{\bar{x}x}^n)\right\|^2 k + \frac{1}{4c_2}\left(\frac{2}{L}\right)^p \left\|x^{\frac{p}{2}}(v_{\bar{x}x}^{m+1}+v_{\bar{x}x}^m)\right\|^2 +$$

$$+\left\{\left(\frac{8}{L}+1\right)T^2+\left(\frac{8}{L}+c_2\right)2T\right\}\sum_{n=1}^{m+1}\left\|v_{\bar{x}\bar{t}}^n\right\|^2 k + R_3 \qquad ,$$

where c_2 is an arbitrarily chosen constant and R_3 is a positive constant independent of h and k . Therefore, we get

$$-\sum_{n=1}^{m}\sum_{j=1}^{J-1} f_j^n v_{\bar{x}x\hat{t},j}^n hk \leqslant$$

$$\leqslant\left\{\frac{3}{2}C_f^2T^2L+\frac{1}{2}C_f^2T^2+\frac{3}{2}C_f^2L+C_r+2+\left(\frac{8}{L}+1\right)T^2+\left(\frac{8}{L}+c_2\right)2T\right\}\cdot$$

$$\cdot\sum_{n=1}^{m+1}\left\|v_{\bar{x}\bar{t}}^n\right\|^2 k +$$

$$+\frac{1}{4}\left(\frac{2}{L}\right)^p \sum_{n=1}^{m-1}\left\|x^{\frac{p}{2}}(v_{\bar{x}x}^{n+1}+v_{\bar{x}x}^n)\right\|^2 k + \frac{1}{4c_2}\left(\frac{2}{L}\right)^p \left\|x^{\frac{p}{2}}(v_{\bar{x}x}^{m+1}+v_{\bar{x}x}^m)\right\|^2 +$$

$$+\sum_{n=1}^{m} f_o^n v_{x\hat{t},o}^n k + R_2 + R_3 \qquad . \tag{19}$$

Substituting (12),(13),(14),(19) into (11), we obtain

$$\left\|v_{\bar{x}\bar{t}}^{m+1}\right\|^2+\left\{\frac{A_o}{4}-\frac{B^2}{4c_1}-\frac{1}{2c_2}\left(\frac{2}{L}\right)^p\right\}\left\|x^{\frac{p}{2}}(v_{\bar{x}x}^{m+1}+v_{\bar{x}x}^m)\right\|^2-\frac{k^2}{4}\left(C+\frac{B^2}{c_1}L^p\right)\left\|v_{\bar{x}x\bar{t}}^{m+1}\right\|^2$$

$$\leqslant G_1\sum_{n=1}^{m+1}\left\|v_{\bar{x}\bar{t}}^n\right\|^2 k + G_2\sum_{n=1}^{m}\left\|x^{\frac{p}{2}}(v_{\bar{x}x}^{n+1}+v_{\bar{x}x}^n)\right\|^2 k +$$

$$+k^2G_3\sum_{n=1}^{m}\left\|v_{\bar{x}x\bar{t}}^{n+1}\right\|^2 k + G_4 \qquad , \tag{20}$$

where G_s (s=1,2,3,4) are constants independent of h and k .

Let ε be a sufficiently small positive constant, and put

$$c_1=\max\left\{4B^2/A_o\ ,\ B^2L^2/(C\varepsilon)\right\}\ ,\quad c_2=8\left(\frac{2}{L}\right)^p/A_o \qquad . \tag{21}$$

Then the above inequality (20) becomes

$$\left\|v_{\bar{x}\bar{t}}^{m+1}\right\|^2+\frac{A_o}{8}\left\|x^{\frac{p}{2}}(v_{\bar{x}x}^{m+1}+v_{\bar{x}x}^m)\right\|^2-\frac{k^2}{4}C(1+\varepsilon)\left\|v_{\bar{x}x\bar{t}}^{m+1}\right\|^2\leqslant$$

$$\leqslant G_1 \sum_{n=1}^{m+1} \left\| v_{\bar{x}\bar{t}}^{n} \right\|^2 k + G_2 \sum_{n=1}^{m} \left\| x^{\frac{p}{2}} (v_{\bar{x}x}^{n+1} + v_{\bar{x}x}^{n}) \right\|^2 k +$$

$$+ k^2 G_3 \sum_{n=1}^{m} \left\| v_{\bar{x}x\bar{t}}^{n+1} \right\|^2 k + G_4 \quad . \tag{22}$$

And for any discrete function $\{y_j\}$ $(j=0,1, \ldots ,J)$, the following inequality holds

$$\| y_{\bar{x}} \|^2 \leqslant \frac{4}{h^2} \| y \|^2 \quad .$$

Hence, we have

$$\left\| v_{\bar{x}x\bar{t}}^{n+1} \right\|^2 \leqslant \frac{4}{h^2} \left\| v_{\bar{x}\bar{t}}^{n+1} \right\|^2 \quad , \qquad n=0,1, \ldots , N .$$

This implies that

$$\left\| v_{\bar{x}\bar{t}}^{m+1} \right\| - \frac{k^2}{4} C(1+\varepsilon) \left\| v_{\bar{x}x\bar{t}}^{m+1} \right\|^2 \geqslant \left\{ 1 - \frac{k^2}{h^2} C(1+\varepsilon) \right\} \left\| v_{\bar{x}\bar{t}}^{m+1} \right\|^2 \quad .$$

Under the assumption that the time step k and the space step h satisfy the following relation

$$\frac{k}{h} \leqslant \left(\frac{1-\varepsilon}{C} \right)^{\frac{1}{2}} \quad , \tag{23}$$

we obtain from the inequality (22)

$$\varepsilon^2 \left\| v_{\bar{x}\bar{t}}^{m+1} \right\|^2 + \frac{A_o}{8} \left\| x^{\frac{p}{2}} (v_{\bar{x}x}^{m+1} + v_{\bar{x}x}^{m}) \right\|^2 \leqslant$$

$$\leqslant (G_1 + 4\frac{1-\varepsilon}{C} G_3) \sum_{n=1}^{m+1} \left\| v_{\bar{x}\bar{t}}^{n} \right\|^2 k + G_2 \sum_{n=1}^{m} \left\| x^{\frac{p}{2}} (v_{\bar{x}x}^{n+1} + v_{\bar{x}x}^{n}) \right\|^2 k + G_4 \quad .$$

Put $M_1 = \max \left\{ \frac{1}{\varepsilon^2} (G_1 + 4\frac{1-\varepsilon}{C} G_3) \; , \; \frac{8}{A_o} G_2 \right\}$. By the Gronwall inequality we have the estimate

$$\varepsilon^2 \left\| v_{\bar{x}\bar{t}}^{m+1} \right\|^2 + \frac{A_o}{8} \left\| x^{\frac{p}{2}} (v_{\bar{x}x}^{m+1} + v_{\bar{x}x}^{m}) \right\|^2 \leqslant G_4 \cdot \exp(2M_1 T) \; , \quad m=1,2, \ldots , N \tag{24}$$

for sufficiently small $k \leqslant 1/(2M_1)$.

Lemma 3. Suppose that the assumptions (I)–(VI) are satisfied and that the steps k and h satisfy the relation (23) and k is sufficiently small. Then for the solution $\{v_j^n\}$ of the difference scheme (6)–(10) there are the estimates

$$\left\| v_{\bar{x}\bar{t}}^{m+1} \right\| \leqslant K_1 \tag{25}$$

$$\left\| v_{\bar{x}}^{m+1} \right\| \leqslant K_2 \tag{26}$$

$$\left\| v^{m+1} \right\| \leqslant K_3 \tag{27}$$

$$\left\| v^{m+1} \right\|_{\infty} \leqslant K_4 \tag{28}$$

$$\left\| v_{\bar{t}}^{m+1} \right\| \leqslant K_5 \tag{29}$$

$$\left\| v_{\bar{t}}^{m+1} \right\|_{\infty} \leqslant K_6 \tag{30}$$

$$\left\| x^{\frac{p}{2}} v_{\bar{x}x}^{m+1} \right\| \leqslant K_7 \tag{31}$$

$$\left\| x^{p} a^{m+1} v_{\bar{x}x}^{m+1} \right\| \leqslant K_8 \tag{32}$$

$$\left\| v_{\bar{t}t}^{m} \right\| \leqslant K_9 \tag{33}$$

where K_s $(s=1,2,\ \ldots\ ,9)$ are constants independent of h and k , and $m=1,2,\ \ldots\ ,N$.

Proof. The estimate (25) is a direct consequence of the inequality (24). Inequality (26) is obtained from (25) by using Lemma 1 . From (26) and Lemma 2 , it can be verified that (27) and (28) hold. Similarly , from (25) and Lemma 2 , we also get the estimates (29),(30). From (24), we have

$$\left\| x^{\frac{p}{2}} (v_{\bar{x}x}^{m+1} + v_{\bar{x}x}^{m}) \right\|^2 \leqslant 8G_4 \exp(2M_1T) \ / \ A_o \quad . \tag{34}$$

On the other hand, there is

$$\left\| x^{\frac{p}{2}} (v_{\bar{x}x}^{m+1} - v_{\bar{x}x}^{m}) \right\|^2 = k^2 \left\| x^{\frac{p}{2}} \ v_{\bar{x}x\bar{t}}^{m+1} \right\|^2 \leqslant$$

$$\leqslant k^2 L^p 4/h^2 \left\| v_{\bar{x}\bar{t}}^{m+1} \right\|^2 \leqslant 4L^p K_1^2 (1-\varepsilon)/C \quad . \tag{35}$$

Combining (34) and (35),we obtain (31). It is easily to be found that (32) also holds. Using the finite difference scheme (6)–(8) and above obtained estimates (26),(28),(30) and (32), we have (33). This completes the proof of Lemma 3 .

Lemma 4. Under the assumptions of Lemma 3 , there are estimates

$$\max_{1\leq j\leq J}\left|v_j^{m+1}-v_{j-1}^{m+1}\right|\leqslant K_2h^{\frac{1}{2}}\quad,\qquad m=1,2,\ \dots\ ,N \tag{36}$$

and

$$\max_{1\leq m\leq N}\left|v_j^{m+1}-v_j^{m}\right|\leqslant K_6k\quad,\qquad j=0,1,\ \dots\ ,J\quad. \tag{37}$$

Proof. Since

$$\left|v_j^{m+1}-v_{j-1}^{m+1}\right|=\left|v_{x,j}^{m+1}\right|h=\sqrt{(v_{\bar{x},j}^{m+1})^2h}\ h^{\frac{1}{2}}\leqslant\left\|v_{\bar{x}}^{m+1}\right\|h^{\frac{1}{2}}\quad,$$

using (26), we obtain (36). Estimate (37) is a version of (30).

Lemma 5. Under the assumptions of Lemma 3 , there are estimates

$$\max_{1\leq j\leq J}\left|v_{\bar{t},j}^{m+1}-v_{\bar{t},j-1}^{m+1}\right|\leqslant K_1h^{\frac{1}{2}}\quad,\ m=1,2,\ \dots\ ,N \tag{38}$$

and

$$\max_{1\leq m\leq N}\left|v_{\bar{t},j}^{m+1}-v_{\bar{t},j}^{m}\right|\leqslant K_{10}k^{\frac{1}{2}}\quad,\ j=0,1,\ \dots\ ,J\quad, \tag{39}$$

where K_{10} is a constant independent of h and k .

Proof. Estimate (38) can be verified from (25) by the same way as estimate (36) has been obtained from (26). Since

$$\left|v_{\bar{t},j}^{m+1}-v_{\bar{t},j}^{m}\right|=\left|v_{\bar{t}t,j}^{m}\right|k\leqslant\left|v_{\bar{t}t,j}^{m}\right|\left(\frac{1-\varepsilon}{C}\right)^{\frac{1}{4}}h^{\frac{1}{2}}k^{\frac{1}{2}}\leqslant\left(\frac{1-\varepsilon}{C}\right)^{\frac{1}{4}}\left\|v_{\bar{t}t}^{m}\right\|k^{\frac{1}{2}}\quad,$$

from (33), we get (39) .

§5.

Now we turn to study the limiting process for the solution of the difference scheme (6)–(10) when the time and space steps k and h, satisfying the relation (23), tend to zero simultaneously.

Following the method, described by Zhou [8], we construct a set of piecewise constant functions as follows: In every grid $Q_j^n=\{x_j\leqslant x<x_{j+1}\ ,\ t^{n-1}<t\leqslant t^n\}$, let

$$v_{hk}^{(0)}(x,t)=v_j^n\ ,\quad v_{hk}^{(+1,0)}(x,t)=v_{x,j}^n\ ,\quad v_{hk}^{(-1,0)}(x,t)=v_{\bar{x},j}^n\ ,$$

$$v_{hk}^{(0,1)}(x,t)=v_{\bar{t},j}^n\ ,\quad v_{hk}^{(0,2)}(x,t)=v_{\bar{t}t,j}^n\ ,\quad v_{hk}^{(1,1)}(x,t)=v_{x\bar{t},j}^n$$

for $j=0,1,\ldots,J-1$, $n=1,2,\ldots,N+1$.

From the Lemma 3 , we have the following estimates for the above constructed functions:

$$\|v_{hk}^{(0)}(\cdot,t)\|_{L_2(0,L)}\leqslant K_3 \quad , \quad \|v_{hk}^{(+1,0)}(\cdot,t)\|_{L_2(0,L)}\leqslant K_2 \quad ,$$

$$\|v_{hk}^{(-1,0)}(\cdot,t)\|_{L_2(0,L)}\leqslant K_2 \quad , \quad \|v_{hk}^{(0,1)}(\cdot,t)\|_{L_2(0,L)}\leqslant K_5 \quad ,$$

$$\|v_{hk}^{(0,2)}(\cdot,t)\|_{L_2(0,L)}\leqslant K_9 \quad , \quad \|v_{hk}^{(1,1)}(\cdot,t)\|_{L_2(0,L)}\leqslant K_1 \quad .$$

Therefore we can select a step sequence $\{h_i,k_i\}$ $(i=1,2,\ldots)$, in which the steps h_i and k_i satisfy the condition (23) for every i, such that as $i\to\infty$ and $h_i\to 0$, the sequences $\{v_{h_ik_i}^{(0)}(x,t)\}$, $\{v_{h_ik_i}^{(+1,0)}(x,t)\}$, $\{v_{h_ik_i}^{(-1,0)}(x,t)\}$, $\{v_{h_ik_i}^{(0,1)}(x,t)\}$, $\{v_{h_ik_i}^{(0,2)}(x,t)\}$ and $\{v_{h_ik_i}^{(1,1)}(x,t)\}$ weakly converge in the space $L_q(0,T;L_2(0,L))$ (for any $q\in[1,\infty)$) respectively to $u(x,t)$, $u^{(+1,0)}(x,t)$, $u^{(-1,0)}(x,t)$, $u^{(0,1)}(x,t)$, $u^{(0,2)}(x,t)$ and $u^{(1,1)}(x,t)$, where all of the functions belong to the space $L_\infty(0,T;L_2(0,L))$. It is easily to be verified that $u^{(+1,0)}(x,t)$ and $u^{(-1,0)}(x,t)$ are the generalized derivative, with respect to x, of the function $u(x,t)$, i.e. ,

$$u^{(+1,0)}(x,t) = u^{(-1,0)}(x,t) = \frac{\partial}{\partial x}u(x,t) \quad ,$$

and $u^{(0,1)}(x,t)$ and $u^{(0,2)}(x,t)$ are the generalized derivatives of $u(x,t)$, with respect to t, of first and second orders respectively, i.e. ,

$$u^{(0,1)}(x,t) = \frac{\partial}{\partial t}u(x,t) \quad , \quad u^{(0,2)}(x,t) = \frac{\partial^2}{\partial t^2}u(x,t) \quad .$$

In addition, it is also easily to be found that $u^{(1,1)}(x,t)$ is the mixed derivative of the function $u(x,t)$ with respect to x and t, i.e. ,

$$u^{(1,1)}(x,t) = \frac{\partial^2}{\partial x\partial t}u(x,t) \quad .$$

Obviously, $u(x,t)$, $\frac{\partial}{\partial t}u(x,t)$, $\frac{\partial}{\partial x}u(x,t)$, $\frac{\partial^2}{\partial t^2}u(x,t)$, $\frac{\partial^2}{\partial x\partial t}u(x,t)$ all belong to the functional space $L_\infty(0,T;L_2(0,L))$. Furthermore, we construct the functions $\bar{v}_{hk}(x,t)$ and $\bar{v}_{hk}^{(0,1)}(x,t)$ as follows: in every domain $\bar{Q}_j^n=\{x_j\leqslant x\leqslant x_{j+1},\ t^{n-1}\leqslant t\leqslant t^n\}$, $\bar{v}_{hk}(x,t)$ and $\bar{v}_{hk}^{(0,1)}(x,t)$ are obtained by the linear expansion in both directions x and t from the values of $\{v_j^n\}$ and $\{v_{\bar{t},j}^n\}$ at four corners of $\bar{Q}_j^n$ respectively. From Lemmas 4 and 5, we know that $\bar{v}_{hk}(x,t)\in C^{(\frac{1}{2},1)}(\bar{Q})$ and $\bar{v}_{hk}^{(0,1)}(x,t)\in C^{(\frac{1}{2},\frac{1}{2})}(\bar{Q})$. Then we can select from

$\{h_i,k_i\}$ a subsequence still denoted by $\{h_i,k_i\}$ (i=1,2, ...) and there exist functions $\bar{u}(x,t)\in C^{(\frac{1}{2},1)}(\bar{Q})$ and $\bar{u}^{(0,1)}(x,t)\in C^{(\frac{1}{2},\frac{1}{2})}(\bar{Q})$, such that as $i\to\infty$, the subsequence $\{\bar{v}_{h_ik_i}(x,t)\}$ and $\{\bar{v}^{(0,1)}_{h_ik_i}(x,t)\}$ uniformly converge to $\bar{u}(x,t)$ and $\bar{u}^{(0,1)}(x,t)$ respectively on the domain $\bar{Q}$. By the definition of the functions $v^{(0)}_{hk}(x,t),\bar{v}_{hk}(x,t),v^{(0,1)}_{hk}(x,t),\bar{v}^{(0,1)}_{hk}(x,t)$ and Lemma 4,5, we have

$$\left|v^{(0)}_{hk}(x,t) - \bar{v}_{hk}(x,t)\right| \leq K_{11}(h^{\frac{1}{2}}+ k) \qquad ,$$

$$\left|v^{(0,1)}_{hk}(x,t) - \bar{v}^{(0,1)}_{hk}(x,t)\right| \leq K_{12}(h^{\frac{1}{2}}+ k^{\frac{1}{2}}) \qquad .$$

Therefore, we can again select a subsequence $\{h_i,k_i\}$, such that the subsequences $\{v^{(0)}_{h_ik_i}(x,t)\}$ and $\{v^{(0,1)}_{h_ik_i}(x,t)\}$ also uniformly converge to the functins $\bar{u}(x,t)$ and $\bar{u}^{(0,1)}(x,t)$ on the domain $\bar{Q}$ as $i\to\infty$. This implies

$$\bar{u}(x,t) = u(x,t) \quad , \qquad \bar{u}^{(0,1)}(x,t) = \frac{\partial}{\partial t}u(x,t) \quad ,$$

and we get

$$\left|u(x,t)\right| \leq K_4 \quad , \qquad \left|\frac{\partial}{\partial t}u(x,t)\right| \leq K_6 \qquad .$$

Then the function $u(x,t)$ satisfies the initial and boundary conditions (2)–(5) in the common sense.

At last, we construct piecewise constant functions as follows: in every grid Q^n_j , put $\widetilde{v}_{hk}(x,t) = x_j^{\frac{2}{2}} v^n_{\bar{x}x,j}$ and $\widetilde{v}^{(2,0)}_{hk}(x,t) = v^n_{\bar{x}x,j}$. From the estimate (31) , we have

$$\left\|\widetilde{v}_{hk}(\cdot,t)\right\|_{L_2(0,L)} \leq K_7 \qquad .$$

Therefore, from the above obtained sequence we can select a subsequence still denoted by $\{h_i,k_i\}$, such that as $i\to\infty$, the sequence $\{\widetilde{v}_{h_ik_i}(x,t)\}$ weakly converges to a function $\widetilde{u}(x,t)$ in the space $L_q(0,T;L_2(0,L))$, where $q\in[1,\infty)$ and

$$\sup_{0\leq t\leq T}\left\|\widetilde{u}(\cdot,t)\right\|_{L_2(0,L)} \leq K_7 \qquad .$$

For arbitrary $\varepsilon>0$, according to (31), in the domain $Q_\varepsilon=\{\varepsilon<x<L, 0<t<T\}$ we have

$$\left\{\sum_{j=J_0}^{J-1}(v^n_{\bar{x}x,j})^2h\right\}^{\frac{1}{2}} \leq \varepsilon^{-\frac{1}{2}}K_7 \qquad ,$$

where $j_0=\left[\frac{\varepsilon}{h}\right]+1$. Therefore, from the above mentioned sequence we can select a subsequence still denoted by $\{h_i,k_i\}$, such that as $i\to\infty$, the sequence $\{\widetilde{v}^{(2,0)}_{h_ik_i}(x,t)\}$ weakly converges to a function $\widetilde{u}^{(2,0)}_\varepsilon(x,t)$ in the space $L_q(0,T;L_2(0,L))$, where $q\in[1,\infty)$ and

$$\sup_{0\le t\le T}\left\|(\cdot)^{\frac{p}{2}}\,\widetilde{u}^{(2,0)}_\varepsilon(\cdot,t)\right\|_{L_2(0,L)}\leqslant K_7 \quad .$$

Since for arbitrary $\varepsilon>0$, the function $\widetilde{u}^{(2,0)}_\varepsilon(x,t)$ is the generalized derivative of the function $u(x,t)$, with respect to x, of second order in the domain Q_ε, we get that there exists a subsequence $\{h_i,k_i\}$, such that as $i\to\infty$, $\widetilde{v}^{(2,0)}_{hk}(x,t)$ weakly converges to $\partial^2u/\partial x^2$ in the domain Q, meanwhile we have

$$\sup_{0\le t\le T}\left\|(\cdot)^{\frac{p}{2}}\,\frac{\partial^2}{\partial x^2}u(\cdot,t)\right\|_{L_2(0,L)}\leqslant K_7 \quad .$$

Let

$$\overset{*}{W}{}^{(2)}_2(Q) = W^{(2)}_\infty(0,T;L_2(0,L))\cap W^{(1)}_\infty(0,T;H^1(0,L))\cap L_\infty(0,T;\overset{*}{H}{}^2(0,L))$$

where

$$\overset{*}{H}{}^2(0,L) = \left\{f(x)\;\middle|\; f\in H^1(0,L),\ \int_0^L x^p(f'')^2\,dx<\infty\right\} \quad .$$

We define a function $U(x,t)$, belonging to the space $\overset{*}{W}{}^{(2)}_2(Q)$, to be the generalized solution of the equation (1), if the following integral relation holds:

$$\iint_Q \varphi\left[\frac{\partial^2U}{\partial t^2} - x^pa\frac{\partial^2U}{\partial x^2} - b\frac{\partial U}{\partial x} - f(x,t,U,\frac{\partial U}{\partial t})\right]dxdt = 0 \qquad \forall\,\varphi\in\Phi ,$$

where $\Phi=\{\varphi(x,t)\mid \varphi(x,t)\in H^1(Q)\}$ is the set of the test functions.

Now we are going to prove that the above obtained limiting function $u(x,t)\in\overset{*}{W}{}^{(2)}_2(Q)$ is the generalized solution of the equation (1).

From the finite difference scheme (6), we have the identity

$$\sum_{n=1}^{N}\sum_{j=1}^{J-1}\varphi^n_j\left[v^n_{\bar{t}t,j} - x^p_j\,a^n_j\,v^n_{\bar{x}x,j} - b^n_j\,v^n_{\overset{\circ}{x},j} - f^n_j\right]hk = 0 \quad .$$

This is evidently identical to the integral relation

$$\iint_Q \varphi_{hk}\left[v^{(0,2)}_{hk} - x^p_h\,a_{hk}\widetilde{v}^{(2,0)}_{hk} - b_{hk}(v^{(-1,0)}_{hk} + v^{(+1,0)}_{hk})/2 - f_{hk}\right]dxdt = 0 \quad ,$$

where $\varphi_{hk}(x,t)=\varphi^n_j$, $x^p_h(x)=x^p_j$, $a_{hk}(x,t)=a^n_j$, $b_{hk}(x,t)=b^n_j$, $f_{hk}(x,t)=f^n_j$ in every small grid Q^n_j ($j=0,1,\ldots,J-1$; $n=1,2,\ldots,N+1$). By the assumptions (I),(II),(III) and the above obtained results that the functional sequences $v^{(-1,0)}_{h_ik_i}$ and $v^{(+1,0)}_{h_ik_i}$ weakly converge to $\partial u/\partial x$, that $v^{(0,2)}_{h_ik_i}$ and $\tilde{v}^{(2,0)}_{h_ik_i}$ weakly converge to $\partial^2u/\partial t^2$ and $\partial^2u/\partial x^2$ respectively, that $v^{(0)}_{h_ik_i}$ and $v^{(0,1)}_{h_ik_i}$ uniformly converge to u and $\partial u/\partial t$ respectively, and that $\varphi_{h_ik_i}$, $x^p_{h_i}$, $a_{h_ik_i}$, $b_{h_ik_i}$, $f_{h_ik_i}$ uniformly converge to $\varphi(x,t),x^p,a(x,t),b(x,t),f(x,t,u,\frac{\partial u}{\partial t})$ respectively, we know that as h_i and $k_i \to 0$ in such a way that the relation (23) is preserved, the above integral relation tends to

$$\iint_Q \varphi\left[\frac{\partial^2 u}{\partial t^2} - x^p a\,\frac{\partial^2 u}{\partial x^2} - b\,\frac{\partial u}{\partial x} - f(x,t,u,\frac{\partial u}{\partial t})\right] dxdt = 0 \quad .$$

Thus the limiting function $u(x,t) \in \overset{*}{W}{}^{(2)}_2(Q)$ is the generalized solution of the degenerate hyperbolic equation (1) and satisfies the initial and boundary conditions (2)–(5) in the common sense.

We will show that the generalized solution u(x,t), which satisfies the initial and boundary conditions (2)–(5) in the common sense, is unique in $\overset{*}{W}{}^{(2)}_2(Q)$. Suppose that u(x,t) and v(x,t) are two different generalized solutions of equation (1) and satisfy the conditions (2)–(5). Put w(x,t) =u(x,t)-v(x,t). Then the function w(x,t) satisfies the following integral relation

$$\iint_Q \varphi\left[\frac{\partial^2 w}{\partial t^2} - x^p a\,\frac{\partial^2 w}{\partial x^2}\right] dxdt =$$

$$= \iint_Q \varphi\left[b\,\frac{\partial w}{\partial x} + f(x,t,u,\frac{\partial u}{\partial t}) - f(x,t,v,\frac{\partial v}{\partial t})\right] dxdt \quad , \quad \forall \varphi \in \Phi \qquad (40)$$

and the initial and boundary conditions

$$w(x,0) = \left.\frac{\partial w}{\partial t}\right|_{t=0} = w(L,t) = 0 \quad , \qquad (41)$$

$$w(0,t) \text{ is finite on } 0 \leqslant t \leqslant T \ . \qquad (42)$$

Put $\varphi(x,t) = e^{-st}x^2\frac{\partial w}{\partial t}$, where s is an arbitrary constant. Then the first term on the left-hand side of (40) is

$$\int_0^T\int_0^L e^{-st}x^2\frac{\partial w}{\partial t}\,\frac{\partial^2 w}{\partial t^2}\,dxdt =$$

$$= \frac{1}{2}\int_0^L e^{-sT}x^2\left(\frac{\partial w}{\partial t}\right)^2\bigg|_{t=T} dx + \frac{1}{2}sL_1^2\int_0^T\int_0^L e^{-st}\left(\frac{\partial w}{\partial t}\right)^2 dxdt \quad , \qquad (43)$$

where $L_1 \in (0,L)$. The second term on the left-hand side of (40) can be estimated as follows

$$-\int_0^T\int_0^L e^{-st}x^{2+p}a\,\frac{\partial w}{\partial t}\,\frac{\partial^2 w}{\partial x^2}\,dxdt = (2+p)\int_0^T\int_0^L e^{-st}x^{1+p}a\,\frac{\partial w}{\partial t}\,\frac{\partial w}{\partial x}\,dxdt +$$

$$+\int_0^T\int_0^L e^{-st}x^{2+p}\frac{\partial a}{\partial x}\,\frac{\partial w}{\partial t}\,\frac{\partial w}{\partial x}\,dxdt + \frac{1}{2}\int_0^T\int_0^L e^{-st}x^{2+p}a\,\frac{\partial}{\partial t}\Big(\frac{\partial w}{\partial x}\Big)^2 dxdt \geqslant$$

$$\geqslant \frac{1}{2}\int_0^L e^{-sT}x^{2+p}a(x,T)\Big(\frac{\partial w}{\partial x}\Big)^2\Big|_{t=T} dx + \frac{1}{2}sA_0L_2^{2+p}\int_0^T\int_0^L e^{-st}\Big(\frac{\partial w}{\partial x}\Big)^2 dxdt -$$

$$-\Big[L^{2+p}C_a + \Big(1+\frac{p}{2}\Big)LC\Big]\int_0^T\int_0^L e^{-st}\Big(\frac{\partial w}{\partial x}\Big)^2 dxdt -$$

$$-\frac{1}{2}\Big[L^{2+p}C_a + (2+p)LC\Big]\int_0^T\int_0^L e^{-st}\Big(\frac{\partial w}{\partial t}\Big)^2 dxdt \qquad , \qquad (44)$$

where $L_2 \in (0,L)$ and C_a is the upper bound of both functions $\left|\frac{\partial a}{\partial x}\right|, \left|\frac{\partial a}{\partial t}\right|$. The first term of the right-hand side of (40) is

$$\int_0^T\int_0^O e^{-st}x^2 b\,\frac{\partial w}{\partial t}\,\frac{\partial w}{\partial x}\,dxdt \leqslant$$

$$\leqslant L^{2+\frac{p}{2}}B\,\frac{1}{2}\left\{\int_0^T\int_0^L e^{-st}\Big(\frac{\partial w}{\partial t}\Big)^2 dxdt + \int_0^T\int_0^L e^{-st}\Big(\frac{\partial w}{\partial x}\Big)^2 dxdt\right\} \quad . \qquad (45)$$

The second term of the right-hand side of (40) is

$$\int_0^T\int_0^L e^{-st}x^2\frac{\partial w}{\partial t}\Big[f(x,t,u,\frac{\partial u}{\partial t}) - f(x,t,v,\frac{\partial v}{\partial t})\Big]\,dxdt =$$

$$=\int_0^T\int_0^L e^{-st}x^2\frac{\partial w}{\partial t}\Big\{\int_0^1\frac{d}{d\lambda}f(x,t,\lambda u+(1-\lambda)v,\frac{\partial u}{\partial t})\,d \;+$$

$$+\Big[f(x,t,v,\frac{\partial u}{\partial t}) - f(x,t,v,\frac{\partial v}{\partial t})\Big]\Big\}\,dxdt \leqslant$$

$$\leqslant \frac{1}{2}L^2C_f\int_0^T\int_0^L e^{-st}w^2\,dxdt + L^2\Big(\frac{1}{2}C_f + C_r\Big)\int_0^T\int_0^L e^{-st}\Big(\frac{\partial w}{\partial t}\Big)^2 dxdt \quad .$$

Since

$$\int_0^T e^{-st}(w(x,t))^2 dt = \int_0^T e^{-st}\Big[\int_0^t\frac{\partial}{\partial\zeta}w(x,\zeta)d\zeta\Big]^2 dt \leqslant$$

$$\leqslant T\int_0^T e^{-st}\int_0^t\Big(\frac{\partial}{\partial\zeta}w(x,\zeta)\Big)^2 d\zeta dt \leqslant \frac{T}{s}\int_0^T e^{-st}\Big(\frac{\partial w}{\partial t}\Big)^2 dt \qquad ,$$

we have

$$\int_0^T\int_0^L e^{-st}x^2\frac{\partial w}{\partial t}\Big[f(x,t,u,\frac{\partial u}{\partial t}) - f(x,t,v,\frac{\partial v}{\partial t})\Big]\,dxdt \leqslant$$

$$\leqslant L^2\left(\tfrac{1}{2}C_f T/s + \tfrac{1}{2}C_f + C_r\right)\int_0^T\int_0^L e^{-st}\left(\frac{\partial w}{\partial t}\right)^2 dxdt \quad . \tag{46}$$

Substituting (43)–(46) into (40), we get

$$\int_0^L e^{-sT}x^2\left(\frac{\partial w}{\partial t}\right)^2\Big|_{t=T} dx + \int_0^L e^{-sT}x^{2+p}a(x,T)\left(\frac{\partial w}{\partial x}\right)^2\Big|_{t=T} dx +$$

$$+ sL_1^2\int_0^T\int_0^L e^{-st}\left(\frac{\partial w}{\partial t}\right)^2 dxdt + sA_0L_2^{2+p}\int_0^T\int_0^L e^{-st}\left(\frac{\partial w}{\partial x}\right)^2 dxdt \leqslant$$

$$\leqslant\left[2L^{2+p}C_a + (2+p)LC + L^{2+\frac{p}{2}}B\right]\int_0^T\int_0^L e^{-st}\left(\frac{\partial w}{\partial x}\right)^2 dxdt +$$

$$+\left[L^{2+p}C_a + (2+p)LC + L^{2+\frac{p}{2}}B + L^2C_f(1+T/s) + 2L^2C_r\right]\cdot$$

$$\cdot\int_0^T\int_0^L e^{-st}\left(\frac{\partial w}{\partial t}\right)^2 dxdt \quad .$$

For sufficiently large s, we have

$$\int_0^T\int_0^L e^{-st}\left[\left(\frac{\partial w}{\partial x}\right)^2 + \left(\frac{\partial w}{\partial t}\right)^2\right] dxdt \leqslant 0 \quad .$$

Since

$$w(x,0) = w(L,t) = 0 \quad ,$$

this implies that

$$w(x,t) = 0 \quad , \qquad \forall\ (x,t)\in\overline{Q} \quad .$$

Thus we obtain the uniqueness of the solution of the equation (1).

Theorem 1. Assume that the conditions (I)–(VI) hold and the time and space steps k and h satisfy the relation (23), then as h tends to zero, the solution of the difference scheme (6)–(10) converges to a limiting function $u(x,t)\in \overset{*}{W}{}_2^{(2)}(Q)$, where $u(x,t)$ is the unique generalized solution of the equation (1) (in which $p\geqslant 1$) and satisfies the initial and boundary conditions (2)–(5) in the common sense.

REMARK 1°. If in the difference scheme (6) and (7) we take

$$f_j^n = f(x_j, t^n, v_j^n, v_{\bar{t},j}^n) \quad ,$$

then the scheme is explicit even for nonlinear f. In this case, if we replace the assumption (IV) by

$$\text{(IV}'\text{)} \qquad \left|\frac{\partial}{\partial r}f(x,t,u,r)\right| \leqslant C_r \quad ,$$

then Lemmas 3–5 and Theorem 1 remain true.

REMARK 2. The existence of the solution of the nonlinear equations (6),(7) can be obtained by the fixed point principle using a prior estimation which may be derived as the estimate (28) is done . The uniqueness of the solution can also be shown easily.

§6.

When the exponent p in the coefficient of $\partial^2 u/\partial x^2$ of the equation (1) is a positive number less than 1, for the mixed initial-boundary problem of degenerate hyperbolic equation (1), the boundary condition at x=0 can also be given, i.e., the boundary condition (4) may be replaced no loss of generality by

$$u(0,t) = 0 \quad . \tag{4'}$$

Then in the assumption (V) the restriction on x=0 may also be added. The assumption (V) may be formulated as follows:

(V′) For x=0 and x=L, the function f(x,t,u,r) is Lipschitz continuous with respect to $t \in [0,T]$.

Besides, in the assumption (VI) we add the consistent condition $\phi(0) = \psi(0) = 0$, and then the assumption (VI) is denoted by (VI′). The difference equation (7) will be replaced by

$$v_0^n = 0 \quad , \qquad n=0,1, \ldots , N+1 \quad . \tag{7'}$$

Thus, starting from the equality (11), we see that the last term of (15) can not be cancelled with the last term of (12). According to the condition (7′), we have

$$\sum_{n=1}^{m} v^n_{\bar{t}t,0} v^n_{x\mathring{t},0} k = 0 \quad .$$

Therefore, we must estimate the sum $\sum_{n=1}^{m} f_0^n \, v^n_{x\mathring{t},0} k$. Summing by parts we have

$$\sum_{n=1}^{m} f_0^n \, v^n_{x\mathring{t},0} k = - \frac{1}{2} \sum_{n=1}^{m-1} f^{n+1}_{\bar{t},0} (v^{n+1}_{x,0} + v^n_{x,0}) k +$$

$$+ \frac{1}{2} f_0^m \, (v^{m+1}_{x,0} + v^m_{x,0}) - \frac{1}{2} f_0^1 \, (v^1_{x,0} + v^0_{x,0}) \quad ,$$

where appear the terms $v^{n+1}_{x,0} + v^n_{x,0}$. Since

$$v_{x,o}^{n+1} + v_{x,o}^{n} = - \sum_{j=1}^{J-1} (v_{\bar{x}x,j}^{n+1} + v_{\bar{x}x,j}^{n})h + v_{\bar{x},J}^{n+1} + v_{\bar{x},J}^{n} \quad ,$$

we get

$$\left|v_{x,o}^{n+1} + v_{x,o}^{n}\right| \leqslant \sum_{j=1}^{J-1} \left|v_{\bar{x}x,j}^{n+1} + v_{\bar{x}x,j}^{n}\right| h + \left|v_{\bar{x},J}^{n+1} + v_{\bar{x},J}^{n}\right| \quad ,$$

where $\left|v_{\bar{x},J}^{n+1} + v_{\bar{x},J}^{n}\right|$ has been estimated by (18) in §4 . Now,we have

$$\sum_{j=1}^{J-1} \left|v_{\bar{x}x,j}^{n+1} + v_{\bar{x}x,j}^{n}\right| h = \sum_{j=1}^{J-1} x_j^{-\frac{p}{2}} x_j^{\frac{p}{2}} \left|v_{\bar{x}x,j}^{n+1} + v_{\bar{x}x,j}^{n}\right| h \leqslant$$

$$\leqslant c \sum_{j=1}^{J-1} x_j^{-p} h + \frac{1}{4c} \sum_{j=1}^{J-1} x_j^{p} (v_{\bar{x}x,j}^{n+1} + v_{\bar{x}x,j}^{n})^2 h \quad ,$$

where $c > 0$ is an arbitrarily chosen constant. When $p \in (0,1)$,we have

$$\sum_{j=1}^{J-1} x_j^{-p} h \leqslant \int_0^L x^{-p} dx = \frac{1}{1-p} L^{1-p} \quad .$$

Therefore, there is

$$\sum_{j=1}^{J-1} \left|v_{\bar{x}x,j}^{n+1} + v_{\bar{x}x,j}^{n}\right| h \leqslant \frac{c}{1-p} L^{1-p} + \frac{1}{4c} \left\| x^{\frac{p}{2}} (v_{\bar{x}x}^{n+1} + v_{\bar{x}x}^{n}) \right\|^2 \quad .$$

At last we obtain

$$\sum_{n=1}^{m} f_o^n v_{x\mathring{t},o}^n k \leqslant \frac{1}{2} \sum_{n=1}^{m-1} \left|f_{\bar{t},o}^{n+1}\right| \left\{ \sum_{j=1}^{J-1} \left|v_{\bar{x}x,j}^{n+1} + v_{\bar{x}x,j}^{n}\right| h + \left|v_{\bar{x},J}^{n+1} + v_{\bar{x},J}^{n}\right| \right\} k +$$

$$+ \frac{1}{2}\left|f_o^m\right| \left\{ \sum_{j=1}^{J-1} \left|v_{\bar{x}x,j}^{m+1} + v_{\bar{x}x,j}^{m}\right| h + \left|v_{\bar{x},J}^{m+1} + v_{\bar{x},J}^{m}\right| \right\} + \frac{1}{2}\left|f_o^1 (v_{x,o}^1 + v_{x,o}^o)\right| \leqslant$$

$$\leqslant \frac{1}{2} C_o \sum_{n=1}^{m-1} \left\{ \frac{1}{1-p} L^{1-p} + \frac{1}{4} \left\| x^{\frac{p}{2}} (v_{\bar{x}x}^{n+1} + v_{\bar{x}x}^{n}) \right\|^2 \right\} k + \frac{1}{4} \sum_{n=1}^{m-1} \left|f_{\bar{t},o}^{n+1}\right|^2 k +$$

$$+ \frac{1}{4} \sum_{n=1}^{m-1} (v_{\bar{x},J}^{n+1} + v_{\bar{x},J}^{n})^2 k + \frac{1}{2} C_o \left\{ \frac{c_3}{1-p} L^{1-p} + \frac{1}{4c_3} \left\| x^{\frac{p}{2}} (v_{\bar{x}x}^{m+1} + v_{\bar{x}x}^{m}) \right\|^2 \right\} +$$

$$+ \frac{1}{4} \left|f_o^m\right|^2 + \frac{1}{4}(v_{\bar{x},J}^{m+1} + v_{\bar{x},J}^{m})^2 + \frac{1}{2} \left|f_o^1\right| \cdot \left|v_{x,o}^1 + v_{x,o}^o\right| \leqslant$$

$$\leqslant \frac{1}{8} C_o \sum_{n=1}^{m-1} \left\| x^{\frac{p}{2}} (v_{\bar{x}x}^{n+1} + v_{\bar{x}x}^{n}) \right\|^2 k + \frac{1}{4} \sum_{n=1}^{m-1} \left\{ (\frac{8}{L} + 1) \left\| v_{\bar{x}}^{n+1} + v_{\bar{x}}^{n} \right\|^2 + \right.$$

$$\left. + (2/L)^p \left\| x^{\frac{p}{2}} (v_{\bar{x}x}^{n+1} + v_{\bar{x}x}^{n}) \right\|^2 \right\} k + \frac{C_o}{8c_3} \left\| x^{\frac{p}{2}} (v_{\bar{x}x}^{m+1} + v_{\bar{x}x}^{m}) \right\|^2 +$$

$$+\frac{1}{4}\left\{(c_2+\frac{8}{L})\left\|v_{\bar{x}}^{m+1}+v_{\bar{x}}^{m}\right\|^2+\frac{1}{c_2}(2/L)^p\left\|x^{\frac{p}{2}}(v_{\bar{x}x}^{m+1}+v_{\bar{x}x}^{m})\right\|^2\right\}+R_4\ , \qquad (47)$$

where the constant C_o is the larger one of the upper bounds of $\left|f_{\bar{t},o}^{n}\right|$ and $\left|f(0,t,0,0)\right|$, and R_4 is a constant independent of h and k.

Considering the boundary condition (7′) and the above inequality (47) we obtain the following inequality, corresponding to the inequality (20),

$$\left\|v_{\bar{x}\bar{t}}^{m+1}\right\|^2+\left\{\frac{A_o}{4}-\frac{B^2}{4c_1}-\frac{3}{4c_2}(2/L)^p-\frac{C_o}{4c_3}\right\}\left\|x^{\frac{p}{2}}(v_{\bar{x}x}^{m+1}+v_{\bar{x}x}^{m})\right\|^2-$$

$$-\frac{k^2}{4}(C+B^2L^p/c_1)\left\|v_{\bar{x}x\bar{t}}^{m+1}\right\|^2\leqslant$$

$$\leqslant G_1'\sum_{n=1}^{m+1}\left\|v_{\bar{x}\bar{t}}^{n}\right\|^2k+G_2'\sum_{n=1}^{m}\left\|x^{\frac{p}{2}}(v_{\bar{x}x}^{n+1}+v_{\bar{x}x}^{n})\right\|^2k+k^2G_3'\sum_{n=1}^{m}\left\|v_{\bar{x}x\bar{t}}^{n+1}\right\|^2k+G_4'\ , \qquad (48)$$

where G_s' $(s=1,2,3,4)$ are constants independent of h and k. Let c_1 , c_2 be constants given in (21) and $c_3=8C_o/A_o$. Then from (48) we obtain

$$\left\|v_{\bar{x}\bar{t}}^{m+1}\right\|^2+\frac{A_o}{16}\left\|x^{\frac{p}{2}}(v_{\bar{x}x}^{m+1}+v_{\bar{x}x}^{m})\right\|^2-\frac{k^2}{4}C(1+\varepsilon)\left\|v_{\bar{x}x\bar{t}}^{m+1}\right\|^2\leqslant$$

$$\leqslant G_1'\sum_{n=1}^{m+1}\left\|v_{\bar{x}\bar{t}}^{n}\right\|^2k+G_2'\sum_{n=1}^{m}\left\|x^{\frac{p}{2}}(v_{\bar{x}x}^{n+1}+v_{\bar{x}x}^{n})\right\|^2k+k^2G_3'\sum_{n=1}^{m}\left\|v_{\bar{x}x\bar{t}}^{n+1}\right\|^2k+G_4'\ .$$

Starting from this inequality, by the same way described in §4 and §5, we obtain the following Lemma and Theorem.

Lemma 6. Suppose that the assumptions (I)-(IV), (V′), (VI′) are satisfied and suppose that the steps k and h satisfy the relation (23) and k is sufficiently small. Then for the solution $\{v_j^n\}$ of the difference scheme (6), (7′), (8), (9), (10) there are the estimates (25)-(33).

Theorem 2. Assume that the conditions (I)-(IV), (V′), (VI′) hold and the time and space steps k and h satisfy the relation (23), then as h tends to zero, the solution of the difference scheme (6), (7′), (8)-(10) converges to a limiting function $u(x,t)\in \overset{*}{W}{}_2^{(2)}(Q)$, where $u(x,t)$ is the unique generalized solution of the equation (1) (in which $p\in(0,1)$) and satisfies the initial and boundary conditions (2), (3), (4′), (5) in the common sense.

References

[1] Zhou Yu-lin, Finite Difference Method for Nonlinear Partial Dif-

ferential Equations, Applied Mathematics and Mathematics of Computation, 5(1983), pp.11 – 21.

[2] Zhou Yu-lin, Finite Difference Method of the Boundary Problems for the Systems of Generalized Schrödinger Type, J. Comput. Math., 1(1983), pp.170 – 181.

[3] Zhou Yu-lin, Guo Bo-ling, Finite Difference Solutions of the Boundary Problems for the Systems of Ferro-Magnetic Chain, J. Comput. Math., 1(1983), pp.294 – 302.

[4] Zhou Yu-lin, Finite Difference Method of the Boundary Problem for the Systems of Sobolev-Galpern Type, J. Comput. Math., 1(1983), pp.363 – 371.

[5] Zhou Yu-lin, Interpolation Formulas of Intermediate Quotients of Discrete Functions and Their Applications — Finite Difference Method and Finite Slice Method for Nonlinear Evolutional Systems of Partial Differential Equations, Proceedings of the 1984 Symposium on Differential Geometry and Differential Equations, Ed. by Feng Kang, Science Press, Beijing,China, 1985.

[6] Zhou Yu-lin, Interpolation Formulas of Intermediate Quotients for Discrete Functions with Several Indices, J. Comput. Math., 2 (1984), pp.376 — 381.

[7] Zhou Yu-lin, Xu Guo-rong, Finite Difference Solutions of the Nonlinear Mutual Boundary Problems for the Systems of Ferro-Magnetic Chain, J. Comput. Math., 2(1984), pp.263 — 271

[8] Zhou Yu-lin, Finite Difference Method for the First Boundary Value Problems of Quasilinear Parabolic Systems, Scientia Sinica (A), 3(1985), pp.206 — 220.

[9] Zhou Yu-lin, Finite Difference Method of Boundary Problem for Nonlinear Pseudo-Hyperbolic Systems, J. of Mathematical Research and Exposition, 5(1985), pp.73 — 79.

[10] Zhou Yu-lin, Some Nonlinear Boundary Problems for the Systems of Nonlinear Wave Equations by Finite Slice Method, J. Comput. Math., 3(1985), pp.50 — 71.

[11] Zhou Yu-lin, The General Nonlinear Mutual Boundary Problems for the Systems of Nonlinear Wave Equations by Finite Difference Method, J. Comput. Math., 3(1985), pp.134 – 160.

[12] Zhou Yu-lin, Guo Bo-ling, Some Boundary Problems of the Spin Systems of Ferro-Magnetic Chain, (I) Nonlinear Boundary Problems, Acta Mathematical Scientia, 6(1986), pp.321 – 337.

[13] Zhou Yu-lin, Guo Bo-ling, Some boundary Problems of the Spin Systems of Ferro-Magnetic Chain, (II) Mixed Problems and others, Acta Mathematical Scientia, (to be published) .

[14] Li De-yuan, Difference Schemes of Degenerate Parabolic Equations,

J. Comput. Math., 1(1983), pp.211 – 222.

[15] Li De-yuan, Difference Method for Semi-Linear Degenerate Parabolic Equations, in "Proceedings of BAIL III", Ed. J. Miller, Boole Press, Dublin, 1984.

[16] Shen Long-jun, Du Ying-yan, The Stability of the Finite Difference Solutions for a Class of Degenerate Parabolic Equations, J. Comput. Math., 2(1985), pp.175 – 187.

[17] Ding Lee, R. M. Kennedy, A Numerical Treatment of a Mixed Type Dynamic Motion Equation Arising from a Towed Acoustic Antenna in the Ocean, Comp. and Maths. with Appls., 11(1985), pp.807 – 816.

[18] M. L. Krasnov, Mixed Boundary Value Problems for Degenerate Linear Hyperbolic Differential Equations of Second Order, Mat. Sb., 49(91) (1959), pp.29 – 84.

ERROR EXPANSIONS FOR FINITE ELEMENT APPROXIMATIONS AND THEIR APPLICATIONS

Lin Qun and Xie Rui-feng
Institute of Systems Science
Academia Sinica, Beijing, China

We review and extend results on the error expansions of finite element approximation developed since 1983. We stress the expansion for the gradient error and the applications to the a posterior error estimate and the Richardson extrapolation and also to the adaptive refinement.

1. INTRODUCTION

In this paper we consider piecewise linear approximations $u^h (h>0)$ on triangulations T^h to the solution u of a model Dirichlet problem. It is known that, for the general (regular) triangulation, we have only the error estimates of low order:

$$|u^h - u| \leq ch^2 \ln \frac{1}{h}, \qquad |\nabla u^h - \nabla u| \leq ch, \tag{1}$$

where the factor $\ln\frac{1}{h}$ cannot be moved (see [6][15]). If we impose certain geometrical condition on T^h, however, we have the superconvergence estimates, [6] [17] [18]

$$|u^h - u| \leq ch^2, \qquad |(\bar{\nabla} u^h - \nabla u)(z)| \leq ch^2, \tag{2}$$

where $z \in T^h$ is the nodal points and $\bar{\nabla}$ the average gradient of two vertical triangles at vertex z. Notice that the constants c are over-conservative even if it can be estimated mathematically. Therefore, it remains to be answered that how a reasonable size of h can be decided for a given accuracy.

One of the approaches for answering this problem is to study the error expansions for the solution and its gradient like the following:

$$(u^h - u)(z) = e(z)h^2 + O(h^4 \ln \frac{1}{h}), \tag{3}$$

$$(\bar{\nabla}u^h - \nabla u)(z) = e'(z)h^2 + O(h^4 \ln \frac{1}{h}). \qquad (4)$$

If (3),(4) hold true for the nodal points $z \in T^h$ (with certain geometrical restriction on T^h), then the following a posterior error estimates

$$(u^{h/2} - u)(z) = -\frac{1}{3}(u^{h/2} - u^h)(z) + O(h^4 \ln \frac{1}{h}), \qquad (5)$$

$$(\bar{\nabla}u^{h/2} - \nabla u)(z) = -\frac{1}{3}\bar{\nabla}(u^{h/2} - u^h)(z) + O(h^4 \ln \frac{1}{h}) \qquad (6)$$

hold also true for $z \in T^h$. Thus, the approximation errors at nodal points $z \in T^h$ can be estimated during the solution process since $O(h^4 \ln \frac{1}{h})$ in (5),(6) can be omitted. Moreover, after the error is small enough, we can use the Richardson extrapolation to generate the fourth order approximations for both of solution and its gradient:

$$\frac{1}{3}(4u^{h/2} - u^h)(z) = u(z) + O(h^4 \ln \frac{1}{h}), \qquad (7)$$

$$\frac{1}{3}\bar{\nabla}(4u^{h/2} - u^h)(z) = \nabla u(z) + O(h^4 \ln \frac{1}{h}) \qquad (8)$$

at nodal points $z \in T^h$. So, for solution u, the extrapolation of linear elements will have the same accuracy as the superconvergence of quadratic elements (see §5). For gradient ∇u, however, the extrapolation of linear elements will have an accuracy one order higher than the superconvergence of quadratic elements (see §5).

Thus, the error expansions (3),(4) contain informations (see (5), (6),(8)) more than the higher order elements and hence is worthy of recommendation. Unfortunately, we can have only the pointwise estimate (1) for the general triangulation T^h and hence we must impose certain geometrical restriction on T^h in order to guarantee the expansions (3), (4).

For the polygonal domain with a piecewise uniform triangulation T^h as shown in Figure 1 the expansion (3) has been established in [6] (see also [19][20][24]). For the curved domain with an interior uniform triangulation (but keeps flexible in the neighborhood of the boundary as shown in Figure 2) an interior expansion with a remainder of reduced order $O(h^3 \ln \frac{1}{h})$ has been derived in [6],[22]. Recently a piecewise almost uniform triangulation as shown in Figure 3 has been introduced for the curved domain in [23] and a global expansion (3) (with a remainder of same order $O(h^4 \ln \frac{1}{h})$) can be recovered.

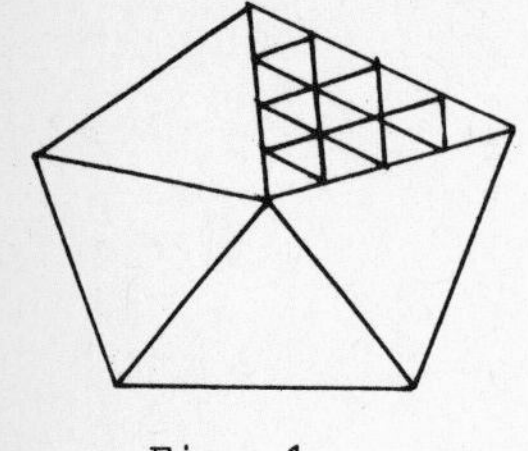

Fig. 1

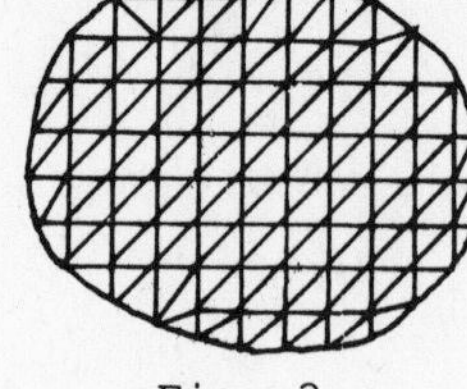

Fig. 2

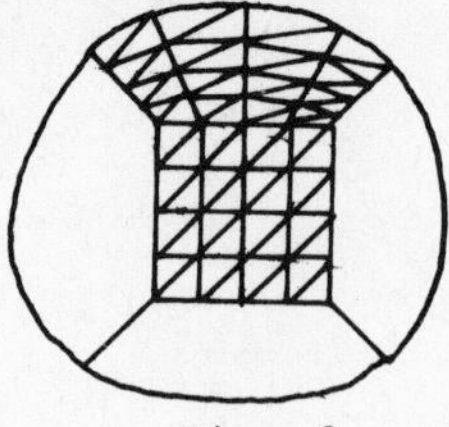

Fig. 3

In §3 we review the proof of solution expansion (3) and we derive in §4 the gradient expansion (4). The gradient superconvergence of quadratic elements will be mentioned in §5. Finally, in §6, we establish an error expansion for nonregular triangulation appeared in the adaptive refinement approach as shown in Figure 4.

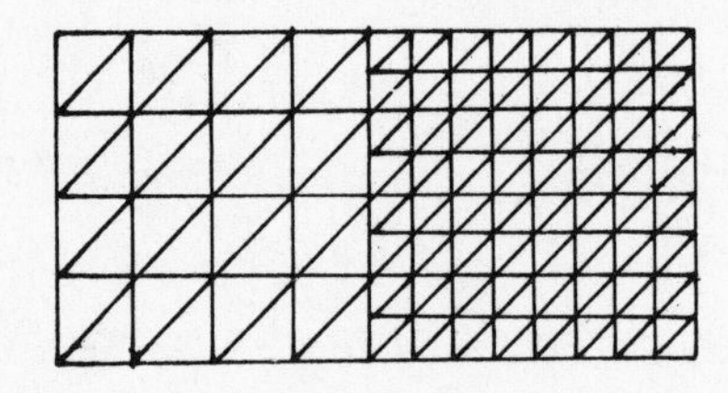

Fig. 4

We guess the error expansion similar to (3) holds also true for quadratic elements. In fact we have established in [22] the following expansion

$$\lambda^h - \lambda = eh^4 + O(h^5)$$

for piecewise quadratic approximation λ^h to the eigenvalue λ.

All of the above results are based on the smoothness assumption of the solution u. For the local smooth solution we can establish only a local expansion with a remainder of reduced order (see [25]). However, Babuska [1], Blum and Rannacher [7], Huang and Lin [11], Huang and Mu [13] have established the error expansion for the corner singularity case.

Rannacher [27] [28] has also established the error expansion for second order problems which are not separable and for second order elliptic systems coming from the fourth order elliptic problems and also for the nonstandard finite element methods. Helfrich [10], Huang and Lin [12],Lin and Wang [21] have established the error expansion for the time-dependent problems. The error expansions have also been established for certain nonlinear problem (see chen [8] Lu [26]), for Neumman problem (see Lin and Xie [22]), for the free boundary problem (see Shen [30]) and also for certain nonconforming finite elements (see [16]).

All of these results show that the study of error expansions (3),(4) on certain realizable triangulation (see Fig. 1-4) will provide a powerful and widely used tool in deriving the a posterior error estmates (5),(6) and the adaptive approach (see Figure 4) and the Richardson extrapolations (7),(8).

2. REVIEW OF THE INTEGRAL EXPANSION

Let us consider the model Dirichlet problem

$$-\Delta u = f \quad \text{in } D, \quad u = 0 \quad \text{on } \partial D, \tag{9}$$

where D is a bounded domain of R^2 with straight or curvilinear boundary ∂D. We approximate (9) by the simplest finite element method, namely, linear elements. For this, let $T^h = \{K\}$ be a finite regular triangulation of D of width h with all its boundary vertices on ∂D. Corresponding to T^h, we define the linear finite element space

$$S^h = \{\ v^h \in H_0^{1,\infty} : v^h \text{ is linear on each } K\ \}$$

and the linear finite element solution $u^h \in S^h$ satisfying

$$(\nabla u^h, \nabla v^h) = (f, v^h) \qquad \forall\ v^h \in S^h.$$

Below, for ease of understanding, we shall consider "uniform" triangulation, namely, the three sides of any element $K \in T^h$ parallel to the three sides of a fixed triangle. We use the notations T_d (d=1,2,3) for the tangent unit vectors along the three edges of the fixed triangle in counter-clockwise ordering.

For any element $K \in T^h$ we introduce the notation:

$$s_d\text{- side of K parallel with } T_d, \quad h_d \text{ - length of } s_d,$$

n_d-outer normal unit vector along s_d, a-area of K,

p_d-vertex of K opposite to s_d, t_d-tangent unit vector along s_d.

We have: either

$$t_d = T_d \quad \text{or} \quad t_d = -T_d \quad (d=1,2,3).$$

Let us introduce the sign function:

$$g(K) = t_d \cdot T_d . \qquad (10)$$

Then

$$g(K) = \pm 1, \quad t_d = g(K) T_d .$$

For any two adjacent elements K and K', we have

$$g(K) = -g(K').$$

We will use the notation:

$$\partial_d = \frac{\partial}{\partial t_d}, \quad D_d = \frac{\partial}{\partial T_d} .$$

Then $\partial_d = g(K) D_d$.

The normal vector can be expressed as a linear combination of tangent vectors along two sides (see [22]):

$$n_1 = \frac{h_1 h_2}{2a} (t_1 \cdot t_2 \ t_1 - t_2). \qquad (11)$$

Furthermore, we can transfer one line integral to another (see [6] [24]):

$$\int_{s_1} v \, ds = \frac{h_1}{h_2} \int_{s_2} v ds + \frac{h_1 h_3}{2a} \int_K \partial_3 v dx. \qquad (12)$$

As usual, the error of finite element approximation u^h is linked to the interpolation $i^h u \in S^h$:

$$(u^h - i^h u)(z) = \int_D \nabla(u^h - i^h u) \nabla g_z^h dx = \int_D \nabla(u - i^h u) \nabla g_z^h dx,$$

where, for any fixed $z \in D$, $g_z^h \in s^h$ is the discrete Green function defined as in [6].

What needs to be expanded is the following integral

$$I(v^h) = \int_D \nabla(u - i^h u) \nabla v^h dx \qquad \forall \, v^h \in S^h.$$

Note that ∇v^h is a constant on K. We have, by integral by parts,

$$I(v^h) = \sum_{K \in T^h} \int_K \nabla(u - i^h u) \nabla v^h dx = \sum_{K \in T^h} \int_{\partial K} (u - i^h u) \frac{\partial v^h}{\partial n} ds$$

$$= \sum_{d=1} \sum_{K \in T^h} \int_{s_d} (u - i^h u) \frac{\partial v^h}{\partial n_d} ds.$$

To make definite we consider the line integral on S_1:

$$I_K^1 = \int_{s_1} (u-i^h u) \frac{\partial v^h}{\partial n_1} ds.$$

Note that $\frac{\partial v^h}{\partial n_1}$ is a constant. We have, by the Euler-Maclaurin formula,

$$I_K^1 = -\frac{h_1^2}{12} \int_{s_1} D_1^2 u \frac{\partial v^h}{\partial n_1} ds + h_1^4 J_K^1, \quad J_K^1 = \int_{s_1} b(s) D_1^4 u \frac{\partial v^h}{\partial n_1} ds. \qquad (13)$$

Since $\frac{\partial v^h}{\partial n_1}$ is discontinuous: $\frac{\partial v^h}{\partial n_1}(K) \neq \frac{\partial v^h}{\partial n_1}(K')$ for two adjacent elements K and K' with common side s_1 but $\partial_1 v^h$ is continuous on s_1, we must transfer normal vector to tangent vectors by using (11):

$$I_K^1 = -\frac{h_1^3 h_2}{24a} (t_1 \cdot t_2 \int_{s_1} D_1^2 u \, \partial_1 v^h ds - \int_{s_1} D_1^2 u \partial_2 v^h ds) + h_1^4 J_K^1.$$

For the second line integral, in view of (12),

$$\int_{s_1} D_1^2 u \partial_2 v^h ds = \frac{h_1}{h_2} \int_{s_2} D_1^2 u \, \partial_2 v^h ds + \frac{h_1 h_3}{2a} \int_K \partial_3 D_1^2 u \partial_2 v^h dx.$$

Hence we have, by using the sign function g defined in (10),

$$I_K^1 = -g(K) \frac{h_1^3 h_2}{24a} T_1 \cdot T_2 \int_{s_1} D_1^2 u D_1 v^h ds + g(K) \frac{h_1^4}{24a} \int_{s_2} D_1^2 u \, D_2 v^h ds$$

$$- \frac{h_1^4 h_2 h_3}{48a^2} \int_K D_3 D_1^2 u D_2 v^h dx + h_1^4 J_K^1.$$

For $s_1 \subset D$ there exist $K, K' \in T^h$ such that $K \cap K' = s_1$ and we have

$$D_1 v^h(K) = D_1 v^h(K'), \qquad g(K) = -g(K').$$

Thus,

$$g(K) \int_{s_1} D_1^2 u D_1 v^h ds + g(K') \int_{S_1} D_1^2 u D_1 v^h ds = 0.$$

For $s_1 \subset \partial D$ we have

$$v^h = 0 \text{ on } S_1, \quad \partial_1 v^h = 0 \text{ on } S_1.$$

Hence

$$\sum_{K \in T^h} g(K) \int_{s_1} D_1^2 u D_1 v^h ds = 0.$$

By the same reason,

$$\sum_{K \in T^h} g(K) \int_{s_2} D_1^2 u D_2 v^h ds = 0.$$

We obtain

$$I^1 = \sum_K I_K^1 = -\frac{h_1^4 h_2 h_3}{48a^2} \int_D D_3 D_1^2 u D_2 v^h dx + h_1^4 J^1.$$

Since $v^h = 0$ on ∂D we have, by integral by parts,

$$I^1 = \frac{h_1^4 h_2 h_3}{48a^2} \int_D D_3 D_2 D_1^2 \; u \; v^h \; ds + h_1^4 J^1 .$$

Let $h = \max h_d$. Since all $K \in T^h$ are similar to a fixed triangle, there exist constants c_d and a_0 such that

$$h_d = c_d h, \qquad a = a_0 h^2$$

and hence

$$I^1 = h^2 \frac{c_1^4 c_2 c_3}{48a_0^2} \int_D D_2 D_3 D_1^2 \; u \; v^h \; dx + h_1^4 J^1,$$

where

$$J^1 = \sum_K \int_{s_1} b(s) D_1^4 u \frac{\partial v^h}{\partial n_1} ds.$$

We can extend the Euler-MacLaurin formula (13) to be

$$I_K^1 = - \frac{h_1^2}{12} \int_{s_1} D_1^2 u \frac{\partial v^h}{\partial n_1} ds - \frac{h_1^4}{b_0} \int_{s_1} D_1^4 u \frac{\partial v^h}{\partial n_1} ds + h_1^6 \int_{s_1} b(s) D_1^6 u \frac{\partial v^h}{\partial n_1} ds$$

Then, by the same argument as above, we obtain the further expansion

$$I^1 = h^2 \frac{c_1^4 c_2 c_3}{48a_0^2} \int_D D_3 D_2 D_1^2 u v^h dx + h^4 \frac{c_1^6 c_2 c_3}{4b_0 a_0^2} \int_D D_2 D_3 D_1^4 u v^h dx + h^6 R^1$$

$$R^1 = \sum_K \int_{s_1} b(s) D_1^6 u \frac{\partial v^h}{\partial n_1} ds.$$

We can have the similar expansions for I^2 and I^3. To sum up, we obtain

$$I(v^h) = h^2 \int_D D^4 u v^h dx + \sum_{d=1}^{3} h_d^4 J^d, \tag{14}$$

$$I(v^h) = h^2 \int_D D^4 u v^h dx + h^4 \int_D D^6 u v^h dx + \sum_{d=1}^{3} h_d^6 R^d, \tag{15}$$

where

$$D^4 = \frac{1}{48a_0^2} \sum_{d=1} c_d^4 c_{d+1} c_{d+2} D_{d+1} D_{d+2} D^2,$$

$$D^6 = \frac{1}{4b_0 a_0^2} \sum_{d=1} c_d^6 c_{d+1} c_{d+2} D_{d+1} D_{d+2} D_d^4,$$

$$J^d = \sum_K \int_{s_d} b(s) D_d^4 u \frac{\partial v^h}{\partial n_d} ds, \quad R^d = \sum_K \int_{s_d} b(s) D_d^6 u \frac{\partial v^h}{\partial n_d} ds.$$

3 SOLUTION EXPANSION

Let us take in (14) $v^h = g_z^h$. We first estimate J^d. For this, we introduce the regularized Green function $\tilde{g}_z \in H^2(D)$. Then, we have

$$J^1 = \sum_K \left(\int_{s_1} b(s) D_1^4 u \frac{\partial \tilde{g}_z}{\partial n_1} ds + \int_{s_1} b(s) D_1^4 u \frac{\partial}{\partial n_1} (g_z^h - \tilde{g}_z) ds \right).$$

For $s_1 \subset D$ there exist K, $K' \in T^h$ such that $K \cap K' = s_1$ and we have

$$n_1(K) = -n_1(K'), \quad \frac{\partial \tilde{g}_z}{\partial n_1} = -\frac{\partial \tilde{g}_z}{\partial n_1'} .$$

For $s_1 \subset \partial D$ we have

$$u = D_1^4 u = 0 \text{ on } s_1 .$$

Hence

$$\sum_K \int_{s_1} b(s) D_1^4 u \frac{\partial \tilde{g}_z}{\partial n_1} ds = 0.$$

It remains to estimate the second term in J^1. By a trace theorem (see [6]),

$$|\int_{s_1} b(s) D_1^4 u \frac{\partial}{\partial n_1}(g_z^h - \tilde{g}_z) ds| \leq c||u||_{4,\infty} \int_{\partial K} |\nabla(g_z^h - \tilde{g}_z)| ds$$

$$\leq c||u||_{4,\infty} (h^{-1} \int_K |\nabla(g_z^h - \tilde{g}_z)| ds + \int_K |\nabla^2 \tilde{g}_z| ds).$$

Hence, by the same reason, $J^d = 0(\ln \frac{1}{h})$. We obtain

$$(u^h - i^h u)(z) = I(g_z^h) = h^2 \int_D D^4 u g_z^h dx + 0(h^4 \ln \frac{1}{h}). \tag{16}$$

We denote by e the solution of the dirichlet problem

$$-\Delta e = D^4 u \quad \text{in } D, \quad e = 0 \text{ on } \partial D \tag{17}$$

and e^h the Ritz projection of e. Then e^h is nothing but the coefficient in the representation (16):

$$e^h(z) = (\nabla e^h, \nabla g_z^h) = (\nabla e, \nabla g_z^h) = \int_D D^4 u g_z^h dx.$$

It remains to estimate $e - e^h$. Let m_j be the vertices of polygon D. Then, for $D_0 \subset D$ and $m_j \bar{\in} \bar{D}_0$, we have, by local estimate (3.18) in [6],

$$||e - e^h||_{0,\infty,D_0} \leq ch^2.$$

Combining with (16) yields the desired expansion (3).

Remark 1. It has been proved in [23] that, for a curved domain D with the triangulation T^h shown as in Figure 3, the error expansion (3) holds also true. The difficulty which rises from the nonuniform triangulation is the different tangent directions T_d (and the different sizes h_d) for different elements $K \in T^h$. All these local quantities, however, have been standardized in [23] by the values of certain function at certain points, namely, by certain united reference quantities.

4. GRADIENT EXPANSION

We first discuss the gradient expansion for interpolants. Normally we have only lower gradient convergence:

$$|\nabla(i^h u-u)| \le ch.$$

We will explain below that the gradient superconvergence can happen when the average gradient of two vertical triangles is introduced at the vertex.

Lemma 1. Let $i^1 u$ be the linear interpolant of u with nodes 0 and 1. Then

$$(i^1 u-u)'(0)=\sum_{j=2}^{4} \frac{1}{j!} u^{(j)}(0)+O(1)\,|u|_{5,\infty}.$$

Lemma 2. For $K\in T^h$ and $u\in C^5(K)$, we have

$$\partial_1(i^h u-u)(p_2)=\sum_{j=2}^{4} \frac{1}{j!} h_1^{j-1} \partial_1^j u(p_2)+O(h^4)|u|_{5,\infty},$$

$$D_1(i^h u-u)(p_2)=g(K)\sum_{j=2}^{4} \frac{1}{j!} h_1^{j-1}(g(K))^j D_1^j u(p_2)+O(h^4).$$

It is easy to see that, for any interior nodal point $p \in D$, there exist K, K' such that

$$p = p_2(K) = p_2(K').$$

Since $g(K) = -g(K')$ we have

$$D_1 i^h u|_K + D_1 i^h u|_{K'} - 2D_1 u(p_2)= \frac{1}{3} h_1^2 D_1^3 u(p_2)+O(h^4).$$

We introduce the notation

$$\bar{D}_1 i^h u(p) = \frac{1}{2}(D_1 i^h u|_K + D_1 i^h u|_{K'}),$$

$$\bar{D}_1 i^h u(p_2)-D_1 u(p_2)= \frac{1}{6} h_1^2 D_1^3 u(p_2) + O(h^4). \tag{18}$$

By the same argument,

$$\bar{D}_3 i^h u(p_2)-D_3 u(p_2)= \frac{1}{6} h_3^2 D_3^3 u(p_2)+O(h^4). \tag{19}$$

We denote by

$$\bar{\nabla} i^h u(p_2)= \frac{1}{2}(\nabla i^h u|_K + \nabla i^h u|_{K'})$$

the average gradient of two vertical triangles K and K'. Then, by (18), (19),

$$|\bar{\nabla} i^h u(p_2) - \nabla u(p_2)| \le ch^2, \tag{20}$$

$$\bar{\nabla} i^h u(p_2)- \nabla u(p_2) = \vec{w}(p_2)h^2 + O(h^4), \tag{21}$$

where $\vec{w}$ is a vector function consisting of the third derivatives of u.

We now turn to the gradient expansion for finite element approximation. For this, we introduce $G_z^h \in S^h$ defined by

$$(\nabla G_z^h, \quad \nabla v) = \quad \partial v(z) \qquad v \in S^h. \tag{22}$$

Then, we have

$$\partial(u^h - i^h u) = \int_D \quad \nabla(u - i^h u) \nabla G_z^h \; dx.$$

Taking $v^h = G_z^h$ in (14) and integrating by parts, we obtain by $b(s) \in C_0^1(s_d)$,

$$\partial(u^h - i^h u) = h^2 \int_D D^3 u D G_z^h dx - \Sigma\, h_d^4 \sum_k \int_{s_d} D_d b(s) D_d^3 u \frac{\partial}{\partial n_d} G_z^h ds.$$

In view of the estimates

$$\|G_z^h\|_{1,1} \leq c \ln \frac{1}{h}, \int_{\partial K} |\nabla G_z^h| ds \leq ch^{-1} \int_K |\nabla G_z^h| ds, \tag{23}$$

there holds

$$|\nabla(u^h - i^h u)| \leq ch^2 \ln \frac{1}{h} \|u\|_{3,\infty}.$$

Combining with (20) implies that, for the nodal point $p \in D$,

$$|\bar{\nabla} u^h(p) - \bar{\nabla} u(p)| \leq ch^2 \ln \frac{1}{h} \|u\|_{3,\infty}. \tag{24}$$

A local form of (24) can also be obtained for $u \in C^3(D_0)$ with $D_0 \subset D$.

We take $v^h = G_z^h$ in (15):

$$\partial(u^h - i^h u) = h^2 \int_D D^4 u G_z^h dx + h^4 \int_D D^6 u G_z^h \; dx + \Sigma\, h_d^6 R^d.$$

It is easy to estimate the last two terms by using (23), and we have

$$\partial(u^h - i^h u) = h^2 \int_D D^4 u \; G_z^h \; dx + O(h^4 \ln \frac{1}{h}) \|u\|_{5,\infty}. \tag{25}$$

We again use the function e defined in (17) and its Ritz projection e^h. Then, by (22), ∂e^h is nothing but the coefficient in the representation (25):

$$\partial e^h(z) = (\nabla G_z^h, \nabla e^h) = (\nabla G_z^h, \nabla e) = \int_D D^4 u \; G_z^h \; dx. \tag{26}$$

By a local form of (24), for $p \in D_0 \subset D$ and $m_j \in \bar{D}_0$,

$$|\bar{\nabla} e^h(p) - \bar{\nabla} e(p)| \leq ch^2 \ln \frac{1}{h}.$$

Combining with (21) (25) (26) yields the following

Theorem 1. For $p \in D_0 \subset D$ and $m_j \in \bar{D}_0$, there holds

$$\bar{\nabla} u^h(p) - \bar{\nabla} u(p) = (\bar{\nabla} e(p) + \vec{w}(p)) h^2 + O(h^4 \ln \frac{1}{h}).$$

Remark 2. the gradient expansion (4) holds also true for a curved domain D with the triangulation T^h shown as in Figure 3. In order to derive the interpolant expansion (21) we again need to expand each local quantity for $K \in T^h$ by certain united reference quantities as show in [23].

5. GRADIENT SUPERCONVERGENCE FOR QUADRATIC ELEMENTS

In this section the symbol S^h will denote the quadratic finite element space. We have proved in [23] the following expansion: for $v^h \in S^h$,

$$I(v^h)=h^4(\int_D D^5uDv^hdx + \sum_K \int_K D^4uD^2v^hdx)+O(h^5)||v^h||'_{2,1}. \quad (27)$$

Taking $v^h=G_z^h$ (see (22)) and observing that

$$||G_z^h||_{1,1} \leq c \ln \frac{1}{h}, \quad ||G_z^h||'_{2,1} \leq ch^{-1} \ln \frac{1}{h},$$

we obtain

$$|\partial(u^h-i^hu)(z)| = |I(G_z^h)| \leq ch^3 \ln \frac{1}{h} . \quad (28)$$

Let q_1^*, q_{-1}^* be the Guass points on S_1:

$$|q_{-1}^* - p_2| = |q_1^*-p_3| = \frac{1}{2} (1- \frac{\sqrt{3}}{3})h_1 .$$

It is easy to verify that

$$|\partial_1 i^hu(q_j^*)- \partial_1 u(q_j^*)| \leq ch^3 \quad (j= -1,1). \quad (29)$$

For any interior nodal point $p \in D$, there exist K, K' such that

$$p = p_2(K) = p_2(K').$$

Let q_1^*, q_{-1}^* be the Guass points on $s_1=s_1(K)$ and q_1', q_{-1}' the Gauss points on $s_1' = s_1(K')$. We denote by i_3v the cubic interpolant of $v \in C(s_1 \cup s_1')$ with nodes q_{-1}^*, q_1^*, q_{-1}', q_1'. Then

$$\begin{aligned} &|v - i_3v| \leq ch^3 ||v||_{3,\infty}, \\ &i_3v(p_2)=(\frac{\sqrt{3}}{6} + \frac{1}{4})(v(q_{-1}^*)+v(q_{-1}'))+(\frac{\sqrt{3}}{6} - \frac{1}{4})[v(q_1^*)+v(q_1')]. \end{aligned} \quad (30)$$

Hence, by (28), (29), (30),

$$\begin{aligned} |i_3\partial_1u^h(p_2)- \partial_1u(p_2)| &\leq |i_3 \partial_1(u^h-i^hu)(p_2)|+|i_3\partial_1 i^hu- \partial_1u| \\ &\leq ||u^h-i^hu||_{1,\infty} + |i_3(\partial_1 i^hu- \partial_1u)| +|i_3\partial_1u-\partial_1u| \leq ch^3 \ln\frac{1}{h}. \end{aligned}$$

By the same argument, there also holds

$$|i_3 \partial_3u^h(p_2) - \partial_3u(p_2)| \leq ch^3 \ln \frac{1}{h}.$$

To sum up, we obtain the following

Theorem 2. For any interior nodal point $p \in D$, there hold

$$|i_3D_1u^h(p)-D_1u(p)| \leq ch^3 \ln \frac{1}{h} ,$$

$$|i_3D_3u^h(p)-D_3u(p)| \leq ch^3 \ln \frac{1}{h} ,$$

where

$$\begin{aligned} i_3D_1u^h(p)=&(\frac{\sqrt{3}}{6} + \frac{1}{4})(D_1u^h(q_{-1}^*)+D_1u^h(q_{-1}')) \\ &+(\frac{\sqrt{3}}{6} - \frac{1}{4})(D_1u^h(q_1^*)+D_1u^h(q_1')), \\ i_3D_3u^h(p)=&(\frac{\sqrt{3}}{6} + \frac{1}{4})(D_3u^h(q_{-3}^*)+D_3u^h(q_{-3}')) \\ &+(\frac{\sqrt{3}}{6} - \frac{1}{4})(D_3u^h(q_3^*)+D_3u^h(q_3')). \end{aligned}$$

Remark 3. Our proof for the gradient superconvergence of quadratic elements is based on the expansion method shown in (27). This kind of expansion method is more powerful than the usual superconvergence proof (see [31] for example) since it provides an approach to prove the asymptotic error expansion for quadratic element approximation which has many applications.

6. EXPANSION ON A NONREGULAR TRIANGULATION

Suppose that the polygonal domain D consists of two subdomains D* and D' and that there exist two uniform triangulations T^h and T^k defined on D* and D' respectively, as shown in Figure 4. We assign $\ell = \partial D^* \cap \partial D'$, and assume that the mesh size of D* is the half of the mesh size of D' along the common boundary ℓ. We use the notation $\ell_h \subset \bar{D}^*$ to denote the line segment parallele to ℓ and having a distance with one layer of elements from ℓ. The nodal point sets of T^h and T^k will be denoted by N* and N' respectively. Then

$$N^* \cap \ell \neq N' \cap \ell .$$

p is nonregular point if $p \in N^* \cap \ell \setminus N'$ and the other points of $N^* \cup N'$ will be defined as the regular points. Let S^h be a linear finite element space defined by

$$S^h = \{ v^h \in C_0(\bar{D}),\ v^h \text{ is linear for each } K \in T^h \cup T^k \} .$$

Then, each $v^h \in S^h$ is defined uniquely by its values at the regular points, and there is no freedom for its value at the nonregular point.

$K \in T^h \cup T^k$ is a nonregular element if at least one of the vertices is nonregular point. Otherwise, K is regular element. For the regular element K, we have

$$|u - i^h u|_{s,p,K} \leq ch^{t-s+\frac{2}{p} - \frac{2}{q}} |u|_{t,q,K} \quad (0 \leq s \leq t \leq 2),$$

where $i^h u \in S^h$ is the interpolant of u having the same values at the regular points.

Note that any nonregular element will belong to a set of elements as shown in Figure 5. Let p be a nonregular point and D_p the union of nonregular elements with the vertex p. Then, $D_d \subset \bar{D}^*$. It is easy to see that D_p can be transformed into a reference region $\tilde{D}_p$ where the Bramble lemma can be used to derive the following estimate:

$$|\tilde{u} - \tilde{i}u|_{s,p,\tilde{D}_p} \leq c |\tilde{u}|_{t,q,\tilde{D}_p},$$

and hence, by the standard analysis,

$$|u - i^h u|_{s,p,D_p} \leq ch^{t-s+\frac{2}{p}-\frac{2}{q}} |u|_{t,q,D_p} ,$$

Fig. 5

$$|u-i^h u|_{s,p,D} \le ch^{t-s+\frac{2}{p}-\frac{2}{q}} |u|'_{t,q,D}.$$

Thus, the interpolant estimate holds true for the nonregular triangulation $T^h \cup T^k$. In addition, S^h is a conforming finite element space. Therefore, all the estimates in the finite element analysis (including the estimate for the discrete Green function) hold also true for our case.

Note that all $K \subset D'$ are of regular elements, hence, by §3,

$$\int_{D'} \nabla(u-i^k u)\, \nabla g_z^k dx = k^2 \int_{D'} D^4 u g_z^k dx + k^2 \int_\ell D^3 u g_z^k ds + O(k^3)$$

$$= k^2 \int_{D'} D^4 u g_z dx + k^2 \int_\ell D^3 u g_z ds + O(k^3 \ln \tfrac{1}{k}).$$

Let D_h^* be the region with boundary consisting of $\partial D^* \setminus \ell$ and ℓ_h. Since all $K \subset D_h^*$ are of regular elements, we have the usual expansion as in §3,

$$\int_{D_h^*} \nabla(u-i^h u) \nabla g_z^h dx = h^2 \int_{D_h^*} D^4 u g_z^h dx + h^2 \int_{\ell_h} D^3 u g_z^h dx + O(h^3).$$

For $D_0 \subset D$, dist $(D_0, \ell) > h$ and $m_j \in \bar{D}_0$, there hold, for $z \in D_0$,

$$|g_z^h - g_z|_{1,\infty,D\setminus D_h^*} \le ch, \qquad ||g_z||_{2,\infty,D\setminus D_h^*} \le c,$$

and hence

$$\left|\int_{D^*\setminus D_h^*} D^4 u g_z dx\right| \le ch|u|_{4,\infty} |g_z|_{0,\infty,D^*\setminus D_h^*},$$

$$\left| \int_{\ell_h} D^3 u g_z ds - \int_\ell D^3 u g_z ds \right| \le \int_{D^*\setminus D_h^*} |\partial(D^3 u g_z)| dx \le ch,$$

which lead to the following expansion

$$\int_{D_h^*} \nabla(u-i^h u) \nabla g_z^h dx = h^2 \int_{D^*} D^4 u g_z dx + h^2 \int_\ell D^3 u g_z + O(h^3 \ln \tfrac{1}{h}).$$

It remains to estimate

$$\int_{D^*\setminus D_h^*} \nabla(u-i^h u) \nabla g_z^h dx = \int_{D^*\setminus D_h^*} \nabla(u-i^h u) \nabla(g_z^h - g_z) dx$$

$$+ \int_{D^*\setminus D_h^*} \nabla(u-i^h u) \nabla g_z dx.$$

The second integral can be estimated as follows:

$$\left| \int_{D^*\setminus D_h^*} \nabla(u-i^h u) \nabla(g_z^h - g_z) dx \right| \le ch |u-i^h u|_{1,\infty} |g_z^h - g_z|_{1,\infty,D^*\setminus D_h^*} \le ch^3.$$

For the last integral, observing $\Delta g_z = 0$ in $D^*\setminus D_h^*$, there holds

$$\int_{D^*\setminus D_h^*} \nabla(u-i^h u) \nabla g_z dx = \int_{\partial(D^*\setminus D_h^*)} (u-i^h u) \frac{\partial g_z}{\partial n} ds$$

$$= \left(\int_\ell - \int_{\ell_h}\right)(u-i^h u) \frac{\partial g_z}{\partial n} ds = \left(\int_\ell - \int_{\ell_h}\right)(u-i^h u) \frac{\partial}{\partial n} g_z^h ds + O(h^3)$$

$$= h^2 \int_\ell D^2 u \frac{\partial}{\partial n} g_z^h ds - \frac{h^2}{2} \int_{\ell_h} D^2 u \frac{\partial}{\partial n} g_z^h ds + O(h^3)$$

$$= h^2 \int_{\ell} D^2u \frac{\partial}{\partial n} g_z ds - \frac{h^2}{2} \int_{\ell_h} D^2u \frac{\partial}{\partial n} g_z ds + O(h^3)$$

$$= \frac{h^2}{2} \int_{\ell} D^2u \frac{\partial}{\partial n} g_z ds + O(h^3).$$

To sum up, we obtain the following

Theorem 3. Suppose that $u \in C^4(\bar{D})$ and that $D_0 \subset D$, dist $(D_0, \ell) > h$ and $m_j \notin \bar{D}_0$. Then, there exist functions e_1 and e_2 such that, for nodal nodal points $z \in D_0$,

$$u^h(z) - u(z) = e_1(z)k^2 + e_2(z)h^2 + O(h^3 \ln \frac{1}{h}).$$

REFERENCES

1. I. Babuska, private communication.

2. I. Babuska and A. Miller, A feedback finite element method with a posteriori error estimation, Tech. Note BN-1031, IPST, University of Maryland, October 1984.

3. I. Babuska and A.K.Noor, Quality assessment and control of finite element solutions, Tech. Note BN-1049, IPST University of Maryland, May 1986.

4. H. Blum, On Richardson extrapolation for linear finite elements on domains with reentrant corners, To appear in Z. Angew. Math. Mech.

5. H. Blum and M.Dobrowolski, On the finite element method for elliptic equations on domains with corners, Computing 28,53-63, 1982.

6. H. Blum, Q. Lin and R.Rannacher, Asymptotic error expansion and Richardson extrapolation for linear finite elements, Numer. Math. 49, 11-37, 1986.

7. H. Blum and R.Rannacher, To appear.

8. C. Chen, To appear.

9. J. Frehse and R.Rannacher, Eine L -Fehlerabschatzung fur diskrete Grundlosungen in der Methode der finiten Elemente, Bonn, Math. Schr. 89, 92-114, 1976.

10. H.-P. Helfrich, Asymptotic expansion for the finite element approximation of parabolic problems, Bonn. Math. Schr. 158, 11-30, 1984.

11. Y. Huang and Q. Lin, To appear.

12. Y. Huang and Q. Lin, To appear.

13. H. Huang and M. Mu, To appear.

14. M. Krizek and P. Neittaanmaki, On superconvergence techniques, Jyvaskyla University, Finland, Preprint No. 34, 1984.

15. N.Levine, pointwise logarithm-free error estimates for finite elements on linear triangles. Numerical Analysis Report Nr. 6/84, University of Reading.

16. Li Bo, To appear,

17. Q. Lin, T.Lu and S. Shen, Maximum norm estimate, extrapolation and optimal point of stresses for the finite element methods, J. Comput. Math. 1, 376-383, 1983.

18. Q. Lin, T. Lu and S. shen, asymptotic expansion for finite element approximations, Research Report IMS-11, Chendu Branch of Academia Sinica, china, 1983.

19. Q. Lin and T. Lu, asymptotic expansions for linear element approximation of elliptic problems on polygonal domains, Computing Methods in Applied sciences and Engineering VI. Proc. 6th Int. Symp., Versailles 1983 (R. Glowinski, J.L.Lions,ed.), 371-321, North-Holland, amsterdam, 1984.

20. Q.Lin and J.Wang, Some expansions of the finite element approximation, Research Report IMS-15, Chengdu Branch of Academia Sinica, China, 1984.

21. Q.Lin and J.Wang, Asymptotic expansion for finite element approximation of time-dependent problems, To appear.

22. Q.Lin and R.Xie, Some advances in the study of error expansion for finite elements, J. Comput. Math. 4, 368-382, 1986.

23. Q.Lin and R.Xie, How to recover the convergence rate for Richardson extrapolation on bounded domains, J. Comput. Math. 5, 1987.

24. Q. Lin and J.Xu, Linear finite elements with high accuracy, J. Comput. Math. 3, 115-133, 1985.

25. Q. Lin and Q. Zhu, Local asymptotic expansion and extrapolation for finite elements, J. Comput. Math. 4, 263-265, 1986.

26. T. Lu, To appear.

27. R. Rannacher, Richardson extrapolation with finite elements, To appear in Notes on Numerical Fluid Mechanics, Vieweg.

28. R.Rannacher, Richardson extrapolation for a mixed finite element approximation of a plate bending problem, GAMM conference, 1986.

29. R. Rannacher and R.Scott, Some optimal error estimates for piecewise linear finite element approximations, Math. Comp. 38, 437-445, 1982.

30. S. Shen, To appear.

31. Q. Zhu and Q. Lin, The theory of superconvergence for Finite Elements, Lectures in Xiangtan University, 1986.

THE FOURIER PSEUDOSPECTRAL METHOD WITH A RESTRAIN OPERATOR FOR THE M.K.D.V. EQUATION

Ma He-ping and Guo Ben-yu
Shanghai University of Science and Technology

Abstract In this paper we develop a new Fourier pseudospectral method with a restrain operator for the modified Korteweg-de Vries (MKDV) equation. We prove the generalized stability of the scheme, from which the convergence follows.

Recent publications on spectral methods for nonlinear partial differential equations provide new potent techniques (See [1-8]). In many of the relevant papers, pseudospectral methods are used, because they are more efficient than spectral methods (See[9-12]). But sometimes pseudospectral methods have nonlinear instability, so some filtering or smoothing techniques are used (See[13,14]). The authors also proposed a restrain operator for KDV and Burgers equation (See[15, 16]).

In this paper a restrain operator R is used to develop a new Fourier pseudospectral method for the modified Korteweg-de Vries (MKDV) equation. We prove the the generalized stability of the method, from which the convergence follows with some assumption.

1. The Scheme

Consider the MKDV equation with periodic boundary condition:

$$\begin{cases} \partial_t u + u^2 u_x + \delta u_{xxx} = 0, & -\infty<x<\infty , \ 0<t\leq T, \\ u(x+1,t) = u(x,t), & -\infty<x<\infty , \ 0\leq t\leq T, \\ u(x,0) = u_0(x), & -\infty<x<\infty , \end{cases} \tag{1.1}$$

where δ is a constant.

Let $x \in I = (0,1)$. The space $L^2(I)$ is equipped with the inner product $(\cdot,\cdot)$ and norm $\|\cdot\|$. For any real $\beta>0$, let

$$H_p^{\beta}(I)= \{u\in L^2(I) \mid \sum_{k=-\infty}^{\infty} (1+|k|)^{2\beta} \, |\hat{u}_k|^2 < \infty, \ \hat{u}_k=(u,e^{2\pi ikx})\} ,$$

then $H_p^{\beta}(I) \subset H^{\beta}(I)$. The semi-norm and the norm of $H^{\beta}(I)$ are denoted by $|\cdot|_{\beta}$ and $\|\cdot\|_{\beta}$ respectively.

For any positive integer N, set

$$V_N = \text{span}\{ e^{2\pi ikx} \mid |k|\leq N \}$$

and let $\dot{V}_N$ be the subspace of V_N of real-valued functions. Let $h=1/(2N+1)$ be the mesh size in the variable x and $x_j = jh$ $(j=0,1,\dots,2N)$. The discrete inner product and norm are defined by

$$(u,v)_N = h\sum_{j=0}^{2N} u(x_j)\overline{v(x_j)}, \quad \|u\|_N = (u,u)_N^{\frac{1}{2}} .$$

Let P_N: $L^2(I) \to V_N$ and P_c: $C(\bar{I}) \to V_N$ be defined in such a way that

$$(P_N u, \varphi) = (u, \varphi), \qquad \forall \varphi \in V_N, \tag{1.2}$$

$$P_c u(x_j) = u(x_j), \qquad 0 \leq j \leq 2N. \tag{1.3}$$

For any $u, v \in C(\bar{I})$, we have

$$(u,v)_N = (P_c u, P_c v)_N = (P_c u, P_c v). \tag{1.4}$$

Let $\gamma \geq 1$ and $u \in V_N$, we define $R = R(\gamma)$ by

$$Ru(x) = \sum_{|k|\leq N} \left(1-\left(\frac{|k|}{N}\right)^{\gamma}\right) a_k e^{2\pi ikx}, \quad a_k = (u, e^{2\pi ikx}). \tag{1.5}$$

In order to approximate the nonlinear term $u^2 u_x$ reasonably, we define the operator $J_c:(V_N)^3 \to V_N$ as follows

$$J_c(u,v,w) = \frac{1}{6}\left\{ P_c(u_x vw) + P_c((P_c(uv))_x w) + (P_c(uvw))_x \right\} .$$

If u, v and $w \in \dot{V}_N$, then

$$(J_c(u,v,v),w) + (J_c(w,v,v),u) = 0. \tag{1.6}$$

Let τ be the mesh size in the variable t. Denote $u^k(x) = u(x, k\tau)$ by u^k. Define

$$u_t^k = \frac{1}{\tau}(u^{k+1} - u^k), \qquad |||u|||_\beta = \max_{k\tau\leq T} \|u^k\|_\beta .$$

The fully discrete pseudospectral method for problem (1.1) is to find u_c^k in $\dot{V}_N$ such that

$$\begin{cases} u_{ct}^k + RJ_c(Ru_c^k + \theta_1\tau Ru_{ct}^k,\ Ru_c^k + \alpha\tau Ru_{ct}^{k-1},\ Ru_c^k + \alpha\tau Ru_{ct}^{k-1}) \\ \qquad + \delta R(u_c^k + \theta_2\tau u_{ct}^k)_{xxx} = 0, \qquad k \geq 1, \\ u_c^1 = u_{1c}, \\ u_c^0 = P_c u_0 , \end{cases} \tag{1.7}$$

where u_{1c} can be obtained by acheme (1.7) with $\alpha=0$. If $\theta_1=\theta_2=1/2$, then $\|u_c^k\| = \|u_c^0\|$, $\forall k \geq 0$.

2. The Main Theoretical Result

Consider the generalized stability of scheme (1.7) (The definition can be found in [17]). If u_c^k and the right term in (1.7) have errors $\tilde{u}^k$ and $\tilde{f}^k \in \dot{V}_N$ respectively, then

$$\tilde{u}_t^k + R\tilde{J}_c^k + \delta R(\tilde{u}^k + \theta_2 \tau \tilde{u}_t^k)_{xxx} = \tilde{f}^k, \tag{2.1}$$

where

$$\tilde{J}_c^k = J_c(R\tilde{u}^k + \theta_1 \tau R\tilde{u}_t^k,\ Rv_c^k + R\tilde{v}^k,\ Rv_c^k + R\tilde{v}^k) + J_c(Ru_c^k + \theta_1 \tau Ru_{ct}^k,$$
$$R\tilde{v}^k,\ Rv_c^k + R\tilde{v}^k) + J_c(Ru_c^k + \theta_1 \tau Ru_{ct}^k, Rv_c^k, R\tilde{v}^k),$$

in which $v_c^k = u_c^k + \alpha\tau u_{ct}^{k-1}$ and $\tilde{v}^k = \tilde{u}^k + \alpha\tau \tilde{u}_t^{k-1}$. Let

$$Q^n = \|\tilde{u}^n\|^2 + \tau\sum_{k=0}^{n-1} c_0 \tau \|\tilde{u}_t^k\|^2,$$

$$\rho^n = c_1(\|\tilde{u}^0\|^2 + \tau\sum_{k=0}^{n-1} \|\tilde{f}^k\|^2),$$

where c_0, c_1 are some positive constants. Let $\varepsilon>0$ be suitably small, and m a suitable constant, we have

Theorem 1. If $\tau N^2 \le \lambda < \infty$ and

(i) $\tau N^6 c\delta^2 (m-2\theta_2)^2 / 4\varepsilon \le \lambda < \infty$,

(ii) $e^{c^*T} \rho^{[T/\tau]} \le \min\{ \frac{1}{\tau N^3}, \sqrt{\frac{\varepsilon}{c\tau N^4 (m-2\theta_1)^2}} \}$,

then for all $n\tau \le T$,

$$Q^n \le \rho^n e^{c^* n\tau},$$

where λ, c are positive constants and c^* is a positive constant depending on $|||u_c|||_{3/2+r}$ and ε $(r>0)$.

Remark If $\theta_1 = \theta_2 > 1/2$, we can put $m = 2\theta_2$ and so the condition (i) in Theorem 1 can be removed.

The convergence estimates can be got from stability following the technique in [15,16].

3. Numerical Results

Consider the MKDV equation

$$\partial_t u + 3u^2 u_x + \delta u_{xxx} = 0, \tag{3.1}$$

which has the solitary wave

$$u^2 = 2c\,\mathrm{sech}^2 \sqrt{\frac{c}{\delta}}(x-ct).$$

We take $c=0.2$, $\delta=10^{-4}$ and run the example using scheme (1.7). The operator R is good for stability in the case of $\theta_1=0$ (See Table 1,2). But if γ is too small, the precision becomes poor. So the value of γ must be chosen suitably. We also use R in the following ways

$$RJ_c(Ru,Ru,Ru)+\delta u_{xxx}, \tag{3.2}$$

$$RJ_c(Ru,u,u)+\delta u_{xxx}, \tag{3.3}$$

$$J_c(u,Ru,Ru)+\delta u_{xxx}, \tag{3.4}$$

The results of scheme (1.7) altered as (3.2), (3.3) or (3.4) are shown in Table 3. The extrapolation scheme (1.7) with $\alpha=\theta_1=\theta_2=1/2$ gives better results (See Table 4).

t	$\gamma=\infty$	$\gamma=6$
0.50	.1763-1	.4299-1
1.00	.1863-1	.5908-1
1.50	.8396-1	.5493-1
2.00	.1837+0	.3204-1

Table 1. The L^2-error of scheme (1.7) with $\alpha=\theta_1=0$, $\theta_2=1/2$, $\tau=0.0025$, $h=1/32$.

t	$\gamma=\infty$	$\gamma=3$
0.2	.4642-2	.1083-1
0.4	.1735-1	.1434-1
0.6	.4259-1	.1328-1
0.8	.8325-1	.8842-2
1.0	.1376+0	.1597-1

Table 2. The L^2-error of scheme (1.7) with $\alpha=\theta_1=0$, $\theta_2=1/2$, $\tau=0.002$, $h=1/64$.

t	case (3.2)	case (3.3)	case (3.4)
0.2	.1585-1	.1238-1	.6880-2
0.4	.2346-1	.1681-1	.7028-2
0.6	.2742-1	.1692-1	.6976-2
0.8	.2692-1	.1255-1	.2088-1
1.0	.2189-1	.1076-1	.4765-1

Table 3. The L^2-error of scheme (1.7) altered as (3.2), (3.3) or (3.4) with $\gamma=3$, $\alpha=\theta_1=0$, $\theta_2=1/2$, $\tau=0.002$, $h=1/64$.

t	$\gamma=\infty,\ \alpha=0$	$\gamma=\infty,\ \alpha=1/2$	$\gamma=25,\ \alpha=1/2$
0.15	.8735-3	.3011-3	.3617-3
0.30	.1276-2	.3373-3	.4145-3
0.45	.1582-2	.3541-3	.4209-3
0.60	.1880-2	.4551-3	.5521-3
0.75	.2125-2	.5070-3	.6396-3

Table 4. The L^2-error of scheme (1.7) with $\theta_1=\theta_2=1/2$, $\tau=0.0025$, $h=1/32$.

4. Some Lemmas

In order to prove the theorem in section 2, we need the following lemmas, the constants c in which are independent of N or the function u and may be different in different cases.

Lemma 1. [13] If $u\in H_p^{\beta}(I)$, then

$$\|P_N u\|_\beta \le c\|u\|_\beta ,$$

and if $\beta>1/2$, then

$$\|P_c u\|_\beta \le c\|u\|_\beta .$$

Lemma 2. [13] If $0\le\mu\le\beta$ and $u\in V_N$, then

$$\|u\|_\beta \le cN^{\beta-\mu}\|u\|_\mu , \qquad \|Ru\|_\beta \le \|u\|_\beta .$$

Lemma 3. If $u\in H^1(I)$, then

$$\|u\|_{L^\infty} \le c\|u\|^{\frac{1}{2}}\|u\|_1^{\frac{1}{2}} .$$

Lemma 4.[5] If $u,v\in V_N$, then

$$\|uv\|^2 \le (2N+1)\|u\|^2\|v\|^2 .$$

Now let

$$u=\sum_{|\ell|\le N} a_\ell e^{2\pi i l x} , \qquad v=\sum_{|\ell|\le N} b_\ell e^{2\pi i l x}$$

and extend the cofficients periodically, i.e.,

$$a_{\ell+2N+1}=a_\ell , \qquad b_{\ell+2N+1}=b_\ell .$$

We define the circle convolution by

$$u*v(x)=\sum_{|\ell|\le N}\sum_{|k|\le N} a_k b_{\ell-k} e^{2\pi i l x} .$$

Lemma 5.[15] If $u,v\in V_N$, $w\in \dot V_N$, then

$$P_c(uv)=u*v, \qquad (u*w,v)=(u,w*v). \tag{4.1}$$

Lemma 6.[15] If $u,v \in V_N$, $w \in H_p^{3/2+r}(I)$ $(r > 0)$, then

$$|(u_x*Rv,w)+(u*Rv_x,w)| \leq c_r\gamma \|w\|_{\frac{3}{2}+r} \|u\| \|v\|,$$

$$|(u_x*Ru,w)-(u*Ru_x,w)| \leq c_r\gamma \|w\|_{\frac{3}{2}+r} \|u\|^2,$$

where $R=R(\gamma)$ $(\gamma \geq 1)$ is defined by (1.5).

Lemma 7. If u,v and $w \in \dot{V}_N$, then

$$|(Ru*Ru_x,w)| \leq c \|w\|_{\frac{3}{2}+r} \|u\|^2,$$

$$|(w*Ru_x-(w*Ru)_x,v)| \leq c\|w\|_{\frac{3}{2}+r} \|u\| \|v\|,$$

$$\|w*Ru_x-(w*Ru)_x\| \leq c\|w\|_{\frac{3}{2}+r} \|u\|,$$

$$|(w*u_x-(w*u)_x,Rv)| \leq c\|w\|_{\frac{3}{2}+r} \|u\| \|v\|, \tag{4.2}$$

$$\|R(w*u_x-(w*u)_x)\| \leq c\|w\|_{\frac{3}{2}+r} \|u\|. \tag{4.3}$$

Proof. From Lemma 6 we have

$$|(w*u_x-(w*u)_x,Rv)| = |(u_x*Rv+u*Rv_x,w)| \leq c\|w\|_{\frac{3}{2}+r} \|u\| \|v\|.$$

We get (4.3) by letting $v=R(w*u_x-(w*u)_x)$ in (4.2). The others can be derived similarly.

Lemma 8. [5] Let ρ, M_1, M_2 and M_3 be positive constants, E^k be nonnegative function. Suppose that

(i) for all $\tau \leq n\tau \leq T$,

$$E^n \leq \rho + M_1\tau \sum_{k=0}^{n-1} [E^k+M_2(E^k)^2 + M_3(E^k)^3],$$

(ii) $E^0 \leq \rho$ and $\rho e^{3M_1T} \leq \min(M_2^{-1}, M_3^{-\frac{1}{2}})$,

then for all $n\tau \leq T$, we have

$$E^n \leq \rho e^{3M_1 n\tau}.$$

5. The Proof of Theorem

By (4.1), rewrite

$$J_c(u,v,w) = \frac{1}{6}\{u_x*v*w + (u*v)_x*w+(u*v*w)_x\}.$$

The proof of Theorem 1.

Taking the inner product of (2.1) with $2\tilde{u}^k+m\tau\tilde{u}_t^k(m>0)$ and denoting $\xi_c^k=Ru_c^k, \tilde{\xi}^k=R\tilde{u}^k$, $\eta_c^k=\xi_c^k+\alpha\tau\xi_{ct}^{k-1}$ and $\tilde{\eta}^k=\tilde{\xi}^k+\alpha\tau\tilde{\xi}_t^{k-1}$ for simplicity, we have

$$\|\tilde{u}^k\|^2_t+\tau(m-1-\varepsilon)\|\tilde{u}^k_t\|^2+\sum_{\ell=1}^{6}\tilde{F}^k_\ell \le \|\tilde{u}^k\|^2+(1+\frac{\tau m^2}{4\varepsilon})\|\tilde{f}^k\|^2, \tag{5.1}$$

where

$$\tilde{F}^k_1 = 2(J_c(\xi^k_c+\theta_1\tau\xi^k_{ct},\tilde{\eta}^k,\eta^k_c+\tilde{\eta}^k),\tilde{\xi}^k),$$

$$\tilde{F}^k_2 = 2(J_c(\xi^k_c+\theta_1\tau\xi^k_{ct},\eta^k_c,\tilde{\eta}^k),\tilde{\xi}^k),$$

$$\tilde{F}^k_3 = m\tau(J_c(\xi^k_c+\theta_1\tau\xi^k_{ct},\tilde{\eta}^k,\eta^k_c+\tilde{\eta}^k),\tilde{\xi}^k_t),$$

$$\tilde{F}^k_4 = m\tau(J_c(\xi^k_c+\theta_1\tau\xi^k_{ct},\eta^k_c,\tilde{\eta}^k),\tilde{\xi}^k_t),$$

$$\tilde{F}^k_5 = \tau(m-2\theta_1)(J_c(\tilde{\xi}^k,\eta^k_c+\tilde{\eta}^k,\eta^k_c+\tilde{\eta}^k),\tilde{\xi}^k_t),$$

$$\tilde{F}^k_6 = \delta\tau(m-2\theta_2)(\tilde{\xi}^k_{xxx},\tilde{\xi}^k_t).$$

Now we are going to estimate $|\tilde{F}^k_\ell|$. By Lemma 7, we have

$$|((w*(Ru)_x*v+(w*Ru*v)_x,Ru)|$$

$$=|(w*(Ru*v)_x-(w*Ru*v)_x,Ru)|\le c\|w\|_{\frac{3}{2}+r}\|u\|\,\|uv\|.$$

Thus we can get

$$|\tilde{F}^k_1|\le\varepsilon\tau\|\tilde{u}^{k-1}_t\|^2+c(|||u_c|||_{\frac{3}{2}+r})\{\|\tilde{u}^k\|^2+(\|\tilde{u}^k\|^4+\|\tilde{u}^{k-1}\|^4)(1+\tau N^3)\},$$

$$|\tilde{F}^k_2|\le\varepsilon\tau\|\tilde{u}^{k-1}_t\|^2+c(|||u_c|||_{\frac{3}{2}+r})(\tau N^2+1)\|\tilde{u}^k\|^2,$$

$$|\tilde{F}^k_3|\le\varepsilon\tau\|\tilde{u}^k_t\|^2+c(|||u_c|||_{\frac{3}{2}+r})\{\tau N^2(\|\tilde{u}^k\|^2+\|\tilde{u}^{k-1}\|^2)+\tau N^3(\|\tilde{u}^k\|^4+\|\tilde{u}^{k-1}\|^4)\},$$

$$|\tilde{F}^k_4|\le\varepsilon\tau\|\tilde{u}^k_t\|^2+c(|||u_c|||_{\frac{3}{2}+r})\tau N^2(\|\tilde{u}^k\|^2+\|\tilde{u}^{k-1}\|^2),$$

$$|\tilde{F}^k_5|\le\varepsilon\tau\|\tilde{u}^k_t\|^2+c(|||u_c|||_{\frac{3}{2}+r})(m-2\theta_1)^2\{\tau N^2\|\tilde{u}^k\|^2$$
$$+\tau N^3(\|\tilde{u}^k\|^4+\|\tilde{u}^{k-1}\|^4)\}+c\varepsilon^{-1}\tau N^4(m-2\theta_1)^2(\|\tilde{u}^k\|^6+\|\tilde{u}^{k-1}\|^6),$$

$$|\tilde{F}^k_6|\le\varepsilon\tau\|\tilde{u}^k_t\|^2+c\delta^2\tau N^6(m-2\theta_2)^2\|\tilde{u}^k\|^2/4\varepsilon.$$

By substituting the above estimates into (5.1), we obtain

$$\|\tilde{u}^k\|^2_t+\tau(m-1-7\varepsilon)\|\tilde{u}^k_t\|^2+2\varepsilon\tau^2\|u^{k-1}_t\|^2_t\le c(|||u_c|||_{\frac{3}{2}+r})\{\|\tilde{u}^k\|^2+\|\tilde{u}^{k-1}\|^2$$
$$+\tau N^3(\|\tilde{u}^k\|^4+\|\tilde{u}^{k-1}\|^4)\}+c\varepsilon^{-1}\tau N^4(m-2\theta_1)^2(\|\tilde{u}^k\|^6+\|\tilde{u}^{k-1}\|^6)$$
$$+(c\delta^2\tau N^6(m-2\theta_2)^2/4\varepsilon)\|\tilde{u}^k\|^2+(1+\tau m^2)/4\varepsilon)\|\tilde{f}^k\|^2,$$

for k=0 the similar estimate holds but the terms related **to k-1 level do not appear.**
Therefore, let $m=1+7\varepsilon+c_0$, we get

$$Q^n \leqq \rho^n + \tau \sum_{k=0}^{n-1} \{ c(|||u_c|||_{\frac{3}{2}+r})(Q^k + \tau N^3 (Q^k)^2)$$

$$+ c\varepsilon^{-1} \tau N^4 (m-2\theta_1)^2 (Q^k)^3 + (c\delta^2 \tau N^6 (m-2\theta_2)^2 / 4\varepsilon) Q^k \} .$$

Finally we use Lemma 8 to complete the proof.

References

[1] K.Abe, O.Inoue, Fourier expansion solution of the KdV equation, J.Comp.Phys., 34(1980), 202-210.

[2] Y.Maday, A.Quarteroni, Legendre and Chebyshev spectral approximations of Burgers' equation, Numer.Math., 37(1981), 321-332.

[3] O.H.Hald, Convergence of Fourier methods for Navier-Stokes equations, J.Comp. Phys., 40(1981), 305-317.

[4] Y.Maday, A.Quarteroni, Spectral and pseudospectral approximations of the Navier-Stokes equations, SIAM J.Numer.Anal., 19(1982), 761-780.

[5] Guo Ben-yu, Error estimations of the spectral method for solving K.D.V.-Burgers equation, Acta Mathematica Sinica, 28(1985), 1-15.

[6] Guo Ben-yu, The convergence of spectral scheme for solving two-dimensional vorticity equation, J.Comp.Math., 1(1983), 353-362.

[7] Guo Ben-yu, Error estimation of spectral method for solving three-dimensional vorticity equation, **Acta Mathematicae Applicatae Sinica, 2(1985), 229-240.**

[8] Guo Ben-yu, Spectral method for solving Navier-Stokes equations, Scientia Sinica, Series A, 8(1985), 715-728.

[9] H.Schamel,K.Elsässer, The application of the spectral method to nonlinear wave propogation, J.Comp.Phys., 22(1976), 501-516.

[10] J.Canosa, J.Gazdag, The Korteweg-de Vries-Burgers equation, J.Comp.Phys., 23(1977), 393-403.

[11] B.Fornberg, G.B.Whitham, A numerical and theoretical study of certain nonlinear wave phenomena, Philos.Trans.Royal Soci.London, 289(1978), 373-404

[12] P.Cornille, A pseudospectral scheme for the numerical calculation of shocks, J.Comp.Phys., 47(1982), 146-159.

[13] H.O.Kreiss, J.Oliger, Stability of the Fourier method,SIAM J.Numer.Anal., 16 (1979), 421-433.

[14] T.A.Zang, M.Y.Hussaini, in "Proceedings, Seventh International Conference on Numerical Methods in Fluid Dynamics", Springer-Verlag, Berlin/New York, 1980.

[15] Ma He-ping, Guo Ben-yu, The Fourier pseudospectral method with a restrain operator for the Korteweg-de Vries equation, J.Comp.Phys., 65(1986), 120-137.

[16] Guo Ben-yu, Ma He-ping, Fourier pseudospectral method for Burgers equation, to appear in Northeastern Math.J..

[17] Guo Ben-yu, Difference Methods for Partial Differential Equations, Science Press, in press.

A NUMERICAL METHOD FOR A CLASS OF
NONLINEAR FOURTH ORDER EQUATIONS

Pan Zhong-xiong
Shanghai University of Science and Technology
Shanghai, China.

Abstract This paper considers the behaviour and numerical method of the solutions of a class of nonlinear fourth order equations. Under some conditions on a(x), r(x,u) and the domain, the problem at least exists three solutions. Based on the difference method the nontrivial solutions can be found numerically.

I. Introduction

Consider the following problem:

$$\begin{cases} \Delta\Delta u + \nabla\cdot a\nabla u = r(x,u) & x \in \Omega \\ u|_{\partial\Omega} = 0 , & \partial_\nu u|_{\partial\Omega} = 0 \end{cases} \quad (1)$$

where Ω is a bounded set in R^N with boundary $\partial\Omega$; $\partial_\nu u$ is the outer normal derivative on $\partial\Omega$. Particularly, if $r(x,u) = \varepsilon u^3$ then(1) is the perturbed problem of Bretherton's type[1].

In order to consider the behaviour of the solutions, let

$E = H_0^2(\Omega) = \{u|\, u\in H^2(\Omega)\ ,\ u|_{\partial\Omega} = \partial_\nu u|_{\partial\Omega} = 0\}$ be a Sobolev space and the following assumptions be satisfied:

1) Assume the $N \leq 3$, $d = \mathrm{diam}(\Omega)$, $a = a(x) \in C^1(\Omega)$.Then there are constants a_1, d_0 such that $a(x) \leq a_1$, $\frac{1}{4}a_1 d^2 \leq d_0 < 1$;

2) $r(x,t) \in C^1(\Omega \times R)$, $\frac{\partial r}{\partial t} \geq 0$, $r(x,-t) = - r(x,t)$;

3) Let $\int_0^z r(x,t)dt = p(x,z)$, for $|z| > 0$. Then there exists $\mu > 2$ such that

$$0 < \mu p(x,z) \leq zr(x,z)$$

hold uniformly for $x \in \overline{\Omega}$;

4) There exist positive constants C_1 , C_2 such that

$$|r(x,t)| \le C_1 + C_2 |t|^s$$

holds for any $(x,t) \in \bar{\Omega}xR$; The limit

$$\lim_{|t|\to\infty} |r(x,t)| \Big/ |t|^s$$

is bounded for $x \in \bar{\Omega}$, where s is a positive integer.

II. Existence of the solutions

We define the functional I(u) on E as follows

$$I(u) = \frac{1}{2}\int_\Omega (\Delta u)^2 - a(x)\,|\nabla u|^2 - 2p(x,u)d\Omega . \tag{2}$$

If its Fréchet derivative is denoted by I' then for any u,V ∈ E

$$(I'(u), V) = \int_\Omega (\Delta u \Delta V - a\nabla u \cdot \nabla V - r(x,u)V)d\Omega .$$

Evidently, the critical points of I(u) in E are the weak solutions of problem(1), so that, we'll discuss the existence of critical points.

<u>Lemma 1</u>. Let

$$|||u|||^2 = \int_\Omega (\Delta u)^2 d\Omega - \int_\Omega a|\nabla u|^2 d\Omega \qquad u \in E$$

and the assumption 1) hold . Then

$$C_3 \|u\|_{H_0^2(\Omega)} \le |||u||| \le C_4 \|u\|_{H_0^2(\Omega)} \tag{3}$$

where

$$C_3 = d_0 \Big/ \frac{d^2}{4}\left(1+ \frac{d^2}{2}\right) \quad ; \qquad C_4 = \max(a_2 , N) ;$$

$$a_2 = \max_{x\in\bar{\Omega}} |a(x)| .$$

<u>Proof</u> Using Green's formula and Poncare's inequality, the result can be obtained.

<u>Lemma 2</u> . Suppose that the assumptions 2), 3), 4) hold. Then the functional I defined by (2) is continuously differentiable and satisfies Pailas- Smale condition.

<u>Proof</u> Based on the continuity of a(x) and r(x,u) and the definition of Fréchet

derivative, we can prove $I(u) \in C^1(E,R)$ directly.

Assume that $u_m \in E$ and

$$|I(u_m)| \leq M \tag{4}$$

$$I'(u_m) \to 0 \tag{5}$$

hold. Then there is a subsequence of $\{u_m\}$ (will still be denoted by $\{u_m\}$) such that

$$\|u_j - u_k\|^2_{H_0^2(\Omega)} \to 0 \qquad j,k \to \infty .$$

First of all, we prove that $\{u_m\}$ is bounded set in E. In fact, referring to (5), we obtain

$$\int_\Omega \Delta u_m \Delta V d\Omega - \int_\Omega a \nabla u_m \cdot \nabla V d\Omega - \int_\Omega r(x,u_m) V d\Omega = o(1) \|V\|_{H_0^2(\Omega)}. \tag*{$(6)_m$}$$

If let $V = u_m$, then

$$\int_\Omega (\Delta u_m)^2 - a|\nabla u_m|^2 - r(x,u_m)u_m d\Omega = o(1)\|u_m\|_{H_0^2(\Omega)}. \tag{7}$$

From (4), we have

$$-M \leq \int_\Omega (\Delta u_m)^2 - a|\nabla u_m|^2 - 2p(x,u_m) d\Omega \leq M. \tag{8}$$

Under the assumption 1), using (7), (8) and lemma 1, we have

$$-M \leq |||u_m|||^2 - 2\int_\Omega p(x,u_m) d\Omega \leq M, \tag{9}$$

$$|||u_m|||^2 - \int_\Omega r(x,u_m)u_m d\Omega = o(1)|||u_m|||. \tag{10}$$

From the assumptions 2), 3), we obtain

$$|||u_m|||^2 - \frac{2}{\mu}\int_\Omega r(x, u_m)u_m d\Omega \leq M,$$

i. e.,

$$|||u_m|||^2 - \frac{2}{\mu}[\,|||u_m|||^2 - o(1)|||u_m|||\,] \leq M ,$$

$$(1 - \frac{2}{\mu})|||u_m|||^2 + \frac{2}{\mu}\, o(1)|||u_m||| \leq M.$$

Because of $\mu > 2$, we obtain that $|||u_m|||$ is bounded. Using lemma 1, the boundedness of $\|u_m\|_{H_0^2(\Omega)}$ is proved.

According to the embedding theorem, for $N \leq 3$, there is a subsequence of $\{u_m\}$ (will still be denoted by $\{u_m\}$) that converges in $C(\Omega)$

Subtracting $(5)_k$ from $(5)_j$, we have

$$\int_\Omega \Delta(u_j - u_k)\Delta V d\Omega - \int_\Omega a\nabla(u_j - u_k)\cdot\nabla V d\Omega - \int_\Omega (r(x,u_j)-r(x,u_k))d\Omega = o(1)\|V\|_{H_0^2(\Omega)} .$$

Putting $V = u_j - u_k$, we obtain

$$|||u_j - u_k|||^2 - \int_\Omega (r(x,u_j) - r(x,u_k))(u_j - u_k)d\Omega = o(1)|||u_j - u_k||| .$$

Because of the convergence of $\{u_m\}$ in $C(\Omega)$, we have

$$|||u_j - u_k||| \to 0 \qquad j,k \to \infty ,$$

i. e.,

$$\|u_j - u_k\|_{H_0^2(\Omega)} \to 0 \qquad j, k \to \infty ,$$

which completes the proof.

<u>Theorem 1</u> Assume that the conditions 1)-4) are satisfied. Then there exist trivial solution and at least two nontrivial solutions of problem (1).

<u>Proof</u> First of all, the assumption $r(x,-t) = -r(x,t)$ implies that $r(x,0) = 0$. Evidently, there is trivial solution of problem(1).

In order to prove that there exist nontrivial solutions, from lemma 2, it is enough to prove that there exist nontrivial critical points of $I(u)$ in E.

According to the mountain pass lemma, on the one hand, we show that there exists an open ball B_ρ with center 0 and radius ρ such that $I(u)|_{\partial B_\rho} \geq C_0 > 0$ holds.

In fact, for any $\eta > 0$ there exists $\delta > 0$ such that

$$|p(x,t)| \leq \eta|t|^2 \qquad \text{as} \quad |t| \leq \delta$$

and

$$|r(x,t)| \leq C_1 + C_2|t|^S \qquad \text{as} \quad |t| > \delta$$

hold.

$$I(u) = \frac{1}{2}\int_\Omega (\Delta u)^2 - a|\nabla u|^2 - 2p(x,u)d\Omega$$

$$= \frac{1}{2}|||u|||^2 - \int_{(x|x\in\Omega,|u|\geq\delta)} p(x,u)d\Omega - \int_{(x|x\in\Omega,|u|\not\geq\delta)} p(x,u)d\Omega$$

$$\geq \frac{1}{2}|||u|||^2 - \eta||u||^2_{L_2(\Omega)} - \frac{1}{\mu}b_2|u|^{s+1}d\Omega$$

from lemma 1 and Pancare's inequality, we have

$$I(u) \geq \{\frac{1}{2}[d_0 \Big/ \frac{d^2}{4}(1+\frac{d^2}{2})] - \eta - \frac{1}{\mu}b_2\rho^{s-1}\} ||u||^2_{H_0^2(\Omega)}$$

Let $\eta > 0$, $\rho > 0$ be small enough so that

$$I(u)|_{\partial B\rho} \geq \frac{d_0}{d^2(1+\frac{d^2}{2})}\rho^2 > 0$$

holds.

On the other hand, the assumption 4) implies that there exist positive constants b_1, b_2 such that $|p(x,t)| \geq b_1|t|^{s+1} - b_2$ holds for $|t|$ large enough. Let $u_1 \neq 0$, $u_1 \in E$ be given and $V = \frac{u_1}{|||u_1|||}$. Then

$$I(\lambda V) = \frac{1}{2}\lambda^2 - \int_\Omega p(x, V)d\Omega \leq \frac{1}{2}\lambda^2 - b_1\int_\Omega|\lambda v|^{s+1}d\Omega + b_2 \,\mathrm{mes}(\Omega)$$

$$= \frac{1}{2}\lambda^2 - b_1\lambda^{s+1}\int_\Omega |V|^{s+1}d\Omega + b_2\,\mathrm{mes}(\Omega) .$$

If V is fixed and $\lambda \to \infty$ then $I(\lambda V) \to \infty$, Hence, there exists $u_0 \in E \setminus \bar{B}_\rho$ such that

$$I(u_0) < \frac{d_0}{d^2(1+\frac{d^2}{2})}\rho^2 \leq \inf_{u\in\partial B\rho} I(u) \text{ holds.}$$

Finally, $I(0) = 0 < \inf_{u\in\partial B\rho} I(u)$ holds. Therefore, $I(u)$ satisfies P.S. condition. Applying mountain pass lemma, we obtain that there exist nontrivial critical points. The assumption $p(x,-t) = p(x,t)$ implies that if u^* is a nontrivial critical point then $-u^*$ is a critical point too, which completes the proof.

Corollary If $r(x,t) = \varepsilon t^3$ then the problem (1) is Bretherton's type. It is easy to see that the theorem holds.

III. The difference method for nontrivial solutions

Let Ω be rectangular domain, $P_{ij} = (x_i, y_j)$ $\quad i = 0,\ldots,N, \quad j = 0,\ldots, M$ be net points. Then $u^h(P_{ij})$ or u^h denotes the net function, for simplicity, it will be denoted by u.

In order to apply difference method to the boundary conditions, we put a layer of imaginary net points outside the rectangle.

Definition

$$\Delta^h u = u_{x\bar{x}} + u_{y\bar{y}}$$

$$\Delta^h \Delta^h u = u_{xx\bar{x}\bar{x}} + 2u_{x\bar{x}y\bar{y}} + u_{yy\bar{y}\bar{y}}$$

$$\nabla^h \cdot a \bar{\nabla}^h u = (a u_{\bar{x}})_x + (a u_{\bar{y}})_y$$

The method of successive iteration is the following:

$$\begin{cases} u^{m+1} = u^m - \alpha[\Delta^h \Delta^h u^{m+1} + \nabla^h \cdot a \bar{\nabla}^h u^{m+1} - r(x, u^m)] & P_{ij} \in \Omega^h \\ u^{m+1} = 0 \qquad \partial_\nu^h u^{m+1} = 0 & P_{ij} \in \partial\Omega^h \end{cases} \tag{11}$$

where $\alpha > 0$ is a parameter.

Lemma 3 Let E be net function defined on Ω^h and $E|_{\partial\Omega^h} = 0$, $\partial_\nu^h E|_{\partial\Omega^h} = 0$. Then[3]

$$\|E\|_2^2 \le \frac{d^2}{4} \|\nabla^h E\|_2^2 \le \frac{d^4}{16} \|\Delta^h E\|_2^2$$

holds, where

$$\| E\|_2^2 = \sum_{P_{ij} \in \Omega^h} (E(P_{ij}))^2 h^2 \quad .$$

Lemma 4 Let net function e satisfy

$$\begin{cases} \Delta^h \Delta^h e + \nabla^h \cdot a \bar{\nabla}^h e = 0 \\ e|_{\partial\Omega^h} = 0 \, , \quad \partial_\nu^h e|_{\partial\Omega^h} = O(h) = \partial_\nu^h U|_{\partial\Omega^h} \, . \end{cases} \tag{12}$$

Then $|e|_m = O(h^2)$.

Theorem 2 Suppose that U is a nontrivial solution of problem (1); u^m is an iterative solution of (11); e is a solution of problem (12).
If

$$\frac{1}{2}\Theta_0 + R_0 \le \left(1 - \frac{d^2 a_1}{4}\right) \frac{16}{d^4}$$

holds, then

$$\|E^{m+1}\|_2^2 \le \frac{1+\alpha R_0}{1+\alpha(R_0+\theta_0)}\|E^m\|_2^2 + O(h^2),$$

where

$$E^m = U - u^m - e;$$

Θ_0 is a positive constant;

$$R_0 = \max_{ij} \frac{\partial r}{\partial U}(x,U).$$

<u>Proof</u> Putting U in (11), we obtain

$$\begin{cases} U = U - \alpha[\Delta^h\Delta^h U + \nabla^h \cdot a\bar{\nabla}^h U - r(x,U) + O(h^2)] \\ U|_{\partial\Omega^h} = 0, \qquad \partial_\nu^h U|_{\partial\Omega^h} = \partial_\nu U|_{\partial\Omega^h} + O(h) = O(h) \end{cases} \tag{13}$$

From (11), (12), (13) and $E^n = U - u^m - e$, we know

$$\begin{cases} E^{m+1} = E^m - \alpha[\Delta^h\Delta^h E^{m+1} + \nabla^h \cdot a\bar{\nabla}^h E^{m+1} - (r(x,U) - r(x,U-E^m-e))] + O(h^2) \\ E^{m+1}|_{\partial\Omega^h} = 0 \quad , \qquad \partial_\nu^h E^{m+1}|_{\partial\Omega^h} = 0 \end{cases} \tag{14}$$

Hence, from (14) we have

$$\sum_{ij}(E^{m+1})^2 h^2 = \sum_{ij} E^m E^{m+1} h^2 - \alpha[\sum_{ij}(\Delta^h E^{m+1})^2 h^2 - \sum_{ij} a|\nabla^h E^{m+1}|^2 h^2]$$

$$+ \alpha \sum_{ij} E^{m+1}(r(x,U) - r(x,U - E^n - e))\, h^2 + \sum_{ij} E^{m+1} \cdot O(h^2) h^2$$

$$= \sum_{ij} E^{m+1} E^m h^2 - \alpha[\sum_{ij}(\Delta^h E^{m+1})^2 h^2 - \sum_{ij} a|\nabla^h E^{m+1}|^2 h^2]$$

$$+ \alpha \sum_{ij} E^{m+1} \int_0^1 \frac{\partial r}{\partial \xi}(x, Ut + (U-E^m-e)(1-t))dt \cdot (E^m+e)h^2$$

$$+ \sum_{ij} E^{m+1} O(h^2) h^2 .$$

Putting $\sum_{ij} E^{m+1} O(h^2)\, h^2$ together with $\sum_{ij} E^{m+1} \cdot e\, h^2$; using the assumption 2) and the boundedness of U on some bounded region, we have

$$\| E^{m+1} \|_2^2 \le \frac{1 + \alpha R_0}{1+\alpha(R_0+\Theta_0)} \| E^m \|_2^2 + O(h^2) \tag{15}$$

Let $q = \dfrac{1 + \alpha R_0}{1+\alpha(R_0+\Theta_0)}$. Then from (15) , we obtain

$$\| E^{m+1} \|_2^2 \le q^{m+1} \| E^0 \|_2^2 + O(h^2).$$

Corollary If $r(x,t) = \varepsilon t^3$ and ε is small enough, then theorem 2 holds.

References

[1] A. H. Hayfeh Perturbation Methods 1973.

[2] Shen Yao-Tian Science Bulletin 19. 1984. (in Chinese)

[3] A. A. Samarskii The Theory of Difference Schemes 1983. (in Russian).

CONVERGENCE CONDITIONS OF THE EXPLICIT AND WEAK IMPLICIT FINITE DIFFERENCE SCHEMES FOR PARABOLIC SYSTEMS

Shen Long-jun
Institute of Applied Physics and
Computational Mathematics
P.O.Box 8009, Beijing

1. Introduction

Parabolic systems of partial differential equations of second order arise in many disciplines, such as plasma physics, nuclear physics, and chemistry. The system of electron temperature equation strongly coupling with magnetic field equation is an important example of parabolic system of second order[2]. In the study of explicit difference methods for Schrödinger equation appear the parabolic systems of higher order[3][6]. Therefore, the study of parabolic systems not only is an interesting subject from the view point of theoretical research, but also is important for practical applications.

The difference methods for parabolic systems have been studied in [1],[5],[7]. The author of [1] considered the finite difference methods of initial boundary value problem for a class of nonlinear higher order parabolic systems:

$$u_t = (-1)^{M+1} A(x,t,u,\cdots,u_{x^{M-1}})\, u_{x^{2M}} + F(x,t,u,\cdots,u_{x^{2M-1}}) \tag{1}$$

$$u_{x^k}(0,t) = u_{x^k}(I,t) \tag{2}$$

$$u(x,0) = \varphi(x) \quad , \tag{3}$$

where (x,t) belongs to the rectangular domain $Q_T = \{0 \le x \le 1,\ 0 \le t \le T\}$, u, F are m-dimensional vector functions, and A is an $m \times m$ positive definite matrix.

For (1),(2),(3), [1] constructed the finite difference scheme:

$$\frac{V_j^{n+1} - V_j^n}{\Delta t} = (-1)^{M+1} A_j^{n+\alpha} \frac{\Delta_+^M \Delta_-^M V_j^{n+\alpha}}{h^{2M}} + F_j^{n+\alpha}, \quad (j = M, \cdots, J-M;\ n = 1, \cdots, N), \tag{1_h}$$

$$\Delta_+^k V_0^n = \Delta_-^k V_J^n = 0 \ , \quad (k = 0, 1, \cdots, M-1) \ ,$$

$$V_j^0 = \bar{\varphi}_j \ , \quad (j = 0, 1, \cdots, J) \ ,$$

where

$$A_j^{n+\alpha} = A(x_j, t^{n+\alpha}, \overset{*}{\delta}{}_h^0 V_j^{n+\alpha}, \cdots, \overset{*}{\delta}{}_h^{M-1} V_j^{n+\alpha})$$

$$F_j^{n+\alpha} = F(x_j, t^{n+\alpha}, \tilde{\delta}_h^0 V_j^{n+\alpha}, \cdots, \tilde{\delta}_h^{2M-1} V_j^{n+\alpha})$$

$$\overset{*}{\delta}{}_h^k V_j^{n+\alpha} = \sum_{i=j-M}^{j+M-k} \left(\alpha \bar{\beta}_{ki} \frac{\Delta_+^k V_i^{n+1}}{h^k} + \bar{\bar{\beta}}_{ki} \frac{\Delta_+^k V_i^n}{h^k} \right), \quad k=0,1,\cdots,M-1 .$$

$$\tilde{\delta}_h^k V_j^{n+\alpha} = \sum_{i=j-M}^{j+M-k} \tilde{\beta}_{ki} \frac{\Delta_+^k V_i^{n+\alpha}}{h^k}, \quad k=0,1,\cdots,2M-1 .$$

$$V^{n+\alpha} = \alpha V^{n+\alpha} + (1-\alpha) V^n, \quad 0 \leq \alpha \leq 1 .$$

Here β, $\bar{\beta}$, $\bar{\bar{\beta}}$, $\tilde{\beta}$ are constants with the relations

$$\sum_{i=j-M}^{j+M-k} (\alpha \bar{\beta}_{ki} + \bar{\bar{\beta}}_{ki}) = 1, \quad k=0,1,\cdots,M-1 .$$

$$\sum_{i=j-M}^{j+M-k} \tilde{\beta}_{ki} = 1, \quad k=0,1,\cdots,2M-1 .$$

$(1)_h$, $(2)_h$, $(3)_h$ is explicit as $\alpha=0$ and implicit as $0<\alpha\leq 1$. For the sake of convenience in the following, the implicit scheme is divided into two cases: $0<\alpha<1/2$ and $1/2\leq\alpha\leq 1$. We call $(1)_h$, $(2)_h$, $(3)_h$ weak implicit scheme if $0<\alpha<1/2$, and strong implicit scheme if $1/2\leq\alpha\leq 1$. [1] proved the convergence for the solutions of system $(1)_h$, $(2)_h$, $(3)_h$ in the case of strong implicit scheme. In this paper, we are going to consider the convergence in the case of the weak implicit and explicit schemes, and the convergence conditions will be presented.

It is well known, for the simplest second order parabolic equation the strong implicit scheme is stable absolutely. But the explicit and weak implicit schemes are conditionally stable. Two cases are very different.

The symboles in this paper are completely the same with [1] . And we suppose that the following conditions are satisfied:

(I) The coefficient matrix A is an m×m continuous and positive definite matrix, i.e., for any $(x,t)\in Q_T$ and $p_0, \cdots, p_{M-1} \in \mathbb{R}^m$ there is a positive constant $a>0$, such that

$$(\xi, A(x,t,p_0,\cdots,p_{M-1})\xi) \geq a|\xi|^2 \tag{4}$$

for any $\xi \in \mathbb{R}^m$.

(II) The m-dimensional vector function $F(x,t,p_0,\cdots,p_{2M-1})$ is continuous for $(x,t)\in Q_T$ and $p_0,p_1,\cdots,p_{2M-1}\in \mathbb{R}^m$ and Lipschitz continuous with respect to $p_0, p_1,\cdots,p_{2M-1}\in\mathbb{R}^m$. Then there is a constant $K_1>0$, such that

$$|F(x,t,p_0,\cdots,p_{2M-1})|\leq K_1\left(\sum_{k=0}^{2M-1}|p_k| + \bar{F}(x,t)\right) \tag{5}$$

for $(x,t)\in Q_T$ and $p_0,\cdots,p_{2M-1}\in\mathbb{R}^m$, where

$$\bar{F}\equiv F(x,t,0,\cdots,0) .$$

(III) The m-dimensional initial vector function $\varphi(x)$ belongs to $C^{(M)}([0,1])$ and satisfies the homogeneous boundary condition (3) .

2. Basic Lemmas

To consider the convergence of solutions for problem $(1)_h$, $(2)_h$, $(3)_h$ we should have some prior estimations of solutions for this problem. In this section we mainly estimate the norm:

$$\left\|\delta_+^M\left(\frac{V^{n+1}-V^n}{\Delta t}\right)\right\|_h .$$

According to the definition, obviously we have

$$\begin{aligned} I &\equiv \left\|\delta_+^M\left(\frac{V^{n+1}-V^n}{\Delta t}\right)\right\|_h^2 \\ &= \sum_{j=0}^{J-M}\left|\delta_+^M\left(\frac{V^{n+1}-V^n}{\Delta t}\right)_j\right|^2 \\ &= \sum_{j=0}^{J-M}\left|\frac{1}{h^M}\sum_{k=0}^{M}(-1)^k\binom{M}{k}\left(\frac{V^{n+1}-V^n}{\Delta t}\right)_{j+M-k}\right|^2 \\ &= \frac{1}{h^{2M}}\sum_{j=0}^{J-M}\left|\sum_{s=j}^{j+M}(-1)^{j+M-s}\binom{M}{j+M-s}\left(\frac{V^{n+1}-V^n}{\Delta t}\right)_s\right|^2 . \end{aligned}$$

From the boundary condition $(2)_h$, we also have

$$\left(\frac{V^{n+1}-V^n}{\Delta t}\right)_j = 0 , \qquad \begin{aligned} j &= 0,1,\cdots,M-1 ; \\ &J-M+1,\cdots,J . \end{aligned}$$

Hence

$$I = \frac{1}{h^{2M}}\Big\{ \sum_{j=0}^{M-1} \Big| \sum_{s=M}^{j+M} (-1)^{j+M-s} \binom{M}{j+M-s} \Big(\frac{V^{n+1}-V^{n}}{\Delta t}\Big)_s \Big|^2$$

$$+ \sum_{j=M}^{J-2M} \Big| \sum_{s=j}^{j+M} (-1)^{j+M-s} \binom{M}{j+M-s} \Big(\frac{V^{n+1}-V^{n}}{\Delta t}\Big)_s \Big|^2$$

$$+ \sum_{j=J-2M-1}^{J-M} \Big| \sum_{s=j}^{J-M} (-1)^{j+M-s} \binom{M}{j+M-s} \Big(\frac{V^{n+1}-V^{n}}{\Delta t}\Big)_s \Big|^2 \Big\} .$$

Substituting $(1)_h$ into this expression, we have the following estimation:

$$I = \frac{1}{h^{2M}}\Big\{ \sum_{j=0}^{M-1} \Big[\sum_{s=M}^{j+M} \binom{M}{j+M-s} \Big| A_s^{n+\alpha} \frac{\Delta_+^M \Delta_-^M V_s^{n+\alpha}}{h^{2M}} + F_s^{n+\alpha} \Big| \Big]^2$$

$$+ \sum_{j=M}^{J-2M} \Big[\sum_{s=j}^{j+M} \binom{M}{j+M-s} \Big| A_s^{n+\alpha} \frac{\Delta_+^M \Delta_-^M V_s^{n+\alpha}}{h^{2M}} + F_s^{n+\alpha} \Big| \Big]^2$$

$$+ \sum_{j=J-2M+1}^{J-M} \Big[\sum_{s=j}^{J-M} \binom{M}{j+M-s} \Big| A_s^{n+\alpha} \frac{\Delta_+^M \Delta_-^M V_s^{n+\alpha}}{h^{2M}} + F_s^{n+\alpha} \Big| \Big]^2 \Big\}$$

$$\leq \frac{1}{h^{2M}}\Big\{ (1+\varepsilon_1) \sum_{j=0}^{M-1} \Big[\sum_{s=M}^{j+M} \binom{M}{j+M-s} \Big| A_s^{n+\alpha} \frac{\Delta_+^M \Delta_-^M V_s^{n+\alpha}}{h^{2M}} \Big| \Big]^2$$

$$+ C(\varepsilon_1) \sum_{j=0}^{M-1} \Big[\sum_{j=M}^{j+M} \binom{M}{j+M-s} \Big| F_s^{n+\alpha} \Big| \Big]^2$$

$$+ (1+\varepsilon_1) \sum_{j=M}^{J-2M} \Big[\sum_{s=j}^{j+M} \binom{M}{j+M-s} \Big| A_s^{n+\alpha} \frac{\Delta_+^M \Delta_-^M V_s^{n+\alpha}}{h^{2M}} \Big| \Big]^2$$

$$+ C(\varepsilon_1) \sum_{j=M}^{J-2M} \Big[\sum_{s=j}^{j+M} \binom{M}{j+M-s} \Big| F_s^{n+\alpha} \Big| \Big]^2$$

$$+ (1+\varepsilon_1) \sum_{j=J-2M+1}^{J-M} \Big[\sum_{s=j}^{J-M} \binom{M}{j+M-s} \Big| A_s^{n+\alpha} \frac{\Delta_+^M \Delta_-^M V_s^{n+\alpha}}{h^{2M}} \Big| \Big]^2$$

$$+ C(\varepsilon_1) \sum_{j=J-2M+1}^{J-M} \Big[\sum_{s=j}^{J-M} \binom{M}{j+M-s} \Big| F_s^{n+\alpha} \Big| \Big]^2 \Big\} ,$$

where ε_1 , $C(\varepsilon_1)$ are positive constants, $C(\varepsilon_1)$ depends on ε_1 and ε_1 will be determined.

LEMMA 1[4]. Suppose $m_i \geqslant 0$ $(i=1,\cdots,n)$, $k>0$. Then the polynomial in p_i

$$\sum_{i=1}^{n} m_i P_i^2 - k\Big(\sum_{i=1}^{n} P_i\Big)^2 \tag{6}$$

is nonnegative definite if and only if

$$k \leq 1 \Big/ \Big(\sum_{i=1}^{n} 1/m_i\Big) .$$

PROOF: See [4].

Let

$$P_{js} = \binom{M}{j+M-s} \left| A_s^{n+\alpha} \frac{\Delta_+^M \Delta_-^M V_s^{n+\alpha}}{h^{2M}} \right| , \qquad \begin{matrix} j = 0, 1, \cdots, J-M \\ s = M, \cdots, J-M \end{matrix} . \tag{7}$$

$$\rho(A) = \sup_{\xi \in \mathbb{R}^m} \frac{|A\xi|}{|\xi|} , \tag{8}$$

$$\sigma(A) = \inf_{\xi \in \mathbb{R}^m} \frac{(A\xi, \xi)}{|\xi|^2} , \tag{9}$$

where $\rho(A)$ is the spectral radius of the matrix A.

We construct the following expressions of second degree for p_{js} :

$$\sum_{s=M}^{j+M} \frac{2 P_{js}^2}{(M+1)\binom{M}{j+M-s}^2 \max\limits_{s} \frac{\rho^2(A_s^{n+\alpha})}{\sigma(A_s^{n+\alpha})}} - \frac{(1+\varepsilon_2)(1-2\alpha)\Delta t}{h^{2M}} \Big(\sum_{s=M}^{j+M} P_{js}\Big)^2 ,$$

$$j = 0, 1, \cdots, M-1 .$$

$$\sum_{s=j}^{j+M} \frac{2 P_{js}^2}{(M+1)\binom{M}{j+M-s}^2 \max\limits_{s} \frac{\rho^2(A_s^{n+\alpha})}{\sigma(A_s^{n+\alpha})}} - \frac{(1+\varepsilon_2)(1-2\alpha)\Delta t}{h^{2M}} \Big(\sum_{s=j}^{j+M} P_{js}\Big)^2 ,$$

$$j = M, \cdots, J-2M .$$

$$\sum_{s=j}^{j-M} \frac{2 P_{js}^2}{(M+1)\binom{M}{j+M-s}^2 \max\limits_{s} \frac{\rho^2(A_s^{n+\alpha})}{\sigma(A_s^{n+\alpha})}} - \frac{(1+\varepsilon_2)(1-2\alpha)\Delta t}{h^{2M}} \Big(\sum_{s=j}^{J-M} P_{js}\Big)^2 ,$$

$$j = J-2M+1, \cdots, J-M ,$$

where ε_2 is a small positive constant to be determined.

From LEMMA1, it is easy to know that the condition, which keeps the all expressions of second degree for p_{js} nonnegative, is

$$\Delta t \leq h^{2M} \Big/ \Big\{ (1+\varepsilon_2)(\tfrac{1}{2}-\alpha)(M+1) \sum_{s=0}^{M} \binom{M}{s}^2 \max_{\substack{t^n \leq t \leq t^{n+1} \\ 0 \leq x \leq l \\ P_j \in \mathbb{R}^m}} \frac{\rho^2(A)}{\sigma(A)} \tag{10}$$

i.e., if (10) holds, then for any p_{js} we have

$$\frac{(1+\varepsilon_2)(1-2\alpha)\Delta t}{h^{2M}}\Big(\sum_{s=M}^{j+M} P_{js}\Big)^2 \leq \sum_{s=M}^{j+M} \frac{2\,P_{js}^2}{(M+1)\binom{M}{j+M-s}^2 \max\limits_s \dfrac{\rho^2(A_s^{n+\alpha})}{\sigma(A_s^{n+\alpha})}}\,, \quad j=0,\cdots,M-1. \tag{11}$$

$$\frac{(1+\varepsilon_2)(1-2\alpha)\Delta t}{h^{2M}}\Big(\sum_{s=j}^{j+M} P_{js}\Big)^2 \leq \sum_{s=j}^{j+M} \frac{2\,P_{js}^2}{(M+1)\binom{M}{j+M-s} \max\limits_s \dfrac{\rho^2(A_s^{n+\alpha})}{\sigma(A_s^{n+\alpha})}}\,, \quad j=M,\cdots,J-2M. \tag{12}$$

$$\frac{(1+\varepsilon_2)(1-2\alpha)\Delta t}{h^{2M}}\Big(\sum_{s=j}^{J-M} P_{js}\Big)^2 \leq \sum_{s=j}^{J-M} \frac{2\,P_{js}^2}{(M+1)\binom{M}{j+M-s}^2 \max\limits_s \dfrac{\rho^2(A_s^{n+\alpha})}{\sigma(A_s^{n+\alpha})}}\,, \quad j=J-2M+1,\cdots,J-M. \tag{13}$$

Substituting (8),(11),(12),(13) into the expression I , we have

$$\begin{aligned}
I \leq \frac{1}{h^{2M}}\Big\{ &(1+\varepsilon_1)\sum_{j=0}^{M-1}\Big(\sum_{s=M}^{j+M} P_{js}\Big)^2 + C(\varepsilon_1)\sum_{j=0}^{M-1}\Big[\sum_{s=M}^{j+M}\binom{M}{j+M-s}\,|F_s^{n+\alpha}|\Big]^2 \\
&+(1+\varepsilon_1)\sum_{j=M}^{J-2M}\Big(\sum_{s=j}^{j+M} P_{js}\Big)^2 + C(\varepsilon_1)\sum_{j=M}^{J-2M}\Big[\sum_{s=j}^{j+M}\binom{M}{j+M-s}\,|F_s^{n+\alpha}|\Big]^2 \\
&+(1+\varepsilon_1)\sum_{j=J-2M+1}^{J-M}\Big(\sum_{s=j}^{J-M} P_{js}\Big)^2 + C(\varepsilon_1)\sum_{j=J-2M+1}^{J-M}\Big[\sum_{s=j}^{J-M}\binom{M}{j+M-s}\,|F_s^{n+\alpha}|\Big]^2\Big\} \\
\leq \frac{1}{h^{2M}}\Big\{ &\sum_{j=0}^{M-1}\sum_{s=M}^{j+M} 2(1+\varepsilon_1)h^{2M}P_{js}^2\Big/\Big[(1+\varepsilon_2)(1-2\alpha)(M+1)\Delta t\binom{M}{j+M-s}^2\max_s\frac{\rho^2(A_s^{n+\alpha})}{\sigma(A_s^{n+\alpha})}\Big] \\
&+C(\varepsilon_1)\sum_{j=0}^{M-1}\Big[\sum_{s=M}^{j+M}\binom{M}{j+M-s}\,|F_s^{n+\alpha}|\Big]^2 \\
&+\sum_{j=M}^{J-2M}\sum_{s=j}^{j+M} 2(1+\varepsilon_1)h^{2M}P_{js}^2\Big/\Big[(1+\varepsilon_2)(1-2\alpha)(M+1)\Delta t\binom{M}{j+M-s}^2\max_s\frac{\rho^2(A_s^{n+\alpha})}{\sigma(A_s^{n+\alpha})}\Big] \\
&+C(\varepsilon_1)\sum_{j=M}^{J-2M}\Big[\sum_{s=j}^{j+M}\binom{M}{j+M-s}\,|F_s^{n+\alpha}|\Big]^2 \\
&+\sum_{j=J-2M+1}^{J-M}\sum_{s=j}^{J-M} 2(1+\varepsilon_1)h^{2M}P_{js}^2\Big/\Big[(1+\varepsilon_2)(1-2\alpha)(M+1)\Delta t\binom{M}{j+M-s}^2\max_s\frac{\rho^2(A_s^{n+\alpha})}{\sigma(A_s^{n+\alpha})}\Big] \\
&+C(\varepsilon_1)\sum_{j=J-2M+1}^{J-M}\Big[\sum_{s=j}^{J-M}\binom{M}{j+M-s}\,|F_s^{n+\alpha}|\Big]^2\Big\}.
\end{aligned}$$

We take ε_1 , ε_2 , such that $(1+\varepsilon_1)/(1+\varepsilon_2) \leq 1-\varepsilon_3$, and $0<\varepsilon_3<\varepsilon_2/2$. Hence we obtain

$$I' \equiv (\tfrac{1}{2}-\alpha)\Delta t \, \| \delta_+^+ (\frac{V^{n+1}-V^n}{\Delta t}) \|_h^2$$

$$\leq \left\{ \sum_{j=0}^{M-1} \sum_{s=M}^{j+M} + \sum_{j=M}^{J-2M} \sum_{s=j}^{j+M} + \sum_{j=J-2M+1}^{J-M} \sum_{s=j}^{J-M} \right\} (1-\varepsilon_3) P_{js}^2 \Big/ \left[(M+1) \binom{M}{j+M-s}^2 \max_s \frac{\rho^2(A_s^{n+\alpha})}{\sigma(A_s^{n+\alpha})} \right]$$

$$+ \frac{\tilde{c}\Delta t}{h^{2M}} \sum_{s=M}^{J-M} | F_s^{n+\alpha} |^2 ,$$

where $\tilde{c}$ is a constant which depends on ε_1, M. Because of the expression (7) and the positive definite condition of A, it is easy to obtain

$$\frac{P_{js}^2}{\binom{M}{j+M-s} \max_s \frac{\rho^2(A_s^{n+\alpha})}{\sigma(A_s^{n+\alpha})}} \leq \left(A_s^{n+\alpha} \frac{\Delta_+^M \Delta_-^M V_s^{n+\alpha}}{h^{2M}} , A_s^{n+\alpha} \frac{\Delta_+^M \Delta_-^M V_s^{n+\alpha}}{h^{2M}} \right) \Big/ \max_s \frac{\rho^2(A_s^{n+\alpha})}{\sigma(A_s^{n+\alpha})}$$

$$\leq \rho^2(A_s^{n+\alpha}) \left(\frac{\Delta_+^M \Delta_-^M V_s^{n+\alpha}}{h^{2M}} , \frac{\Delta_+^M \Delta_-^M V_s^{n+\alpha}}{h^{2M}} \right) \Big/ \max_s \frac{\rho^2(A_s^{n+\alpha})}{\sigma(A_s^{n+\alpha})}$$

$$\leq \frac{\rho^2(A_s^{n+\alpha})}{\max_s \frac{\rho^2(A_s^{n+\alpha})}{\sigma(A_s^{n+\alpha})}} \cdot \frac{\left(A_s^{n+\alpha} \frac{\Delta_+^M \Delta_-^M V_s^{n+\alpha}}{h^{2M}} , \frac{\Delta_+^M \Delta_-^M V_s^{n+\alpha}}{h^{2M}} \right)}{\sigma(A_s^{n+\alpha})}$$

$$\leq \left(A_s^{n+\alpha} \frac{\Delta_+^M \Delta_-^M V_s^{n+\alpha}}{h^{2M}} , \frac{\Delta_+^M \Delta_-^M V_s^{n+\alpha}}{h^{2M}} \right) .$$

Hence

$$I' \leq \frac{(1-\varepsilon_3)}{M+1} \left[\sum_{j=0}^{M-1} \sum_{s=M}^{j+M} + \sum_{j=M}^{J-2M} \sum_{s=j}^{j+M} + \sum_{j=J-2M+1}^{J-M} \sum_{s=j}^{J-M} \right] \left(A_s^{n+\alpha} \frac{\Delta_+^M \Delta_-^M V_s^{n+\alpha}}{h^{2M}} , \frac{\Delta_+^M \Delta_-^M V_s^{n+\alpha}}{h^{2M}} \right)$$

$$+ \frac{\tilde{c}\Delta t}{h^{2M}} \sum_{s=M}^{J-M} | F_s^{n+\alpha} |^2 .$$

Changing the summing order for these double series, we have

$$I' \leq (1-\varepsilon_3) \sum_{s=M}^{J-M} \left(A_s^{n+\alpha} \frac{\Delta_+^M \Delta_-^M V_s^{n+\alpha}}{h^{2M}} , \frac{\Delta_+^M \Delta_-^M V_s^{n+\alpha}}{h^{2M}} \right) + \bar{c} \sum_{s=M}^{J-M} | F_s^{n+\alpha} |^2 ,$$

where $\bar{c}$ is a constant depending on ε_1, M and $\max(\rho^2(A_s^{n+\alpha})/\sigma(A_s^{n+\alpha}))$. In above derivation we have used the condition (10). So there is

LEMMA 2. Suppose $0 \leq \alpha < 1/2$. If (10) holds, then there is the estimation:

$$(\tfrac{1}{2}-\alpha)\Delta t \left\| \delta_+^M \left(\frac{V^{n+1}-V^n}{\Delta t} \right) \right\|_h^2$$

$$\leq (1-\varepsilon_3) \sum_{s=M}^{J-M} \left(A_s^{n+\alpha} \frac{\Delta_+^M \Delta_-^M V_s^{n+\alpha}}{h^{2M}}, \frac{\Delta_+^M \Delta_-^M V_s^{n+\alpha}}{h^{2M}} \right) + \bar{c} \sum_{s=M}^{J-M} \left| F_s^{n+\alpha} \right|^2 \tag{14}$$

for the solutions of problem $(1)_h, (2)_h, (3)_h$. Here the factor $1+\varepsilon_2$ in (10) can be canceled. However, the restriction $\Delta t \leq \cdots$ has to be replaced by $\Delta t < \cdots$, i.e., (10) can be written in this form:

$$\Delta t < \frac{h^{2M}}{(\frac{1}{2}-\alpha)(M+1) \sum_{s=0}^{M} \binom{M}{s}^2 \max_s \frac{\rho^2(A_s^{n+\alpha})}{\sigma(A_s^{n+\alpha})}} \tag{10'}$$

3. Prior Estimations and Convergence

First we estimate the solution of problem $(1)_h, (2)_h, (3)_h$. Taking the scalar product of the vector $\Delta_+^M \Delta_-^M V_j^{n+\alpha}/h^{2M}$ with the corresponding vector equation of the system and then summing up the resulting relations for $j=M,\cdots,J-M$, we have

$$\sum_{j=M}^{J-M} \left(\frac{V_j^{n+1}-V_j^n}{\Delta t}, \frac{\Delta_+^M \Delta_-^M V_j^{n+\alpha}}{h^{2M}} \right) h$$

$$= \sum_{j=M}^{J-M} \left((-1)^{M+1} A_j^{n+\alpha} \frac{\Delta_+^M \Delta_-^M V_j^{n+\alpha}}{h^{2M}}, \frac{\Delta_+^M \Delta_-^M V_j^{n+\alpha}}{h^{2M}} \right) h + \sum_{j=M}^{J-M} \left(F_j^{n+\alpha}, \frac{\Delta_+^M \Delta_-^M V_j^{n+\alpha}}{h^{2M}} \right) h. \tag{15}$$

For the first term, we get

$$I_1 \equiv \sum_{j=M}^{J-M} \left(\frac{V_j^{n+1}-V_j^n}{\Delta t}, \frac{\Delta_+^M \Delta_-^M V_j^{n+\alpha}}{h^{2M}} \right) h$$

$$= (-1)^M \sum_{j=0}^{J-M} \left(\delta_+^M \left(\frac{V_j^{n+1}-V_j^n}{\Delta t} \right), \alpha \delta_+^M V_j^{n+1} + (1-\alpha) \delta_+^M V_j^n \right) h$$

$$=(-1)^M\left[\frac{\|\delta_+^M V^{n+1}\|_h^2-\|\delta_+^M V^n\|_h^2}{2\,\Delta t}-\left(\frac{1}{2}-\alpha\right)\Delta t\left\|\delta_+^M\left(\frac{V^{n+1}-V^n}{\Delta t}\right)\right\|_h^2\right]. \tag{16}$$

The second term can be written in this form

$$I_2=(-1)^{M+1}\sum_{j=M}^{J-M}\left(A_j^{n+\alpha}\frac{\Delta_+^M\Delta_-^M V_j^{n+\alpha}}{h^{2M}},\ \frac{\Delta_+^M\Delta_-^M V_j^{n+\alpha}}{h^{2M}}\right)h\ . \tag{17}$$

For the third term, we have

$$|I_3|=\left|\sum_{j=M}^{J-M}\left(F_j^{n+\alpha},\ \frac{\Delta_+^M\Delta_-^M V_j^{n+\alpha}}{h^{2M}}\right)h\right|$$

$$\leqslant \varepsilon_4\|\delta_+^M\delta_-^M V^{n+\alpha}\|_h^2+\frac{1}{4\varepsilon_4}\|F^{n+\alpha}\|_h^2\ . \tag{18}$$

Hence (15) becomes

$$\frac{1}{2\Delta t}\left(\|\delta_+^M V^{n+1}\|_h^2-\|\delta_+^M V^n\|_h^2\right)-\left(\frac{1}{2}-\alpha\right)\Delta t\left\|\delta_+^M\left(\frac{V^{n+1}-V^n}{\Delta t}\right)\right\|_h^2$$

$$\leqslant-\sum_{j=M}^{J-M}\left(A_j^{n+\alpha}\frac{\Delta_+^M\Delta_-^M V_j^{n+\alpha}}{h^{2M}},\ \frac{\Delta_+^M\Delta_-^M V_j^{n+\alpha}}{h^{2M}}\right)h$$

$$+\varepsilon_4\|\delta_+^M\delta_-^M V^{n+\alpha}\|_h^2+\frac{1}{4\varepsilon_4}\|F^{n+\alpha}\|_h^2\ . \tag{19}$$

Applying the LEMMA2, when (10) or (10′) holds, we have

$$\frac{1}{2\,\Delta t}\left(\|\delta_+^M V^{n+1}\|_h^2-\|\delta_+^M V^n\|_h^2\right)$$

$$\leqslant-\varepsilon_3\sum_{s=M}^{J-M}\left(A_s^{n+\alpha}\delta_+^M\delta_-^M V_s^{n+\alpha},\ \delta_+^M\delta_-^M V_s^{n+\alpha}\right)$$

$$+\varepsilon_4\|\delta_+^M\delta_-^M V^{n+\alpha}\|_h^2+\left(\bar{c}+\frac{1}{4\varepsilon_4}\right)\|F^{n+\alpha}\|_h^2\ . \tag{20}$$

Using assumption (II), [1] has proved

$$\|F^{n+\alpha}\|_h^2\leqslant\varepsilon_5\|\delta_+^M\delta_-^M V^{n+\alpha}\|_h^2+c_1\|\delta_+^M V^{n+\alpha}\|_h^2+C_2\ .$$

Substituting this inequality into (20) and applying (4), we get

$$\frac{1}{2\,\Delta t}\left(\|\delta_+^M V^{n+1}\|_h^2-\|\delta_+^M V^n\|_h^2\right)$$

$$\leqslant\left[-\varepsilon_3 a+\varepsilon_4+\varepsilon_5\left(\bar{c}+\frac{1}{4\varepsilon_4}\right)\right]\|\delta_+^M\delta_-^M V^{n+\alpha}\|_h^2+c_3\|\delta_+^M V^{n+\alpha}\|_h^2+C_4\ . \tag{21}$$

We take ε_4, ε_5, such that

$$-\varepsilon_3 a+\varepsilon_4+\varepsilon_5\left(\bar{c}+1/4\varepsilon_4\right)<0\ . \tag{22}$$

And we have

$$\frac{\|\delta_+^M v^{n+1}\|_h^2-\|\delta_+^M v^n\|_h^2}{\Delta t}\leq 2c_3\|\delta_+^M v^{n+\alpha}\|_h^2+2c_4$$

$$\leq c_5\|\delta_+^M v^{n+1}\|_h^2+c_5\|\delta_+^M v^n\|_h^2+c_6$$

or

$$\|\delta_+^M v^{n+1}\|_h^2\leq\frac{1+c_5\Delta t}{1-c_5\Delta t}\|\delta_+^M v^n\|_h^2+\frac{c_6\Delta t}{1-c_5\Delta t}\quad, \tag{23}$$

where $1-c_5\Delta t>0$. From (23) we have

$$\|\delta_+^M v^n\|_h^2\leq e^{3c_5 n\Delta t}(\|\delta_+^M v^0\|_h^2+c_6)\ .$$

Therefore we get the following lemma:

LEMMA 3. Suppose $0\leq\alpha<1/2$. If (I),(II),(III) and (10) or (10′) hold, then

$$\max_{n=0,1,\cdots,N}\|\delta_+^k v^n\|_h\leq K_1\ ,\quad k=0,1,\cdots,M\ , \tag{24}$$

where constant K_1 is independent of h, Δt.

COROLLARY. If the assumptions in LEMMA3 are satisfied, then there is the estimation

$$\max_{n=0,1,\cdots,N}\|\delta_+^k v^n\|_\infty\leq K_2\ ,\quad k=0,1,\cdots,M-1\ , \tag{25}$$

where constant K_2 is independent of h, Δt.

Using the same way as [1], we have

LEMMA 4. Under the assumptions of LEMMA3, there is the estimation

$$\sum_{n=0}^{m}\|\delta_+^k v^{n+\alpha}\|_k^2\Delta t\leq K_3\ ,\quad k=0,1,\cdots,2M\ , \tag{26}$$

where $0\leq m\leq n-1$, constant K_3 is independent of h, Δt.

LEMMA 5. Under the assumptions of LEMMA3, there is the estimation

$$\sum_{n=0}^{m}\left\|\frac{v^{n+1}-v^n}{\Delta t}\right\|_h^2\Delta t\leq K_4\ , \tag{27}$$

where $0\leq m\leq n-1$, constant K_4 is independent of h, Δt.

LEMMA 6. Under the assumptions of LEMMA3, there are the estimations

$$\max_{j=0,\cdots,J-k} |\Delta_+^k V_j^n| \leq K_5 h^k, \quad k=0,1,\cdots,M-1. \tag{28}$$

$$\max_{j=0,\cdots,J-M} |\Delta_+^M V_j^n| \leq K_6 h^{M-\frac{1}{2}}. \tag{29}$$

$$\max_{n=0,\cdots,N-1} |\Delta_+^k V_j^{n+1} - \Delta_+^k V_j^n| \leq K_7 h^k \Delta t^{\frac{1}{2}-\frac{k+1/2}{2M}}, \quad k=0,1,\cdots,M-1. \tag{30}$$

$$\left(\sum_{n=1}^{N-1} \|\Delta_+^k V^{n+\alpha} - \Delta_+^k V^{n-1+\alpha}\|_h^2 \Delta t\right)^{\frac{1}{2}} \leq K_8 h^k \Delta t^{1-\frac{k}{2M}}, \quad k=0,1,\cdots,2M-1, \tag{31}$$

where K_5, K_6, K_7, K_8 are independent of h, Δt, and K_5, K_6 are independent of $n(=0,1,\cdots,N)$, and K_7 is independent of $j(=0,1,\cdots,J)$.

So far all the prior estimations for the solution V_j^n $(j=0,1,\cdots,J;$ $n=0,1,\cdots,N)$ have completed. About the convergence for solution V_j^n we should consider the limiting process as $h^2 + \Delta t^2 \longrightarrow 0$. However, this discussion is almost the same as the sixth section in [1]. Their different point is just that our discussion is under the restriction (10). Therefore, we just describe the convergence theorem:

THEOREM (Convergence Theorem). Suppose that the conditions (I), (II), (III) are satisfied. When $h^2 + \Delta t^2 \longrightarrow 0$ with restriction (10) or (10′), the m-dimensional discrete vector solution V_j^n $(j=0,1,\cdots,J;$ $n=0,1,\cdots,N)$ of the weak implicit scheme or the explicit scheme for (1), (2), (3) converges to the generalized solution $u(x,t) \in W_2^{(2M,1)}(Q_T)$ of the boundary value problem (1), (2), (3).

ACKNOWLEDGEMENT. The author would like to thank Prof. Zhou Yulin for his many valuable technical discussions.

References

[1] Zhou Yulin, Finite Difference Method of First Boundary Problem for Quasilinear Parabolic Systems, SCIENTIA SINICA (Series A), Vol. XXVIII, NO.4 (1985), p368-385.

[2] Shen Longjun et al., A Magnetohydrodynamic Numerical Method for a Laser-Irradiated Target, J. On Numerical Methods and Computer Applications, Vol.7, No.1 (1986), p1-7.

[3] Tony F. Chan, Ding Lee, Longjun Shen, Stable Explicit Schemes for Equations of the Schrödinger Type, Research Report YALEU/DCS/RR-305, Yale Univ., March 1984.
Also see SIAM J.Numer. Anal., Vol.23, No.2, April 1986.

[4] Shen Longjun and Du yingyan, The Stability of the Finite Difference Solutions for a Class of Degenerate Parabolic Equations, J.Comp.Math., No.2 (1985). p175-187.

[5] Li Deyuan, The Finite Difference Schemes for the System of Parabolic Equations in One Dimension, J. Comp. Math., No. 4 (1982), p.80-89.

[6] Fu Hongyuan, Explicit Difference Schemes for Generalized Nonlinear Schrodinger System , Proceedings of the 1984 Beijing Symposium on Differential Geometry and Differential Equations, p153-158.

[7] Chen Guangnan, Alternate Finite Difference Schemes for the System of Parabolic Equations in One Dimension, J. Comp. Math., No. 7 (1985), p164-174.

FINITE ELEMENT APPROXIMATIONS FOR A VARIATIONAL INEQUALITY WITH A NONLINEAR MONOTONE OPERATOR

Shen Shu-min
Department of Mathematics
Suzhou University
Suzhou, China

1. Introduction

Let Ω be a bounded convex domain in R^2, with a reasonably smooth boundary $\partial\Omega$. For a positive integer m, we will denote by $W^{m,p}(\Omega)$ the Sobolev space defined for $p \in [1, +\infty]$ by

$$W^{m,p}(\Omega) = \left\{ u \in L^p(\Omega) \mid D^\alpha u \in L^p(\Omega), \ \forall |\alpha| \le m \right\}.$$

It is well known that these spaces are Banach spaces with the norm

$$\|u\|_{m,p} = \sum_{|\alpha| \le m} \|D^\alpha u\|_{L^p(\Omega)} .$$

(If $p = 2$, we write $\|u\|_m$ instead of $\|u\|_{m,p}$.)

In the case $m = 1$, we will also denote by $W_0^{1,p}(\Omega)$, defined for $1 \le p < +\infty$, the closure of $\mathcal{D}(\Omega)$ in $W^{1,p}(\Omega)$, by $(W_0^{1,p}(\Omega))'$ the dual space of the space $W_0^{1,p}(\Omega)$, i.e. the space $W^{-1,p'}(\Omega)$, where $1/p + 1/p' = 1$, and by $\langle \cdot , \cdot \rangle$ duality pairing between $(W_0^{1,p}(\Omega))'$ and $W_0^{1,p}(\Omega)$.

Consider the following variational inequality:

$$\left.\begin{array}{l} \text{Find } u \in K \text{ such that} \\ \int_\Omega |\nabla u|^{p-2} \nabla u \cdot \nabla(v - u)dx \ge \int_\Omega f(v - u)dx \quad \forall v \in K \end{array}\right\} \tag{1}$$

where

$$K = \left\{ v \in W_0^{1,p}(\Omega) \mid v(x) \ge \phi(x) \quad \text{a.e. on } \Omega \right\} \tag{2}$$

and $2 < p < +\infty$, $f \in (W_0^{1,p}(\Omega))'$, $\phi(x) \in W_0^{1,p}(\Omega)$, $\phi(x)|_{\partial\Omega} \le 0$, and here

$$|\nabla v| = \left(\sum_{i=1}^{2} (\partial_i v)^2\right)^{\frac{1}{2}} .$$

Let A be the operator defined from $W_0^{1,p}(\Omega)$ into $(W_0^{1,p}(\Omega))'$ by

$$Au = -\nabla \cdot (|\nabla u|^{p-2} \nabla u), \tag{3}$$

then problem (1) can be written as

$$\left.\begin{array}{l} \langle Au, v - u\rangle \ge \langle f, v - u\rangle \quad \forall v \in K , \\ u \in K, \end{array}\right\} \tag{4}$$

with K still defined by (2).

From the monotonicity of the mapping $x \to |x|^{p-2} \cdot x$ we have

$$<Au - Av, u - v> = \int_\Omega (|\nabla u|^{p-2}\nabla u - |\nabla v|^{p-2}\nabla v)(\nabla u - \nabla v)dx \geq 0 ,$$

which is an equality only for $v = u$. Thus the nonlinear operator A is monotone and even strictly monotone.

The finite element methods for the variational problem with the above nonlinear monotone operator A are considered by Glowinski, Marroco [5]. Wang [6] has studied numerical analysis of the finite element methods for some variational inequalities with nonlinear monotone operators in Hilbert space case. However, it can not be applied to the problem (1) (or (4)). In this paper the finite element approximation of the problem (1) (or (4)) is given. The convergence result and the error estimates of the approximate solution are shown.

For computational convenience, we adopt the symbol:

$$\|v\| \equiv \|\nabla v\|_{L^p(\Omega)} \quad \forall v \in W^{1,p}(\Omega) .$$

It is easy to see that $\|v\|$ is a norm over the space $W_0^{1,p}(\Omega)$, which is equivalent to the standard norm $\|\cdot\|_{1,p}$.

2. Existence and uniqueness results

We have

Lemma 1. The set K defined by (2) is a closed convex and non-empty subset of $W_0^{1,p}(\Omega)$.

For the operator A defined by (3), we also have

Lemma 2. The operator A is coercive on K, i.e. there exists $v_0 \in K$ such that

$$<Au - Av_0, u - v_0> / \|u - v_0\| \to \infty \tag{5}$$

for all $u \in K$ when $\|u\| \to +\infty$.

Proof. By Holder's inequality, we have

$$|<Au, v>| = |\int_\Omega |\nabla u|^{p-2}\nabla u \cdot \nabla v dx|$$
$$\leq C(\int_\Omega |\nabla u|^{(p-1)p'}dx)^{1/p'} \cdot (\int_\Omega |\nabla v|^p dx)^{1/p} .$$

Hence

$$|<Au, v>| \leq C\|u\|^{p-1} \cdot \|v\| \quad \forall u, v \in W_0^{1,p}(\Omega). \tag{6}$$

Let v_0 be an element in K (or in $W_0^{1,p}(\Omega)$), then we obtain, from (6),

$$<Au - Av_0, u - v_0> = <Au, u> - <Av_0, u - v_0> - <Au, v_0>$$
$$\geq \|u\|^p - c_1\|u - v_0\| - c_2\|u\|^{p-1}$$

$$\geq \|u\|^p - c_1\|u\| - c_2\|u\|^{p-1} - c_1\|v_0\|, \qquad \forall u \in W_0^{1,p}(\Omega),$$

where c_1 , c_2 are some constants. Thus

$$\langle Au - Av_0, u - v_0 \rangle / \|u\| \longrightarrow +\infty \text{ as } \|u\| \longrightarrow +\infty ,$$

and the result (5) follows.

Obviously , it follows from (6) that the operator A is continuous on finite dimensional subspaces. Therefore, in view of Lemmas 1, 2 and the result in [6] , we have

Theorem 1. The problem (1) (or (4)) has a unique solution $u \in K$.

The following properties of the operator A will be useful .

Lemma 3. Let A be the operator defined by (3) with $2 < p < +\infty$, then there exist two constants $\alpha > 0$, $M > 0$, such that

$$\langle Aw - Av, w - v \rangle \geq \alpha \|w - v\|^p \quad \forall w, v \in W_0^{1,p}(\Omega), \tag{7}$$

$$\| Aw - Av\|^* \leq M(\|w\| + \|v\|)^{p-2}\|w - v\| \quad \forall w, v \in W_0^{1,p}(\Omega), \tag{8}$$

where

$$\|Av\|^* = \sup_{w \in W_0^{1,p}(\Omega)} \frac{\langle Av, w \rangle}{\|w\|} .$$

For the proof of lemma 3, may see Glowinski et al. [5] or Ciarlet [2].

3. Finite element approximation

Let $\mathcal{T}_h$ be a standard triangulation made up of triangles $e \in \mathcal{T}_h$, in such a way that all the vertices situated on the boundary $\partial\Omega_h$ of the set $\overline{\Omega}_h = \bigcup_{e \in \mathcal{T}_h} \bar{e}$ also belong to the boundary $\partial\Omega$ of the set Ω . The inclusion $\overline{\Omega}_h \subset \overline{\Omega}$ and $\text{dist}(\partial\Omega_h, \partial\Omega) = O(h^2)$ hold, where h is the largest length of the edges of $\mathcal{T}_h$.

With such triangulation $\mathcal{T}_h$ of the set $\overline{\Omega}_h$ we define the piecewise linear finite element space by

$$X_{oh} = \left\{ v_h \mid v_h|_e \in P_1(e),\ v_h|_{\partial\Omega_h} = 0, \quad \forall e \in \mathcal{T}_h \right\} . \tag{9}$$

We also denote by V_h the space formed by the extensions of the functions of the space X_{oh} which vanish over the set $\overline{\Omega} - \overline{\Omega}_h$. (In fact we will not distinguish between the functions in X_{oh} and their corresponding extensions in the space V_h below). Then the inclusion

$$V_h \subset W_0^{1,p}(\Omega)$$

holds [5] .

Denote by N_h the set of the nodes of the space V_h, i.e. the set of all the vertices. It is quite natural to approximate K by

$$K_h = \left\{ v_h \in V_h \,\middle|\, \forall b \in N_h ,\ v_h(b) \geq \Phi(b) \right\} . \tag{10}$$

Then, we may define the finite element approximation of problem (1) (or (4)) as follows:

$$\left.\begin{array}{l} \text{Find } u_h \in K_h \text{, such that} \\ \langle Au_h, v_h - u_h \rangle \geq \langle f, v_h - u_h \rangle \quad \forall v_h \in K_h . \end{array}\right\} \tag{11}$$

Because K_h is clearly a closed convex nonempty subset of v_h (in general $K_h \not\subset K$) and the operator A is coercive on K_h, we have

<u>Theorem 2.</u> The discrete problem (11) has a unique solution $u_h \in K_h$.

<u>4. Convergence result</u>

Assume that we are given a parameter h converging to 0 and a family $\{K_h\}_h$, which K_h is defined by (10), then $\{K_h\}_h$ satisfies the following two conditions.

<u>Lemma 4.</u>

(i) If $\{v_h\}_h$ is such that $v_h \in K_h$, $\forall h$, and $\{v_h\}_h$ is bounded in $W_0^{1,p}(\Omega)$, then the weak cluster points of $\{v_h\}_h$ belong to K.

(ii) There exist $\chi \subset W_0^{1,p}(\Omega)$, $\bar{\chi} = K$ and $r_h : \chi \rightarrow K_h$ such that $r_h v$ converges strongly to v as $h \rightarrow 0$ for any $v \in \chi$.

The proof of lemma 4 is similar to that of the corresponding results in Glowinski [4].

Now we give the convergence result as follows.

<u>Theorem 3.</u> With the above assumptions on K and $\{K_h\}_h$, we have

$$\lim_{h \rightarrow 0} \| u_h - u \| = 0, \tag{12}$$

where u_h and u are the solutions of (11) and (1) (or (4)), respectively.

<u>Proof.</u> We divide the proof into three parts.

1 <u>Estimates for u_h</u>. We will show that there exist three positive constants c_1, c_2 and c_3 independent of h such that

$$\|u_h\|^p \leq c_1 \|u_h\|^{p-1} + c_2 \|u_h\| + c_3 , \quad \forall h . \tag{13}$$

Since u_h is the solution of (11), we have

$$\langle Au_h , v_h \rangle \leq \langle Au_h, v_h \rangle - \langle f, v_h - u_h \rangle ,$$

and hence

$$\|u_h\|^p = \langle Au_h, u_h \rangle \leq \|Au_h\|^* \|v_h\| + \|f\|^*(\|v_h\| + \|u_h\|).$$

Then, setting $w = v_h$, $v = 0$ in (8), we obtain

$$\|u_h\|^p \leq M\|u_h\|^{p-1}\|v_h\| + \|f\|^*(\|v_h\| + \|u_h\|) , \quad \forall v_h \in K_h \quad .$$

Let $v_0 \in \chi$ and $v_h = r_h v_0 \in K_h$. By condition (ii) on K_h we have $r_h v_0 \longrightarrow v_0$ strongly in $W_0^{1,p}(\Omega)$ and hence $\|v_h\|$ is uniformly bounded by a constant m . Thus from the above relation we obtain

$$\|u_h\|^p \leq mM\|u_h\|^{p-1} + \|f\|^*(\|u_h\| + m) = c_1\|u_h\|^{p-1} + c_2\|u_h\| + c_3 .$$

Moreover, the relation (13) implies

$$\|u_h\| \leq c(\|f\|^*), \quad \forall h , \tag{14}$$

where constant $c(\|f\|^*)$ is independent of h .

2 <u>Weak convergence of $\{u_h\}_h$.</u> Since the space $W_0^{1,p}(\Omega)$ is reflexive, the uniform boundedness of u_h , as shown in (14), implies that there exists a sequence, denoted by $\{u_{h_i}\}$, which weakly converges to some element $u^* \in W_0^{1,p}(\Omega)$. By condition (i) on $\{K_h\}_h$, we have $u^* \in K$.

We will prove that u^* is a solution of (4). Since

$$\langle Au_{h_i} , v_{h_i} - u_{h_i} \rangle \geq \langle f, v_{h_i} - u_{h_i} \rangle, \quad \forall v_{h_i} \in K_{h_i} ,$$

if set $v_0 \in \chi$ and $v_{h_i} = r_{h_i} v_0$, we then have

$$\langle Au_{h_i} , v - u_{h_i} \rangle \geq \langle Au_{h_i} , v - r_{h_i} v \rangle + \langle f, r_{h_i} v - u_{h_i} \rangle .$$

By the monotonicity of the operator A , we obtain

$$\langle Av , v - u_{h_i} \rangle \geq \langle Au_{h_i} , v - r_{h_i} v \rangle + \langle f, r_{h_i} v - u_{h_i} \rangle .$$

It follows clearly from (8) and (14) that the sequence $\{Au_h\}_h$ is bounded. Moreover, since $r_{h_i} v$ converges strongly to v and u_{h_i} converges to u^* weakly as $h_i \rightarrow 0$, and by taking the limit in above relation, we deduce that

$$\langle Av , v - u^* \rangle \geq \langle f, v - u^* \rangle \quad \forall v \in \chi , \quad u^* \in K .$$

By (8) and the density of χ in K , we also have

$$\langle Av , v - u^* \rangle \geq \langle f, v - u^* \rangle , \quad \forall v \in K , \quad u^* \in K .$$

Using Minty's lemma [6], thus we obtain

$$\langle Au^* , v - u^* \rangle \geq \langle f, v - u^* \rangle , \quad \forall v \in K, \quad u^* \in K,$$

i.e. u^* is a solution of (4) .

By theorem 1, the solution of (4) is unique, hence $u^* = u$ is the unique solution, and this shows the whole $\{u_h\}_h$ converges to u weakly.

3 Strong convergence of $\{u_h\}_h$. In view of (7), we have

$$\alpha \|u - u_h\|^p \leq \langle Au - Au_h , u - u_h \rangle .$$

Thus it suffices to show that

$$\langle Au - Au_h , u - u_h \rangle \longrightarrow 0 \quad \text{as} \quad h \to 0 . \tag{15}$$

We first have

$$\langle Au - Au_h, u - u_h \rangle = \langle Au_h, u_h - u \rangle + \langle Au , u - u_h \rangle . \tag{16}$$

For any $v \in \chi$ and $r_h v \in K_h$, it follows clearly from (11) that

$$\langle Au_h, u_h - r_h v \rangle \leq \langle f, u_h - r_h v \rangle ,$$

i.e.

$$\langle Au_h , u_h - v \rangle \leq \langle Au_h, r_h v - v \rangle - \langle f, r_h v - u_h \rangle .$$

Since $r_h v$ converges strongly to v, u_h converges to u weakly and Au_h is bounded, we deduce that

$$\operatorname{Lim}_h \sup \langle Au_h, u_h - v \rangle \leq -\langle f, v - u \rangle, \ \forall v \in \chi.$$

By the density of χ in K and the uniform boundedness of Au_h, the above relation also holds for $\forall v \in K$. Then set $v = u$, we obtain

$$\operatorname{Lim}_h \sup \langle Au_h, u_h - u \rangle \leq 0 ,$$

and hence from (16),

$$\operatorname{Lim}_h \sup \langle Au - Au_h , u - u_h \rangle \leq 0.$$

Using the monotonicity of the operator A, (15) is obtained. This proves the theorem.

5. Error estimates.

In order to obtain error estimates for the finite element solution, we will assume that

$$f \in L^p(\Omega), \quad u, \phi \in W^{2,p}(\Omega), \quad 2 < p < +\infty . \tag{17}$$

First, we have the following abstract error estimate.

Lemma 5. Let u_h and u be the solutions of (11) and (4) respectively. If the assumptions (17) hold, then there exists a constant C independent of the set K_h and of the set K, such that

$$\|u - u_h\|^p \leq C\Big(\inf_{v_h \in K_h} \Big\{ \|Au - f\|_0 \|u - v_h\|_0 + (\|u\| + \|u_h\|)^{(p-2)p'} \|u - v_h\|^{p'} \Big\} + \inf_{v \in K} \|Au - f\|_0 \|u_h - v\|_0 \Big), \tag{18}$$

where $\|\cdot\|_0$ is the L^2-norm.

Proof. We have from (7)

$$\alpha\|u - u_h\|^p \leq \langle Au - Au_h , u - u_h\rangle$$
$$= \langle Au, u\rangle + \langle Au_h, u_h\rangle - \langle Au, u_h\rangle - \langle Au_h, u\rangle,$$

and using (4) and (11),

$$\langle Au, u\rangle \leq \langle Au, v\rangle + \langle f, u - v\rangle, \quad \forall v \in K ,$$
$$\langle Au_h, u_h\rangle \leq \langle Au_h, v_h\rangle + \langle f, u_h - v_h\rangle , \quad \forall v_h \in K_h.$$

Hence we obtain that, for all $v \in K$ and all $v_h \in K_h$,

$$\alpha\|u - u_h\|^p \leq \langle Au, v - u_h\rangle + \langle Au , v_h - u\rangle$$
$$+ \langle Au_h - Au, v_h - u\rangle - \langle f, v - u_h\rangle - \langle f, v_h - u\rangle$$
$$= \langle f - Au, u - v_h\rangle + \langle f - Au, u_h - v\rangle + \langle A(u - u_h), u - v_h\rangle.$$

Since $f - Au \in L^2(\Omega)$, we then have

$$\alpha\|u - u_h\|^p \leq \|Au - f\|_0\|u - v_h\|_0 + \|Au - f\|_0\|u_h - v\|_0$$
$$+ \|A(u - u_h)\|^*\|u - v_h\| , \quad \forall v \in K, \forall v_h \in K_h .$$

In view of (8) and the inequality $ab \leq a^p/p + b^{p'}/p'$, we have

$$\|A(u - u_h)\|^*\|u - v_h\| \leq M(\|u\| + \|u_h\|)^{p-2}\|u - u_h\|\cdot\|u - v_h\|$$

$$\leq \frac{\alpha}{2}\|u - u_h\|^p + \frac{1}{p'}(\frac{\alpha p}{2})^{-p'/p} M^{p'}(\|u\| + \|u_h\|)^{(p-2)p'}\|u - v_h\|^{p'} ,$$

Therefore,

$$\frac{\alpha}{2}\|u - u_h\|^p \leq \|Au - f\|_0\|u - v_h\|_0 + \|Au - f\|_0\|u_h - v\|_0$$
$$+ c_1(\|u\| + \|u_h\|)^{(p-2)p'}\|u - v_h\|^{p'} ,$$

where $c_1 = \frac{1}{p'}(\frac{\alpha p}{2})^{-p'/p} M^{p'}$, from which the inequality (18) follows.

To apply this result, we need estimates for the quantities $\|u - v_h\|_0$, $\|u - v_h\|$ and $\|u_h - v\|_0$, for some $v \in K$ and some $v_h \in K_h$, and the prior estimate for Au .

Lemma 6. We have

$$\inf_{v_h \in K_h}\|u - v_h\|_0 \leq ch^2\|u\|_2 , \tag{19}$$

$$\inf_{v_h \in K_h}\|u - v_h\| \leq ch\|u\|_{2,p} , \tag{20}$$

$$\inf_{v \in K}\|u_h - v\|_0 \leq ch^2\|\phi\|_2 . \tag{21}$$

Proof. Let $\pi_h u$ denote as usual the piecewise linear interpolant of u , then $\pi_h u \in V_h$. Since

$$\forall b \in N_h \ , \ \Pi_h u(b) = u(b) \geq \Phi(b) \ ,$$

it is also an element of the set K_h . Thus, using the standard interpolation estimates, the estimates (19) and (20) are proved.

To prove (21), we introduce the function

$$w = \max_{\Omega} \{ u_h, \Phi \} = u_h + (\Phi - u_h)^+ \ , \tag{22}$$

so that the inequality $w \geq \Phi$ holds in Ω . It follows from $u_h \in W_0^{1,p}(\Omega)$, $\Phi \in W^{2,p}(\Omega)$ and $\Phi|_{\partial\Omega} \leq 0 = u_h|_{\partial\Omega}$ that $(\Phi - u_h)^+ \in W^{1,p}(\Omega)$ [6] , and hence $w \in W_0^{1,p}(\Omega)$. Thus the function w is an element of the set K.

Let us introduce the piecewise linear interpolant $\Pi_h \Phi$ of the function Φ . Since $u_h \in K_h$,

$$\forall b \in N_h \ , \ u_h(b) \geq \Phi(b) = \Pi_h \Phi(b) \ .$$

We have, for all $x \in \bar{\Omega}_h$,

$$|u_h(x) - u(x)| \leq |\Phi(x) - \Pi_h \Phi(x)| \ ,$$

and hence

$$\|u_h - w\|_{0,\Omega_h} \leq \|\Phi - \Pi_h \Phi\|_{0,\Omega_h} \leq ch^2 \|\Phi\|_2 \ .$$

Moreover, since

$$\|u_h - w\|_{0,\Omega}^2 \leq \|\Phi - \Pi_h \Phi\|_{0,\Omega_h}^2 + \|\max(0, \Phi)\|_{0,\Omega-\Omega_h}^2 \ ,$$

we obtain

$$\|u_h - w\|_{0,\Omega} \leq ch^2 \|\Phi\|_2 \ .$$

Thus the estimate (21) is proved.

Lemma 7. If $u \in W^{2,p}(\Omega) \cap W_0^{1,p}(\Omega)$, then

$$\|Au\|_0 \leq c\|u\|_{2,p}^{p-1} \ . \tag{23}$$

Proof. Notice that

$$Au = -\nabla\cdot(|\nabla u|^{p-1}\nabla u) = -|\nabla u|^{p-2}\cdot\Delta u + \nabla(|\nabla u|^{p-2})\cdot\nabla u.$$

By the imbedding theorem for $2 < p < +\infty$, we have $u \in C_0^1(\bar{\Omega})$ and $\|\nabla u\|_{C(\bar{\Omega})} \leq c\|u\|_{2,p}$. It follows from some calculation that

$$\|Au\|_p \leq c\|\nabla u\|_{C(\bar{\Omega})}^{p-2} \cdot \|u\|_2 \leq c\|u\|_{2,p}^{p-1} \ .$$

In view of Lemmas 5, 6, 7 and (14), we then obtain

Theorem 4. Let u_h and u be the solutions of (11) and (1) (or (4)) respectively. If the assumptions (17) hold, then there exists a constant $C(u, f, \Phi)$ independent of h, such that

$$\| u - u_h \| \leq C(u, f, \Phi) h^{\frac{1}{p-1}} .$$

The above error estimate is the same as the error estimate of the finite element solution for the corresponding variational problem in [5].

References

[1] M. Chipot, Variational Inequalities and Flows in Porous Media, Springer-Varlag, New York, 1984.

[2] G. Ciarlet, The Finite Element Methods for Elliptic Problems, North-Halland, Amsterdam, 1978.

[3] R. S. Falk, Error estimates for the approximation of a class of Variational Inequalities, Math. Comp., 28, 963-971, 1974.

[4] R. Glowinski, Numerical Methods for Nonlinear Variational Problems, Springer-Varlag, New York, 1984.

[5] R. Glowinski, A. Marroco, Surl' approximation-dual:tè, dûne classe de problèmes. de Dirichlet non linèaires, Rev. Fr. Autom. Inf. Rech. Oper. Anal. Numèr., R-2, 41-76, 1975.

[6] D. Kinderlehrer, G. Stampacchaia, An Introduction to Variational Inequalities and Their Applications, Academic Press, New York, 1980.

[7] Wang Lie-heng, Error estimate for the finite element solutions of some variational inequalities with nonlinear monotone operators, J. Comput. Math., V. 1, No. 2, 99-105, 1983.

CONTOUR DYNAMICS METHODS FOR DISCONTINUOUS VORTICAL FLOWS

Wu Hua-mo and Wu Yu-hua
Computing Center, Academia Sinica
Beijing, China

ABSTRACT

In this paper the recent results of numerical analysis on the contour dynamics methods (CD) for simulation of two dimensional incompressible inviscid vortex flows are reviewed and developed. The error estimations and the convergence proof of the fully discretized CD methods are presented.

1. INTRODUCTION.

The vortex motion being a universal phenomenon in nature still remains to be an important object of active study [1-4,11-14,16-39]. The detailed historical surveys of observations and speculations on vortices are given in [6,8]. In the case of two dimensional inviscid incompressible vortex motion the solution of the Cauchy problems for the Euler equations has attracted attention of numerous mathematicians. For smooth initial vorticity density distributions Kato [5] proved the existence and uniqueness of the classical solutions. Yudovich [30] proved the well-posedness in Hilbert spaces of the Cauchy problems with piecewise smooth initial conditions. Nevertheless, up to now, it is still a tough task to understand and describe completely the motions of vortices by means of the pure mathematics because of the nonlinearities of the problems.

The progress of new generations of supercomputers casts new lights on the study of the vortex motions. One may use computers to simulate the vortical flows numerically in order to get a deeper understanding of the physical phenomena and the mechanism of the vortex motion. A variety of physical models and numerical methods are considered in [3,4,7,10-14,16-28,30-38]. Among all these models the motion of finite piecewise constant vorticity patches (FPCVPs) was classical. Kirchhoff [6] analytically got a class of steady state solutions of elliptical vortex patches with uniform rotation (the so called rotating V-states). Recently the Kirchhoff vortices have been generalized in [39].

The contour dynamics methods (referred as the CD methods) proposed by N.J. Zabusky [36] are the most convenient procedure for numerical simulation of the evolution of FPCVPs. The methods resolve the motion of the vortex patches exactly and reduce the two-dimensional computations to one-dimensional. Hence, they rouse great interest [1-4,7,10-14,16-29,31-38]. Since the late 70's, using the CD methods Zabusky et al [4,23-26,36,38] have found several new stationary solutions of the Euler equations for FPCVPs such as the rotating m-fold symmetric V-states ($m = 3,4,5,6$), the translating vortex patch pairs and the vortex patch streets (collected in Figs.1,2,3,4), and successfully simulated the dynamics of vortex patches and the interactions between them.

Due to the complexity of the vortex motion the work on the numerical analysis of the CD methods is very little. In this paper we summerize and develop the results of [27-29] obtained since 1983 on the error estimates and the convergence proof of the CD methods. In section 2 the CD methods are described briefly. In section 3 certain regularity properties of the physical models under consideration are assumed. Section 4 is devoted to the error estimations and the convergence proof of the Euler explicit method of the CD methods. Lastly, a discussion on the results is given in section 5.

2. CONTOUR DYNAMICS METHODS

Consider the Euler equations for inviscid incompressible vortical flows in two dimensions

$$
\begin{aligned}
&\frac{\partial\omega}{\partial t} + u\frac{\partial\omega}{\partial x} + v\frac{\partial\omega}{\partial y} = 0, \\
&\Delta\psi = -\omega, \qquad\qquad (2.1) \\
&\omega = -\frac{\partial u}{\partial y} + \frac{\partial v}{\partial x},
\end{aligned}
$$

$$U = (u,v)' = (\partial\psi/\partial y,\ -\partial\psi/\partial x)'.$$

Let

$$G(Z-\zeta) = -(2\pi)^{-1} \ln r,$$

$$K(Z-\zeta) = (2\pi)^{-1} r^{-2}(-y+\eta, x-\xi)'$$

where $Z=(x,y)'$, $\zeta=(\xi,\eta)'$, $r^2=(x-\xi)^2+(y-\eta)^2$,
then we have

$$\psi(Z,t)=\iint_{R^2} G(Z-\zeta)\,\omega(\zeta,t)\,d\zeta = G(Z)*\omega(Z)$$

and

$$U(Z,t)=\iint_{R^2} K(Z-\zeta)\,\omega(\zeta,t)\,d\zeta = K(Z)*\omega(Z). \tag{2.2}$$

Consider the plane motion of N_c FPCVPs with boundaries $\Gamma_j=\partial D_j$ and vorticity density jumps $[\omega]_j=\omega_j^{(out)}-\omega_j^{(in)}$, where $\omega_j^{(out)}$ and $\omega_j^{(in)}$ are the constant vorticity densities outside and inside the support D_j of the j-th patch respectively, $(j=1,\ldots,N_c)$. In this case, the integral (2.2) for U can be rewritten as follows:

$$U(Z) = (2\pi)^{-1} \sum_{j=1}^{N_c} [\omega]_j \oint_{\partial D_j} \ln r\,(\cos\theta_j,\sin\theta_j)'ds_j$$

$$= (2\pi)^{-1} \sum_{j=1}^{N_c} [\omega]_j \oint_{\partial D_j} \ln r\,(d\xi_j,d\eta_j)', \tag{2.3}$$

where θ_j is the tangential angle of the contour of j-th patch to x-axis, ξ and η are the local Cartesian coordinates. The CD methods are discrete codes. The methods consist of two main parts: spatial discretization and time stepping. Here, the methods will be described for a single patch. The extention to general case of N_c patches is straightforward

A SPATIAL DISCRETIZATION

In this step, the approximate velocities are calculated. Choose N points $Z_1,\ldots,Z_N$ on the contour Γ. The contour Γ is approximated by an N-polygon with vertices in these nodes, then the line integration (2.3) is performed along the segments of the polygon. One can get the expressions for $u(Z)$, $v(Z)$ [12]:

$$U(Z)=(u(Z),\ v(Z))'=(2\pi)^{-1}\sum_{n=1}^{N} G_n(Z)\ (\Delta x_n,\Delta y_n)' \tag{2.4}$$

where $G_n(Z)=\frac{1}{2}(1+A_n)\cdot\ln r_{n+1}^2-\frac{1}{2}A_n\cdot\ln r_n^2-1$

$$+|B_n|\,[\arctan(2|B_n|/C_n)+\pi H(-C_n)],$$

$$r_n^2=(x-x_n)^2+(y-y_n)^2,$$

$$A_n=((x_n-x)\Delta x_n+(y_n-y)\Delta y_n)/h_n^2,$$

$$B_n=((x_n-x)\Delta y_n+(y_n-y)\Delta x_n)/h_n^2,$$

$$C_n=(r_n^2+r_{n+1}^2-h_n^2)/h_n^2,$$

$$\Delta x_n=x_{n+1}-x_n,\quad \Delta y_n=y_{n+1}-y_n,$$

$$Z_{N+1}=Z_1.\quad h_n^2=\Delta x_n^2+\Delta y_n^2$$

$H(x)$ is the Heaviside step function defined as

$$H(x)=\begin{cases}1 & (x\geqslant 0),\\ 0 & (x<0).\end{cases}$$

Let $(u_m,\ v_m)'=U_m=U(Z_m)$, $G_{mn}=G_n(Z_m)$, $\Delta Z_n=Z_{n+1}-Z_n$,

then we have

$$U_m=1/(2\pi)\sum_{n=1}^{N} G_{mn}\Delta Z_n.$$

The discretization above is denoted by LG2.

It is easy to see that the calculation of u, v using (2.4)includes computations of logarithm and tangent functions which are expensive and may cause instability in computation. In order to simplify the calculation the line integral (2.3) is transformed as

$$U(Z)=1/(2\pi\ell)\sum_{j=1}^{N_c}[\omega]_j\oint_{\partial D_j} r^{-\ell}(x-\xi,\ y-\eta)'\,dr^{\ell}\ ,\ell=1,2. \tag{2.5}$$

In the case of a single patch, after replacing the contour by an N-polygon and then calculating the line integral (2.5) along the edges of this polygon approximately according to the midpoint rule, we have the following improved formulae of velocities [22-26,38]

$$U_m=(2\pi\ell)^{-1}[\omega]\sum_{n=1}^{N}\Delta U_{mn}\Delta r_{mn}^{\ell}, \tag{2.6}$$

$$\Delta U_{mn} = (Z_m - Z_{n+\frac{1}{2}}) \cdot r^{\ell}_{m,n+\frac{1}{2}},$$

$$Z_{n+\frac{1}{2}} = (Z_n + Z_{n+1})/2,$$

$$r^2_{m,n} = |Z_m - Z_n|^2,$$

$$r^2_{m,n+\frac{1}{2}} = |Z_m - Z_{n+\frac{1}{2}}|^2,$$

$$\Delta r^{\ell}_{m,n} = r^{\ell}_{m,n+1} - r^{\ell}_{m,n},$$

$$|Z|^2 = x^2 + y^2.$$

This method is denoted by R1 or R2 for $\ell=1$ or $\ell=2$ respectively.

Recently the CD methods have been improve to be of third order accuracy by using local cubic polynomial approximation of the arc segment between two adjacent nodes Z_j and Z_{j+1} [38]. The method is called LG3, and it is accurate but time consuming.

B. TIME STEPPING

The leap-frog scheme for time stepping was used in the original paper [36]. This method is replaced by the predictor-corrector [22,23, 25,26] and the Runge-Kutta methods afterwards [25,26,38]. In this paper we also use the single step explicit Euler method for numerical analysis.

It should be kept in mind that the LG2 calculation for velocities is equivalent to the formula (2.2) with polygonal supports of the vorticity density functions. Hence, in this paper we analyze the LG2 algorithm of the CD methods using the integral representation of the velocities (2.2).

3. PHYSICAL MODELS

Consider the plane vortical motion of vortex patches with piecewise constant vorticity densities. As is well known, the regularity of the boundaries of vortex patches in motion severely depends on their initial states. Take a single elliptical vortex patch for example. If the aspect ratio (the ratio of the major and minor axes) of the ellipse exceeds 3, the motion of the patch will be unstable. Singularities such as corners, cusps, tails etc. on the contour can be generated during the motion. Our goal in this paper is to study the properties of the CD methods, therefore, it is necessary to restrict the type of the problems under consideration, i.e., to make several assumptions on the regularity properties of the physical models.

In this paper, M_i, C_i and K_i always denote some positive constants independent of $t \leq T$ with fixed T.

ASSUMPTION A. The vorticity distribution $\omega(Z,t)$ is finite with a compact support in an open circle B_R with a sufficiently large radius R for $0 \leq t \leq T < +\infty$.

Let $Z = G(s^t, t)$ be the equation of the contour Γ^t of the patch at time t, where s^t is the arc parameter on Γ^t.

ASSUMPTION B. The contour Γ^t has the following properties for $0 \leq t \leq T$.

a. $G(s,t)$ is differentiable up to 3-rd order with respect to s and

$$\left|\frac{\partial^k G(s,t)}{\partial s^k}\right| \leq M_1, \quad k=0,1,2,3 \tag{3.1}$$

b. $G(s,t)$ is continuously differentiable with respect to t and

$$\left|\frac{\partial G(s,t)}{\partial t}\right| \leq M_2 \tag{3.2}$$

c. Γ^t has a finite length

d. the motion of vortex points on Γ^t is uniform in t in the following sense. Define the ratio $R^t = R^t(Z_1, Z_2; W_1, W_2) = |Z_1^t - Z_2^t| / |W_1^t - W_2^t|$ for any two pairs of vortex points Z_1^t, Z_2^t and W_1^t, W_2^t on Γ^t. Then, the following inequalities

$$M_3^{-1} \leq R^t \leq M_3 \tag{3.3}$$

hold for $0 \leq t \leq T$ with some constant M_3.

Lemma 3.1. By the assumption A, we have

$$|U| \leq M_1 R, \qquad (0 \leq t \leq T) \tag{3.4}$$

Lemma 3.2. Using Lemma 3.1, we can get

$$|U(Z_1) - U(Z_2)| \leq K_1 \delta + K_2 \delta \ln(1/\delta), \quad (0 \leq t \leq T) \tag{3.5}$$

$\delta = |Z_1 - Z_2|$.

Lemma 3.3. Suppose that assumption A is true. Let $Z(t)$ be the solution of the following Cauchy problem

$$\frac{dZ(t)}{dt} = K(Z) * \omega(Z),$$
$$Z(0) = Z_0, \tag{3.6}$$

then for any $0 \leq t_1 \leq t_2 \leq T$ the inequality

$$|Z(t_1) - Z(t_2)| \leq M_1 R(t_2 - t_1) \tag{3.7}$$

holds.

Lemma 3.4. Suppose assumption A is true, $Z_1(0), Z_2(0) \in \operatorname{supp} \overset{o}{\omega}(Z) \subset B_R$. Let $\delta(t) = |Z_1(t) - Z_2(t)| < R$, then the following estimation is true,

$$R \exp\left[-K_1 K_2^{-1} (\exp(K_2 t) - 1)\right] (\delta(0)/R)^{\exp(K_2 t)} \leq \delta(t)$$
$$\leq R \exp\left[K_1 K_2^{-1} (1 - \exp(-K_2 t))\right] (\delta(0)/R)^{\exp(-K_2 t)} . \tag{3.8}$$

Remark 3.1 Lemma 3.4 implies that the solution of (3.6) is unique.

Remark 3.2. Lemmas 3.1-3.4 are true for the general problem (3.6) when $\omega(Z)$ is not a piecewise constant function.

4. ERROR ESTIMATION OF CD METHODS

Before we turn to analyze the error of fully discretized CD methods we have to estimate the error of velocity calculation after replacing the smooth contour Γ by another contour $\bar{\Gamma}$. Suppose that two vorticity functions $\omega(Z), \bar{\omega}(Z)$ have supports D, $\bar{D}$ with boundaries $\Gamma = \partial D$, $\bar{\Gamma} = \partial \bar{D}$ respectively, and

$$\omega(Z)\big|_{Z \in D} = \omega(Z)\big|_{Z \in \bar{D}} = \omega^{(in)},$$
$$\omega(Z)\big|_{Z \bar{\in} D} = \omega(Z)\big|_{Z \bar{\in} \bar{D}} = \omega^{(out)}. \tag{4.1}$$

Let $B_\varepsilon(Z) = \{\zeta \mid |Z - \zeta| < \varepsilon\}$ be a circle, and $D_\varepsilon(\Gamma) = \bigcup_{Z \in \Gamma} B_\varepsilon(Z)$ be an ε-neighbourhood of Γ. The velocities $U(Z)$ and $\bar{U}(Z)$ induced by ω and $\bar{\omega}$ are defined as

$$U(Z) = \iint_{R^2} K(Z - \zeta)\, \omega(\zeta)\, d\zeta \tag{4.2}$$

and

$$\overline{U}(Z)=\iint_{R^2} K(Z-\zeta)\ \overline{\omega}(\zeta)\ d\zeta . \tag{4.3}$$

We have the following basic lemma.

Lemma 4.1. Let Γ, $\overline{\Gamma}$, D, $\overline{D}$, ω, $\overline{\omega}$ be what we have defined above. If Γ is sufficiently smooth, then the estimation

$$|U(Z)-\overline{U}(Z)| \leq M_4\, \varepsilon(\, C_1 + C_2 |\ln(1/(M_4\varepsilon))|\,) \tag{4.4}$$

holds with constants M_4, C_1, C_2 for any $Z \in R^2$ and $t \leq T$.

The LG2 algorithm of the CD methods includes replacing the contour Γ by an N-polygon $\overline{\Gamma}$. By the assumption B, it is not difficult to prove the existence of such an $\varepsilon > 0$ that $\overline{\Gamma} \subset D_\varepsilon(\Gamma)$, $\varepsilon \leq C_3 H^2$, where $H = \max_i |Z_{i+1} - Z_i|$, $Z_i \in \overline{\Gamma}$.

Hence, we have

Theorem 4.1. If the assumptions A and B are valid, then for the error in velocity calculations in LG2 we have the following estimation

$$|U-\overline{U}| = O(\, H^2 \ln(1/H)). \tag{4.5}$$

Remark. If the order of the approximation of the contour Γ is m and $m \geq 2$, then an estimation in the form

$$|U-\overline{U}| = O(\, H^m \ln(1/H))$$

can be derived. A version of third order approximation of the contour (the so called LG3 CD method) may be found in [38].

Now we turn to study the convergence problem of the fully discretized LG2 code of the CD methods for the equations

$$\frac{dZ_j}{dt} = \overline{U}_j = K(Z_j) * \overline{\omega}(Z),\ 1 \leq j \leq N, \tag{4.6}$$

where $\overline{\omega}(Z)$ is the vorticity function induced by the N-polygon with the nodes $Z_1, \ldots, Z_N$ on Γ.

The explicit Euler method for (4.6) is

$$\hat{Z}_j^{n+1} = \hat{Z}_j^n + \Delta t\ K(\hat{Z}_j^n) * \hat{\omega}(\hat{Z}^n),\quad (n+1)\Delta t \leq T,$$

$$\hat{Z}_j^0 = Z_j^0\ ,\qquad 1 \leq j \leq N, \tag{4.7}$$

where $\hat{\omega}(\hat{Z}^n) = \hat{\omega}^n$ is the vorticity function induced by the N-polygon with vertices $\hat{Z}_j^n, \ldots, \hat{Z}_N^n$.

Define the intermediate value $\bar{Z}_j^{n+1}$ as

$$\bar{Z}_j^{n+1} = Z_j^n + \Delta t\ K(Z_j^n) * \bar{\omega}(Z^n) , \qquad (4.8)$$

$$Z_j^n = Z_j(n\Delta t) ,$$

where $\bar{\omega}(Z^n) = \bar{\omega}^n$ is the vorticity function induced by the N-polygon with the nodes $Z_1^n, \ldots, Z_N^n$ on the exact contour Γ (see 4.6).

Consider the error $|\hat{Z}_j^1 - Z_j(\Delta t)|$ in the first time step. We have

$$\hat{\omega}\ (\hat{Z}^0) = \bar{\omega}(\bar{Z}^0) \ , \ \hat{Z}_j^0 = Z_j^0,$$

$$\hat{Z}_j^1 = \bar{Z}_j^1 = Z_j^0 + \Delta t \iint_{R^2} K(Z_j^0 - \zeta)\bar{\omega}^0(\zeta)\ d\zeta .$$

The exact value Z_j^1 is

$$Z_j^1 = Z_j^0 + \int_0^{\Delta t} \iint_{R^2} K(Z_j - \zeta)\,\omega(\zeta)\ d\zeta\, dt.$$

Using the expressions for Z_j^1 and $\bar{Z}_j^1$ we can obtain the following estimate

$$|\bar{Z}_j^1 - Z_j^1| \leq ((C_4 + C_5|\ln(1/\Delta t)|)\Delta t + (C_6 + C_7|\ln(1/H_0)|)H_0)\Delta t. \qquad (4.9)$$

Let $g_0\Delta t$ denote the RHS of the above inequality. We have

$$\max_j |\hat{Z}_j^1 - Z_j^1| = \max_j |\bar{Z}_j^1 - Z_j^1| \leq \Delta t \cdot g_0 . \qquad (4.10)$$

Similarly we can get the estimate

$$|\bar{Z}_j^{n+1} - Z_j^{n+1}| \leq \Delta t\ g_n , \qquad (4.11)$$

where

$$g_n = (C_4 + C_5\ |\ln(1/\Delta t)|)\Delta t + (C_6 + C_7\ |\ln(1/H_n)|)H_n ,$$

$$H_n = \max_j |Z_j^n - Z_{j+1}^n| .$$

By Lemma 3.4 we have

$$H_n \leq C_8\ H_0^{\exp(-K_2 T)} \equiv H_T .$$

Let $g_T = (C_4 + C_5\ |\ln(1/\Delta t)|)\Delta t + (C_6 + C_7\ |\ln(1/H_0)|)H_T$, hence

$$g_n \leq g_T .$$

If Δt and H_0 are sufficiently small, we have

$$H_n \ll 1, \quad g_n \ll 1, \quad 0 \leq n\Delta t \leq T$$

Denote the upper bound for $\max_j |\hat{Z}_j^n - Z_j^n|$ by f_n. We will define f_n inductively. Evidently, for $n=1$, we can adopt $f_1 = \Delta t\, g_0$. For sufficiently small Δt and H_0, we have $f_1 \ll 1$. Suppose that we have defined all f_n for $n \leq k$, and that $f_n \ll 1$. From (4.7) and (4.8) we have for any k the following.

$$\begin{aligned}|\hat{Z}_j^{k+1} - \bar{Z}_j^{k+1}| &\leq |\hat{Z}_j^k - Z_j^k| + \Delta t \iint_{R^2} |K(\hat{Z}_j^k - \zeta) - K(Z_j^k - \zeta)|\, |\omega^k(\zeta)|\, d\zeta \\ &\quad + \Delta t \iint_{R^2} |K(Z_j^k - \zeta)|\, |\hat{\omega}^k(\zeta) - \bar{\omega}^k(\zeta)|\, d\zeta \\ &\leq f_k + \Delta t R_1 + \Delta t R_2 .\end{aligned}$$

For R_1 and R_2 we can get the following estimations

$$\begin{aligned}|R_1| &\leq K_1 |\hat{Z}_j^k - Z_j^k| + K_2 |\bar{Z}_j^k - Z_j^k| \ln(K_3 \exp(M_1 T)/|\bar{Z}_j^k - Z_j^k|) \\ &\leq (K_1 + K_2 \ln(K_3 \exp(M_1 T)))\, f_k + K_2\, f_k \ln 1/f_k ,\end{aligned}$$

$$|R_2| \leq K_4\, (C_9 + C_{10} |\ln(1/K_4)| + C_{10} \ln(1/f_k))\, f_k .$$

At last, we get

$$\begin{aligned}|\hat{Z}_j^{k+1} - Z_j^{k+1}| &\leq |\hat{Z}_j^{k+1} - \bar{Z}_j^{k+1}| + |\bar{Z}_j^{k+1} - Z_j^{k+1}| \\ &\leq \Delta t\, g_k + f_k + \Delta t (C_{11} + C_{12} \ln(1/f_k))\, f_k\end{aligned}$$

Therefore we can define f_k inductively by setting

$$f_{k+1} = f_k + \Delta t (C_{11} + C_{12} \ln(1/f_k)) f_k + \Delta t\, g_k . \tag{4.12}$$

By a careful consideration of the nonlinear recursive formula (4.12) for f_k we can obtain the following

Theorem 4.2. Suppose the assumptions A and B are true. $\hat{Z}_j^n$ is the solution of (4.7). Z_j^n is the exact solution of the Cauchy problem:

$$\frac{dZ_j(t)}{dt} = K(Z_j(t) * \omega(Z(t)))$$

$$Z_j(0) = Z_j^0 , \quad j = 1, \ldots, N$$

where $\omega(Z(t))$ is the exact contour of the vortex patch. Let $T_1 \leq T$, $C_{12} T_1 < 1$, $K = \max(1, (1/T_1)^{C_{12} T_1}) \exp(C_{11} + C_{12} T_1)$. Suppose Δt and H_0 are sufficiently small such that $K \cdot T_1\, g_{T_1}^{1 - C_{12} T_1} \leq 1/e$. Then the error estimation

$$\max_j \left|\hat{Z}_j^n - Z_j^n\right| \leq f_n \leq K\, T_1\, g_{T_1}^{1-C_{12}T_1} \qquad (4.13)$$

holds for any $n\Delta t \leq T_1 \leq T$.

This theorem shows that for a sufficiently small strip $t \leq T_1$ ($C_{12}\, T_1 < 1$) the CD method is convergent with a convergence rate of $O((\Delta t \ln(1/\Delta t))^{1-C_{12}T_1})$.

Remark In the same way we can prove the convergence theorem and obtain an estimation similar to (4.13) for the predictor-corrector CD methods [25,26,38].

5. DISCUSSION

We have proved the convergence properties of the CD methods for sufficiently small T and got the convergence rate in the form

$$\left|\hat{Z} - Z\right| = O((\Delta t \ln(1/\Delta t))^{1-C_{12}T}), \; (C_{12}\, T < 1). \qquad (5.1)$$

The estimations obtained so far are the same for explicit Euler method and predictor-corrector methods although these CD methods are expected to be of first order and second order accuracys. We think that the exponent $(1-C_{12}T)$ in the estimations (4.13) and (5.1) could be removed. This exponent appears mainly because of the natural but rather rough estimation (4.4) of $|U-\bar{U}|$. The term $\ln(1/\varepsilon)$ in (4.4) has led to the appearance of the factor $\ln(1/f_k)$ in the basic recursive formula (4.12). We believe that the more careful estimation of $|U-\bar{U}|$ will delete this factor and make us able to prove that the Euler method of the CD methods possesses a first order of accuracy. We also believe that in this way we can obtain the second order accuracy estimates for the predictor-corrector CD methods.

REFERENCES

1. J. Burbea, Invariants of vortex motions in the plane, Trends in theory and practice of nonlinear differential equations, V. Laksshmikanthan ed. (Marcel Deckker, New York and Basel), 91-98
2. J. Burbea, Motions of vortex patches, Lett. Math. Phys., (1982), 1-16
3. J. P. Christiansen and N. J. Zabusky, Instability, coalescence and fission of finite area vortex structures, J. Fluid Mech., 61(1973) 219
4. G. S. Deem and N. J. Zabusky, Stationary V-states: Interactions, recurrence and breaking, Phys. Rev. Lett., 40(1978), 859
5. T. Kato, On the classical solution of the two dimensional non-stationary Euler equations, Arch. Rat. Mech. Anal., 25(1967), 302
6. H. Lamb, Hydrodynamics, 6th ed. (Dover, New York, 1932)
7. A. Leonard, Vortex methods for flows simulation, J. Comp. Phys., 37(1980), 289-335
8. H. J. Lugt, Vortex flow in nature and technology, John Wiley & Sons, 1983
9. J. Marsden and A. Weinstein, Coadjoint orbits, vortices, and Clebsch variables for incompressible fluids, Physica 7D(1983), 305
10. M. V. Melander, N. J. Zabusky and A. S. Styczek, A moment model for vortex interactions, Part 1. Computational validation of a Hamiltonian elliptical representation, J. Fluid Mech., 167(1986), 95-115
11. E. A. Overman and N. J. Zabusky, Evolution and Merger of isolated vortex structures, Phys. Fluids, 25(1982), 1297-1305
12. E. A. Overman and N. J. Zabusky, Coaxial scattering of Euler equations, translating V-states via contour dynamics, J. Fluid Mech., 125(1982), 187-202
13. E. A. Overman, N. J. Zabusky and S. L. Ossakow, Ionospheric plasma clouds dynamics via regularized contour dynamics, I. Stability and nonlinear evolution of one contour models, Phys. Fluids, 26(1982), 1139-1153
14. E. A. Overman II, Steady-state solutions of the Euler equations in two dimensions II. Local analysis of limiting V-states, SIAM J. Appl. Math., 46(1986), 756-800
15. B. G. Pachpatte, On some integral inequalities similar to Bellman - Bihari inequalities, J. Math. Anal. Appl., 49(1975), 794-807
16. R. T. Pierrehumbert, A family of steady, translating vortex pairs with distributed vorticity, J. Fluid Mech., 99(1980), 129-144
17. R. T. Pierrehumbert and S. E. Widnall, The structure of organized vortices in a free shear layer, J. Fluid Mech., 102(1981), 301-313
18. P. G. Saffman and R. Szeto, Structure of a linear array of uniform vortices, Stud. Appl. Math., 65(1981), 223-248
19. P. G. Saffman and R. Szeto, Equilibrium shapes of a pair of equal uniform vortices,Phys. Fluids, 23(1980), 2339-2347
20. P. G. Saffman and J. C. Schatzman, Properties of a vortex street of finite vortices, SIAM J. Sci. Statist. Comp., 2(1981), 285-295
21. C. H. Su, Motion of fluid with constant vorticity in a singly - connected region, Phys. Fluids, 22(1979), 2032-2033
22. H. M. Wu, The computational methods for incompressible vortex flows in two dimensions, Proceedings of the second national conference on the computational methods in fluid mechanics, Xi'an, China (1983),1-21 (in chinese)
23. H. M. Wu, E. A. Overman and N. J. Zabusky, Vortex and contour dynamics in two dimensions: Steady state and dynamical solutions of the Euler equations, Proc. 2nd Asian Congress of Fluid Mech.,

Beijing, China (1983), 623-628
24. H. M. Wu, E. A. Overman and N. J. Zabusky, Steady state solutions of the Euler equation in two dimensions: rotating and translating V-states with limiting cases I. Numerical results, J. Comp. Phys., 53 (1984), 42-71
25. H. M. Wu, E. A. Overman and N. J. Zabusky, Fast contour dynamics algorithms, Bull. Amer. Phys. Soc., 27(1982), 1194
26. H. M. Wu, L. E. Wang, E. A. Overman, N. J. Zabusky, New contour dynamical algorithms with node adjustment: A sensitivity approach for optimal algorithm selection (in preparation)
27. H. M. Wu, Y. H. Wu, The propagation of initial small disturbance in discrete computations of the contour dynamics (in preparation)
28. Y. H. Wu, Contour dynamics methods, M.S. degree dissertation, Computing Center, Academia Sinica, Beijing, China (1985)
29. Y. H. Wu, H. M. Wu, The convergence of contour dynamics methods (to appear)
30. V. I. Yudovich, Nonstationary flows of the ideal incompressible liquids, Zh. Vych. Mat. Phys., 36(1963), 1032-1066.
31. N. J. Zabusky, Recent development in contour dynamics for the Euler equations, Ann. of the N. Y. Acad. Sci., 373(1981), 160-170.
32. N. J. Zabusky, Contour dynamics: A method for inviscid and nearly inviscid two dimensional flows, Proc. of the IUTAM symposium on turbulence and chaotic phenomena, T. Tatsumi ed. (North-Holland, Amsterdam, 1984), 251-257.
33. N. J. Zabusky, Visualizing mathematics: Evolution of vortical flows, Physica 18D (1986), 15-25
34. N. J. Zabusky, Computational synergetics and exploration of nonlinear science, Lett. Math. Phys., 10(1985), 143-147
35. N. J. Zabusky, Computational synergetics and mathematical inovation, J. Comp. Phys., 43(1981), 195-249
36. N. J. Zabusky, M. H. Hughes and K. V. Roberts, Contour dynamics methods for the Euler equations in two dimensions, J. Comp. Phys., 30(1979), 96-106
37. N. J. Zabusky and E. A. Overman, Regularization of contour dynamical algorithms, I. Tangential regularization, J. Comp. Phys., 52(1983), 350-373
38. Q. Zou, E. A. Overman, H. M. Wu, and N. J. Zabusky, Contour dynamics for the Euler equations: Curvature controlled initial node placement and accuracy,(preprint, 1987)
39. L. M. Polvani and G. R. Flierl, Generalized Kirchhoff vortices, Phys. Fluids, 29(1986), 2376-2379

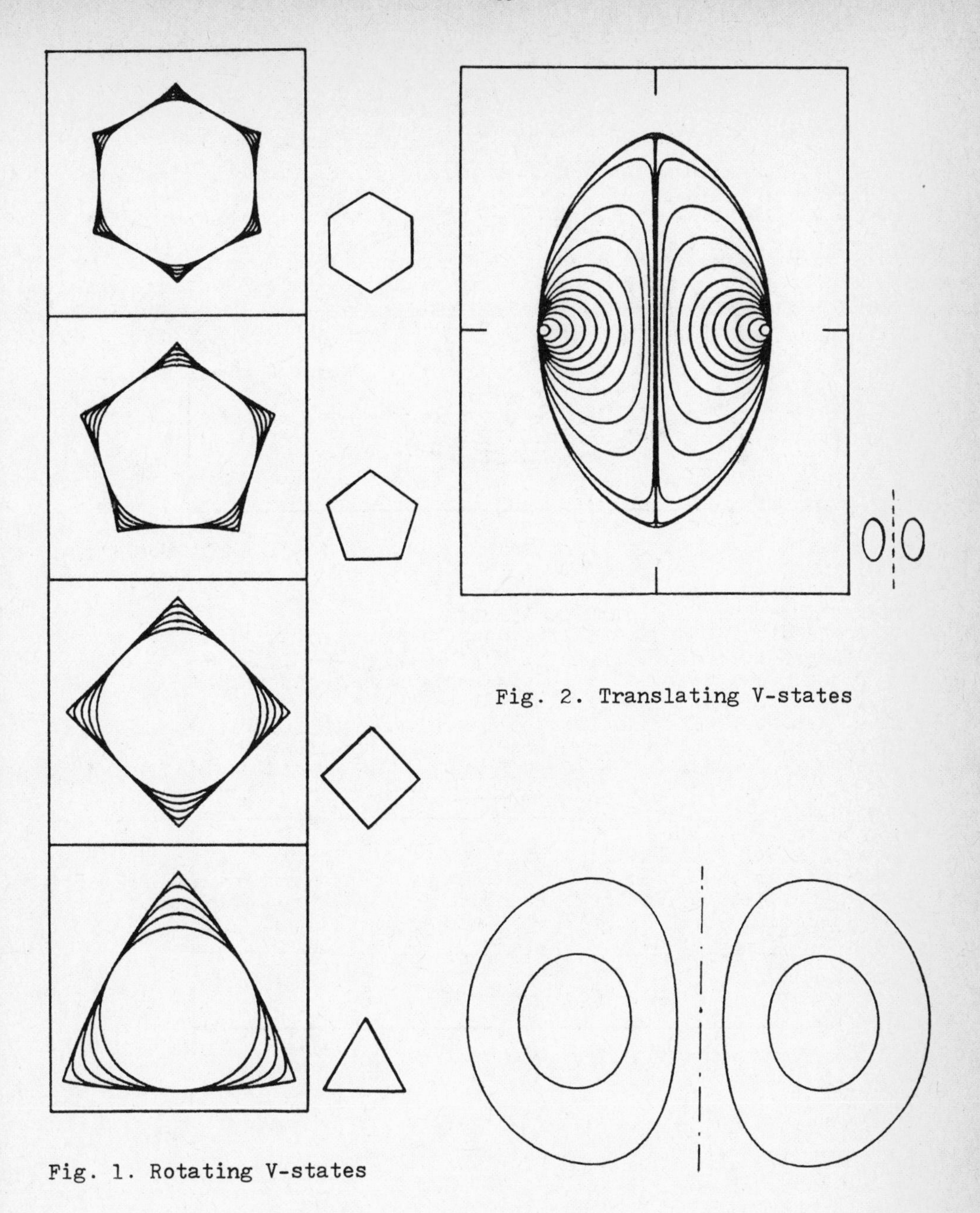

Fig. 1. Rotating V-states

Fig. 2. Translating V-states

Fig. 3. Multi-contour vortex pairs

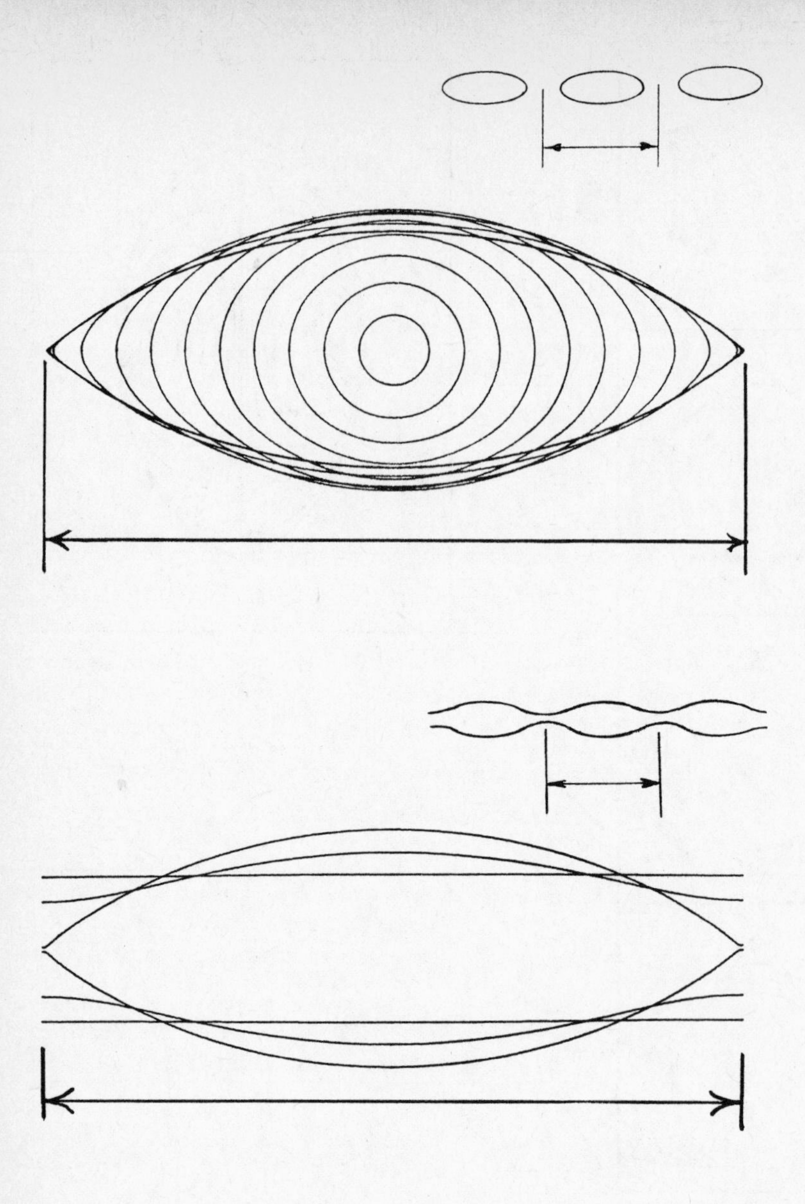

Fig. 4. Vortex patch streets

OPTIMUM DOMAIN PROBLEMS
GOVERNED BY A CLASS OF PDE

Xu Jia-mo
Dept. of Mathematics, Wuhan University

Lu Jun-an
Dept. of Mathematics, Wuhan Institute of
Hydrautic and Electric Engineering

1. Introduction

The optimum domain problems have important practical significance. They are arised from, for example, optimum design of the electrode and aircraft outline contours in electrostatic and gasdynamics respectively. This paper discusses only a simple model for convenience. However the argumentation methods remain effective for general elliptic operators.

Our model problem is to find the pair $\{s, u_s\}$ such that

$$\Delta u_s=0 \quad \text{in } D_s^o \tag{1.1}$$

$$u_s|_\Gamma = u_o(x) \tag{1.2}$$

$$u_s|_s = c \quad \text{(some constant)} \tag{1.3}$$

$$J_1(u_s)=\inf_{s'\in F} J_1(u_{s'}) \tag{1.4}$$

where $J_1(u_{s'})=\sup\limits_{x\in D_{s'}^o} |\nabla u_{s'}(x)|$, (1.5)

Γ and s are boundaries of open sets D and D_s respectively, $D_s\subset D\subset R^2$ and $D_s^o=D\setminus\overline{D}_s$. F is a set of closed contours in D and its definition will be given later.

2. Existence of the classical solution

Definition 2.1. Let F be a set of s which satisfy the following conditions:

i) s is non-degenerate, simple closed curve and has finite length.

ii) s can be expressed by

$$\begin{cases} x_i=x_i^s(t), & t\in[a,b], \\ x_i^s(a)=x_i^s(b), & \end{cases} \qquad i=1,2$$

or

$$s(t)=\begin{pmatrix} x_1^s(t) \\ x_2^s(t) \end{pmatrix}, \quad t\in[a,b), \quad s(a)=s(b).$$

iii) $s(t) \in C^4$ [a,b], there exists a constant M>0 such that $|s^{(4)}(t)| \leq M$ and $x_i^{s'}(a)=x_i^{s'}(b)$, $x_i^{s''}(a)=x_i^{s''}(b)$, i=1,2.

iv) There exists a constant k>0, such that for any $s \in F$,

$$\inf_{\substack{x \in s \\ y \in \Gamma}} \|x-y\|_{R^2} \geq k.$$

Lemma 2.1. F is a compact set with the distance $d(s_1, s_2) = \|s_1(t) - s_2(t)\|_{C^2[a,b]}$.

Proof. Evidently, F is closed. One divides equally [a,b] by nodes $a=t_0<t_1<\ldots<t_n=b$. For each $s \in F$ and these nodes, there exists a cubic spline interpolation, which is denoted by s_p. The interpolation error bound is a well-known result

$$\|s-s_p\|_{C^2[a,b]} \leq \max_{j=0,1,2}(c_j h^{4-j}) \tag{2.1}$$

where $h=(b-a)/n$, and c_j are positive constants. Let F_p denote the set of all s_p, then F_p is a finite dimensional, closed, bounded set. In terms of (2.1), F_p can be taken as a compact ε-net of F, so F is compact itself.

Definition 2.2. $U = \{ u_s \mid u_s \in C^2(D_s^0)$, satisfying (1.1)-(1.3), $s \in F\}$.

Lemma 2.2. U is a compact set.

Proof. Let us consider the mapping: $\begin{cases} F \to U. \\ s \mapsto u_s \end{cases}$ Two elements s_1, s_2 in F are called $s_1 \subset s_2$ if and only if $D_{s_1}^0 \supset D_{s_2}^0$.

Case 1. $s_1 \subset s_2$. For every $x \in D_{s_2}^0$, let $v_s = u_{s_1} - u_{s_2}$, $u_{s_1}, u_{s_2} \in U$, then

$$\Delta v_s(x) = 0, \quad x \in D_{s_2}^0$$

$$v_s|_\Gamma = 0,$$

$$v_s|_{s_2} = u_{s_1}|_{s_2} - c = u_{s_1}|_{s_2} - u_{s_1}|_{s_1}.$$

Because the solution v_s continuously depends on boundary values, $\|v_s\|_{C^2(D_{s_2}^0)} \leq \varepsilon$ if $d(s_1, s_2) \leq \delta$. Thus $\|u_{s_1} - u_{s_2}\|_{C^2(D_{s_2}^0)} \leq \varepsilon$.

Case 2. s_1, s_2 have no inclusion relation. If $d(s_1, s_2) \leq \delta$, then there exists some $s' \in F$ in $D_{s_1}^0 \cap D_{s_2}^0$, such that $s' \subset s_1$, $s' \subset s_2$ and $d(s', s_1)$, $d(s', s_2) \leq 2\delta$. Thus the problem reduces to the above case.

Lemma 2.3. The functional J_1: $U \to R$ is continuous, where

$$J_1(u_s) = \sup_{x \in D_s^0} |\nabla u_s(x)|.$$

Proof. This conclusion is obtained from the obvious expression

$$|J_1(u_{s_1})-J_1(u_{s_2})| \leq c_1 \|u_{s_1}-u_{s_2}\|_{C^2(D^o_{s_1}\cap D^o_{s_2})} \quad \text{for } s_1, s_2 \in F.$$

We have obtained the principal result for this section from above lemmas.

Theorem 2.1. Problem (1.1)-(1.4) has at least a solution.

3. Variational form of the problem

The classical form of the problem in the preceding section is unfit for computation. For this reason we shall give a weak form of the problem.

We begin by recalling the definition of F. The condition iii) in the definition 2.1 will be relaxed by Lipschitz continuous condition. We denote it still by F.

If $v \in C(\bar{D})$, then $\|v\|_{L^\infty(D)}=\max_{x\in\bar{D}}|v(x)|$. Thus if a proposition holds in the L^{2m} norm for an integal number m large enough, then it holds in the L^∞ norm approximately.

We consider the variational problem: Find the pair $\{s, u_s\}\in F\times H^1(D^o_s)$ such that

$$\begin{cases} u_{s'}-f \in H^1_0(D^o_{s'}) & (3.1)\\ \int_{D^o_{s'}} \nabla u_{s'}\cdot\nabla v\,dx=0 \quad \text{for all } v\in H^1_0(D^o_{s'}) & (3.2)\end{cases}$$

$$\text{and } J(u_s)=\inf_{s'\in F} J(u_{s'}) \qquad (3.3)$$

where $f\in H^1(D^o_{s'})$ is a given function satisfying $f|_\Gamma=u_o$ and $f|_{s'}=c$;

$$J(u_{s'})=\int_{D^o_{s'}}|\nabla u_{s'}|^{2m}dx \qquad (3.4)$$

In view of the regularity of solutions (3.1), (3.2), we can ensure $u_{s'}\in W^{1,2m}(D^o_{s'})$. In the following, we shall extend u_s to D_s with value c, and still denote it by u_s.

The continuity of the functional J is obvious. We can prove some results similar to Lemmas 2.1 and 2.2. Thus we have

Theorem 3.1. There exists $\{s, u_s\}$ at least satisfying (3.1)-(3.3).

In fact, we can obtain the following result.

Theorem 3.2. The solution of (3.1)-(3.3) is unique.

Proof. It is sufficient to prove the strict convexity of the functional J. According to (3.4), the G-derivative of J is

$$J'(u, v-u)=2m\int_{D^o_s}(\nabla u\cdot\nabla(v-u))|\nabla u|^{2(m-1)}dx.$$

If $u\neq v$, then

$$J'(u, v-u) < m \int_{D_s^o} |\nabla v|^2 |\nabla u|^{2(m-1)} dx - mJ(u).$$

By the Hölder inequality, a simple computation yields

$$J'(u, v-u) + J(u) < J(v),$$

namely, J is strictly convex [3].

4. Algorithm

The following algorithm is more or less alike with [1], [2]. We shall omit the deduction and give only the convergence proof of the method.

Let us introduce some notation. For any $s_i \in F$, denote $u_i = u_{s_i}$, $J_i = J(u_{s_i})$, $D_i = D^o_{s_i}$, n_i=the outward normal of s_i.

Step 1. Choose arbitrarily $s_o \in F$; set i=0.

Step 2. Let $s=s_i$, slove (3.1), (3.2) to obtain u_i.

Step 3. Find $g_i \in H'(D_i)$ such that

$$g_i - u_i \in H_0^1(D_i)$$

$$\int_{D_i} \nabla g_i \cdot \nabla v dx = 2m \int_{D_i} |\nabla u_i|^{2(m-1)} \nabla u_i \cdot \nabla v dx \qquad (4.1)$$

for all $v \in H_0^1(D_i)$.

Step 4. Let $d_i = -|\nabla u_i|^{2m} + \nabla u_i \cdot \nabla g_i$, if $d_i=0$ then stop; else let $s_{i+1} = s_i + d_i n_i$, set i=i+1 and go to Step 2.

In Step 4, ωd_i may be used instead of d_i, where ω is the relaxation factor. According to the expression of d_i in Step 4, we obtain

$$J_{i+1} - J_i = -\int_{s_i} (|\nabla u_i|^{2m} - \nabla u_i \cdot \nabla g_i)^2 ds \qquad (4.2)$$

so $J_{i+1} < J_i$. The algorithm is a descent method.

Theorem 4.1. The algorithm either terminates with finite steps or generates an infinite sequence $\{s_i, u_i\}$ which converges to solution of (3.1)-(3.3).

Proof. Since grad $J(u_i) = 2m(|\nabla u_i|^2)^{m-1} \nabla u_i$ and ∇u_i with n_i have the same direction on s_i, the method is practically a gradient descent method.

i) If the computation stops in the ith step, then $d_i=0$. In this case, either $\nabla u_i = 0 \Rightarrow$ grad $J(u_i)=0 \Rightarrow J'(u_i, v)=0$

or $\nabla u_i \cdot \nabla g_i = |\nabla u_i|^{2m}$, but

$$\int_{D_i} \nabla g_i \cdot \nabla u_i dx = \int_{D_i} \text{grad } J(u_i) \cdot \nabla u_i dx,$$

so

$$\int_{D_i} |\nabla u_i|^{2m} dx = \int_{D_i} 2m\, |\nabla u_i|^{2m} dx.$$

We obtain grad $J(u_i)=0$ a.e. $D_i \Rightarrow J'(u_i, v)=0$.

ii) For the case of an infinite sequence, $J_{i+1} < J_i$ and $\{J_i\}$ is lower bounded, so

$$\lim_{i\to\infty} J_i \equiv r \tag{4.3}$$

First, there exists some $p \in W^{1,2m}(D)$, such as $p|_\Gamma = u_0$. Let $w_i = u_i - p$, then $w_i \in W_0^{1,2m}(D)$.

By the Clarkson inequality [4], we have

$$\left\|\nabla\left(\frac{u-v}{2}\right)\right\|_{L^{2m}(D)}^{2m} \leqslant \frac{1}{2}(J(u)+J(v))-J\left(\frac{u+v}{2}\right) \qquad \text{for } u,\ v \in W^{1,2m}(D) \tag{4.4}$$

By the algorithm, we obtain

$$r \leqslant J(u_i) \leqslant r+\frac{1}{i} \tag{4.5}$$

From (4.4) and (4.5), we have

$$\left\|\nabla w_i - \nabla w_j\right\|_{L^{2m}(D)}^{2m} \leqslant 2^{2m-1}\left(\frac{1}{i}+\frac{1}{j}\right) \to 0, \quad \text{as } i,j \to \infty .$$

By the Poincaré inequality [4], we thus obtain

$$\|(w_i - w_j)\|_{W^{1,2m}(D)}^{2m} \to 0, \quad \text{as } i,j \to \infty .$$

i.e.

$$\|u_i - u_j\|_{W^{1,2m}(D)} \to 0, \quad \text{as } i,j \to \infty .$$

This implies that $\{u_i\}$ is a Cauchy sequence in $W^{1,2m}(D)$. Therefore there exists $u \in W^{1,2m}(D)$, such that

$$\|u_i - u\|_{W^{1,2m}(D)} \to 0, \quad \text{as } i \to \infty .$$

Next, it is obvious that

$$\lim_{i\to\infty} (J(u_{i+1})-J(u_i))=0.$$

This implies that

$$\lim_{i\to\infty} J'(u_i,\ u_{i+1}-u_i)=0.$$

It follows from the algorithm that $u_{i+1}-u_i$ characterizes the gradient direction. Therefore

$$\lim_{i \to \infty} J'(u_i, v) = 0 \qquad \text{for all } v \in W^{1,2m}(D).$$

By the continuity of $J'(u, v)$ with respect to u, we have

$$J'(u, v) = 0 \qquad \text{for all } v \in W^{1,2m}(D).$$

This completes the proof of theorem 4.1.

The authors wish to thank Mr. X.Y.Huang for his valuable discussions on the background of the problem.

REFERENCES

[1] F. Angrand, R. Glowinski, J. Perioux, O. Pironneau, Optimum Design for Potential Flows, Proceedings of the Third International Conference on Finite Elements in Flow Problems, Banff, Alberta, Canada 10-13 June, 1980 (Vol. 1).

[2] O. Pironneau, Variational Methods for the Numerical Solutions of Free Boundary Problems and Optimum Design Problems, Control Theory of Systems Governed by P D E, Editors: A. K. Aziz, J. W. Wingate, M. J. Balas. Academic Press, Inc. 1977.

[3] J. Céa, Optimisation: théorie et algorithmes, Dunod, Paris, 1971.

[4] Adams, R. A. Sobolev Spaces. Academic Press, New York, London and San Francisco, 1976.

FOLDS OF DEGREE 4 AND SWALLOWTAIL CATASTROPHE

Yang Zhong-hua
Department of Mathematics
Shanghai University of Science and Technology
Shanghai, China

Abstract

In this paper we consider the computation for the folds of degree 4 which corresponds to the swallowtail catastrophe. A regular extended system for the swallowtail catastrophe is proposed. Finally the numerical results in the tubular nonadiabatic reaction problem are given.

1. Introduction

We consider a nonlinear problem with 3 parameters

$$f(\lambda,\mu,\nu,x) = 0 \tag{1.1}$$

where λ,μ,ν are the real parameters, $x \in X$, a Banach space, and f is a C^4 mapping from $\mathbb{R} \times \mathbb{R} \times \mathbb{R} \times X$ to X.

In many applications some loss of criticality in nonlinear problems with several parameters which corresponds to the singularity of higher order at the particular values of parameters is concerned. For example, the loss of criticality in the thermal ignition described by a two-parameters nonlinear problem corresponds to the particular values of two parameters called the folds of degree 3 which is also a cusp catastrophe ([5], [6]).

In this paper we want to compute the swallowtail catastrophe in (1.1), which corresponds to the folds of degree 4.

The plan of the paper is as follows. In Section 2 we define the degree of the folds in brief. Using Liapunov-Schmidt method we get a sufficient and necessary condition for the folds of degree 4. In Section 3 an extended system for the folds of degree 4 is given. The regularity of the extended system is proved. Also we show that the fold of degree 4 actually is the swallowtail catastrophe under the same condition as in the proof of the regularity of extended system. The numerical procedure for computing the swallowtail catastrophe in the tubular nonadiabatic reaction problem is given in Section 4.

2. One-parameter case and fold points

We consider a one-parameter nonlinear problem in a Banach space X

$$f(\lambda,x) = 0 \qquad (2.1)$$

where $\lambda \in R$, $x \in X$ and f is a C^{n+1} ($n \geq 1$ is a suitable positive integer) mapping from $\mathbb{R} \times X$ to X.

The notations $f_\lambda(a)$, $f_{\lambda\lambda}(a)$, $f_x(a)$, $f_{xx}(a)$, $f_{\lambda x}(a)$, $f_{xxx}(a)$,... are used to denote the partial Frechet-derivatives of f at $a=(\lambda,x) \in \mathbb{R} \times X$. We denote the dual pairing of $x \in X$ and $\psi \in X^*$ by ψx where X^* is the conjugate space of X.

Definition 2.1 A point $a_0 = (\lambda_0, x_0) \in \mathbb{R} \times X$ is a fold point of (2.1) with respect to λ if:

$$f(a_0) = 0, \qquad (2.2)$$

$$\mathrm{Ker}\, f_x(a_0) \neq 0, \qquad (2.3)$$

$$f_\lambda(a_0) \notin \mathrm{Range}\, f_x(a_0). \qquad (2.4)$$

Definition 2.2 A fold point a_0 is a simple fold of (2.1) with respect to λ if in addition to (2.2)-(2.4)

$$\dim \mathrm{Ker}\, f_x(a_0) = \mathrm{Codim}\, \mathrm{Range}\, f_x(a_0) = 1 \qquad (2.5)$$

In this case there exist nontrivial $\phi_0 \in X$ and $\psi_0 \in X^*$ such that

$$N \equiv \mathrm{Ker}\, f_x(a_0) = \{ \alpha\phi_0 \mid \alpha \in \mathbb{R}\} \qquad (2.6)$$

$$M \equiv \mathrm{Range}\, f_x(a_0) = \{ x \in X \mid \psi_0 x = 0 \}. \qquad (2.7)$$

As is well known, the zero set of $f(\lambda, x)$ near a simple fold a_0 is a smooth curve

$$\Gamma: \qquad f^{-1}(0) \cap U = \{(\lambda(s),x(s)) \mid |s-s_0| \leq \delta\}$$

where U is an open neighborhood of the simple fold point a_0, δ is a small positive and $\lambda(s), x(s)$ are C^{n+1} mappings satisfying $\lambda_0(s_0) = \lambda_0$, $x(s_0) = x_0$, $|\lambda (s)|^2 + \|x'(s)\|^2 > 0$ and $x(s) = (s-s_0)\phi_0 + v(s)$, $v(s) \in V_0$ where V_0 is a complement of N, i.e. $X = N \oplus V_0$.

Along Γ we have the identity $f(\lambda(s), x(s)) \equiv 0$. Instead of $f_\lambda(\lambda(s), x(s))$, $f_x'(\lambda(s), x(s))$,... we shall write f_λ, f_x,... and denote $\overset{0}{f_\lambda} = f_\lambda(\lambda(s_0), x(s_0))$, $f_x^0 = f_x(\lambda(s_0), x(s_0))$,.... Differentiating $f(\lambda(s), x(s)) \equiv 0$ with respect to s yields

$$f_\lambda \lambda'(s) + f_x x'(s) \equiv 0, \ |s - s_0| \leq \delta . \tag{2.8}$$

Obviously

$$\lambda'(s_0) = 0 \ , \tag{2.9}$$

$$x'(s_0) = \phi_0 . \tag{2.10}$$

Difinition 2.3 A simple fold point $a_0 \in \mathbb{R} \times X$ is said to have degree n+1 if:

$$\lambda''(s_0) = \ldots = \lambda^{(n)}(s_0) = 0, \ \lambda^{(n+1)}(s_0) \neq 0 \tag{2.11}$$

where $\lambda^{(i)}(s_0)$ is the i-th derivative of $\lambda(s)$ with respect to s at $s = s_0$. We also call it the (n+1)th degree fold.

Before working out a sufficient and necessary condition for the 4th degree fold we introduce

$$P_2(\phi_0;f) = f^0_{xx}\phi_0\phi_0 \ , \tag{2.12}$$

$$P_3(\phi_0\phi_1 ; f) = 3f^0_{xx}\phi_0\phi_1 + f^0_{xxx}\phi_0\phi_0\phi_0, \tag{2.13}$$

$$P_4(\phi_0 \ \phi_1 \ \phi_2 f) = 4f^0_{xx}\phi_0\phi_2 + 3f^0_{xx}\phi_1\phi_1 + 6f^0_{xxx}\phi_0\phi_0\phi_1 + f^0_{xxxx}\phi_0\phi_0\phi_0\phi_0 . \tag{2.14}$$

Differentiating $f(\lambda(s),x(s))$ with respect to s at $s=s_0$ we easily obtain

Lemma 2.1

If $\lambda'(s_0) = 0$, $\lambda''(s_0)=0$, $\lambda^{(3)}(s_0)=0$, $\lambda^{(4)}(s_0)\neq 0$ and $x'(s_0)= \phi_0$, $x''(s_0) = \phi_1$, $x^{(3)}(s_0)=\phi_2$, $x^{(4)}(s_0)=\phi_3$, then for n=1,2,3

$$P_{n+1}(\phi_0,\ldots,\phi_{n-1};f) = \frac{d^{n+1}f(\lambda(s),x(s))}{ds^{n+1}}\bigg|_{s=s} - f^0_x\phi_n - f^0_\lambda \lambda^{(n+1)}(s_0).$$

Next we propose a sufficient and necessary condition for a 4th degree fold of (2.1).

Theorem 2.2

The fold $a_0=(\lambda(s_0),x(s_0))$ of (2.1) is to have degree 4 if and only if there exist $\phi_0 \in N$, $\phi_1,\phi_2 \in V_0$ such that

$$\phi_0 = x'(s_0), \ \phi_1 = x''(s_0), \ \phi_2 = x^{(3)}(s_0)$$

which are uniquely determined by

$$f^0_x\phi_1 = - P(\phi_0;f) \tag{2.15}$$

$$f^0_x \phi_2 = -P_3(\phi_0,\phi_1;f) \tag{2.16}$$

and $\phi_3 = x^{(4)}(s_0)$ which satisfies

$$P_4(\phi_0,\phi_1,\phi_2;f) + f_x^0\phi_3 + f_\lambda^0\lambda^{(4)}(s_0) = 0. \tag{2.17}$$

Furthermore

$$\begin{aligned} &\psi_0[P_2(\phi_0;f)] = 0, \\ &\psi_0[P_3(\phi_0,\phi_1;f)] = 0, \\ &\psi_0[P_4(\phi_0,\phi_1,\phi_2;f)] \neq 0. \end{aligned} \tag{2.18}$$

Proof. Differentiating $f(\lambda(s),x(s))=0$ twice with respect to s at $s=s_0$ yields

$$\begin{aligned} &f_x^0 x'(s_0) + f_\lambda^0\lambda'(s_0)=0, \\ &f_{xx}^0 x'(s_0)x'(s_0)+f_x^0 x''(s_0)+2f_{\lambda x}^0\lambda'(s_0)x'(s_0)+f_{\lambda\lambda}^0\lambda'(s_0)^2+f_\lambda^0\lambda''(s_0) = 0 \end{aligned} \tag{2.19}$$

Substituting $\lambda'(s_0)=0$, $x'(s_0)=\phi_0$ and $\lambda''(s_0)=0$ into (2.19) we get

$$f_{xx}^0\phi_0\phi_0 + f_x^0 x''(s_0) = 0 \tag{2.20}$$

Here $x''(s_0)= \phi_1 \in V_0$ is uniquely given by (2.20). Multiplying ψ_0 on (2.20) yields $\psi_0[P_2(\phi_0;f)]=0$. Differentiating (2.19) with respect to s at $s=s_0$ and substituting $\lambda'(s_0)=0$, $\lambda''(s_0)=0$, $\lambda^{(3)}(s_0)=0$, $x'(s_0)=\phi_0$, $x''(s_0)= \phi_1$ into it we obtain

$$P_3(\phi_0,\phi_1;f) + f_x^0 x^{(3)}(s_0) = 0. \tag{2.21}$$

$x^{(3)}(s_0) = \phi_2 \in V_0$ is uniquely determined by (2.21). Applying ψ_0 on (2.21) yields

$$\psi_0[P_3(\phi_0,\phi_1; f)] = 0.$$

Finally differentiating (2.19) with respect to s at $s=s$ once more and substituting $\lambda'(s_0)=\lambda''(s_0)=\lambda^{(3)}(s_0)=0$, $x'(s_0)=\phi_0$, $x''(s_0)=\phi_1$, $x^{(3)}(s_0)=\phi_2$ into it we have

$$P_4(\phi_0,\phi_1,\phi_2;f)+f_x^0 x^{(4)}(s_0)+f_\lambda^0\lambda^{(4)}(s_0) = 0 \tag{2.22}$$

Multiplying ψ_0 on (2.22) and noticing $\lambda^{(4)}(s_0)\neq 0$, we have

$$\psi_0[P_4(\phi_0,\phi_1,\phi_2; f)]\neq 0$$

and

$$\lambda^{(4)}(s_0)= -\psi_0[P_4(\phi_0,\phi_1,\phi_2; f)]/ \psi_0[f_\lambda^0].$$

So far the necessary condition is concluded.

On the other hand, differentiating $f(\lambda(s), x(s)) = 0$ with respect to s at $s=s_0$ once up to four times, the inverse procedure leads the sufficient condition of the theorem.

3. Extended system for folds of degree 4

At first we introduce the extended system of (1.1) for folds of degree 4 as follows

$$
\begin{aligned}
& f(\lambda,\mu,\nu,x)=0 \\
& f_x\phi_0 =0 \\
& e\phi_0 -1=0 \\
& f_x\phi_1 + f_{xx}\phi_0\phi_0 =0 \\
& e\phi_1 =0 \\
& f_x\phi_2 + 3f_{xx}\phi_0\phi_1 + f_{xxx}\phi_0\phi_0\phi_0 =0 \\
& e\phi_2 =0
\end{aligned}
\tag{3.1}
$$

where $e \in X^*$, the conjugate space of X. In other words e is a linear functional on X. We can decompose $X = N \oplus V_0$ such that $e\phi_0 =1$ for ϕ_0 in (2.6) and ex=0 means $x \in V_0$.

Let

$$g_1(\phi_0;\ f) = f^0_x \ , \tag{3.2}$$

$$g_2(\phi_0\ \phi_1;\ f) = f^0_x\phi_1 + f^0_{xx}\phi_0\phi_0 \ , \tag{3.3}$$

$$g_3(\phi_0,\phi_1,\phi_2;\ f)=f^0_x\phi_2+ P_3(\phi_0,\phi_1;\ f) . \tag{3.4}$$

The straight calculation leads to the formulae

$$[g_1(\phi_0;\ f)]_x\phi_0= P_2(\phi_0;\ f), \tag{3.5}$$

$$[g_2(\phi_0;f)]_x\phi_0 +[g_2(\phi_0;f)]_{\phi_0} \phi_1 = P_3(\phi_0,\phi_1;\ f), \tag{3.6}$$

$$[g_3(\phi_0,\phi_1;\ f)]_x\phi_0+[g_3(\phi_0,\phi_1;\ f)]_{\phi_0}\phi_1 + [g_3(\phi_0,\phi_1;\ f)]_{\phi_1}\phi_2 \tag{3.7}$$
$$= P_4(\phi_0,\phi_1,\phi_2;\ f)$$

where $[\cdot]_x$, $[\cdot]_{\phi_0}$, $[\cdot]_{\phi_1}$ are respectively the Frechet-derivative with respect to x, ϕ_0 and ϕ_1 .

Next we are going to discuss the regularity of the extended system (3.1) at the fold of degree 4.

The order of the variables is arranged as $(x,\phi_0,\phi_1,\phi_2,\lambda,\mu,\nu)$. Rearranging the order of the equations in (3.1) we can get the Jacobian matrix of (3.1) at the fold $a_0 =(x^0,\phi_0^0,\phi_1^0,\phi_2^0,\lambda^0,\mu^0,\nu^0)$ as follows

$$J=\begin{pmatrix} f_x & 0 & 0 & 0 & f_\lambda & f_\mu & f_\nu \\ f_{xx}\phi_0 & f_x & 0 & 0 & f_{\lambda x} & f_{\mu x} & f_{\nu x} \\ [g_2]_x & 2f_{xx}\phi_0 & f_x & 0 & [g_2]_\lambda & [g_2]_\mu & [g_2]_\nu \\ [g_3]_x & [g_3]_{\phi_0} & [g_3]_{\phi_1} & f_x & [g_3]_\lambda & [g_3]_\mu & [g_3]_\nu \\ 0 & e & 0 & 0 & 0 & 0 & 0 \\ 0 & 0 & e & 0 & 0 & 0 & 0 \\ 0 & 0 & 0 & e & 0 & 0 & 0 \end{pmatrix}$$

where the superscript °, which means that the Jacobian is valued at the fold of degree 4, is omitted for simplicity of the notations.

We consider

$$JU = 0 \tag{3.8}$$

where $U=(u_0,u_1,u_2,c_0,c_1,c_2)^T$, $u_i \in X$, $c_i \in \mathbb{R}$. If (3.8) only has the trivial solution U=0, injectivity will be proved. Expanding (3.8) u_i, c_i should satisfy

$$f_x u_0 + c_0 f_\lambda + c_1 f_\mu + c_2 f_\nu = 0 \tag{3.9}$$

$$f_{xx}\phi_0 u_0 + f_x u_1 + c_0 f_{\lambda x}\phi_0 + c_1 f_{\mu x}\phi_0 + c_2 f_{\nu x}\phi_0 = 0 \tag{3.10a}$$

$$eu_1 = 0 \tag{3.10b}$$

$$[g_2]_x u_0 + 2f_{xx}\phi_0 u_1 + f_x u_2 + c_0[g_2]_\lambda + c_1[g_2]_\mu + c_2[g_2]_\nu = 0 \tag{3.11a}$$

$$eu_2 = 0 \tag{3.11b}$$

$$[g_3]_x u_0 + [g_3]_{\phi_0} u_1 + [g_3]_{\phi_1} u_2 + f_x u_3 + c_0[g_3]_\lambda + c_1[g_3]_\mu + c_2[g_3]_\nu = 0 \tag{3.12a}$$

$$eu_3 = 0 \tag{3.12b}$$

Projecting (3.9) on M (i.e.Range $f_x(a_0)$) yields

$$u_0 = c\phi_0 + c_0 w_1^{(\lambda)} + c_1 w_1^{(\mu)} + c_2 w_1^{(\nu)} \tag{3.13}$$

Here $w_1^{(\lambda)}$, $w_1^{(\mu)}$, $w_1^{(\nu)} \in V_0$ are uniquely determined by

$$\begin{aligned} f_x w_1^{(\lambda)} &= -Q[f_\lambda] \\ f_x w_1^{(\lambda)} &= -Q[f_\mu] \\ f_x w_1^{(\nu)} &= -Q[f_\nu] \end{aligned} \tag{3.14}$$

where Q is the project operator on M . Applying Ψ_0 on (3.9) we know that c_i must satisfy

$$c_0\psi_0[f_\lambda]+c_1\psi_0[f_\mu]+c_2\psi_0[f_\nu]=0 \tag{3.15}$$

Substituting (3.13) into (3.10a) we obtain

$$f_{xx}\phi_0(c\phi_0+c_0w_1^{(\lambda)}+c_1w_1^{(\mu)}+c_2w_1^{(\nu)}) + f_xu_1 +c_0f_{\lambda x}\phi_0+c_1f_{\mu x}\phi_0+c_1f_{\mu x}\phi_0+c_2f_{\nu x}\phi_0=0$$

Projecting above expression on M and noticing (3.10b), we have

$$u_1= c\phi_1+c_0w_2^{(\lambda)}+c_1w_2^{(\mu)}+c_2w_2^{(\nu)}. \tag{3.16}$$

Here $\phi_1, w_2^{(\lambda)}, w_2^{(\mu)}, w_2^{(\nu)} \in V_0$ are uniquely determined by

$$\begin{aligned} &f_x\phi_1+ f_{xx}\phi_0\phi_0 = 0, \\ &f_xw_2^{(\lambda)} = -Q[f_{xx}\phi_0w_1^{(\lambda)}+f_{\lambda x}\phi_0], \\ &f_xw_2^{(\mu)}= -Q[f_{xx}\phi_0w_1^{(\mu)}+ f_{\mu x}\phi_0] \\ &f_xw_2^{(\nu)} = -Q[f_{xx}\phi_0w_1^{(\nu)}+f_{\nu x}\phi_0] \end{aligned} \tag{3.17}$$

and c_i must satisfy

$$c_0\psi_0[F_1^\lambda]+c_1\psi_0[F_1^\mu]+c_2\psi_0[F_1^\nu] = 0 \tag{3.18}$$

where

$$\begin{aligned} &F_1^\lambda=f_{xx}\phi_0w_1^{(\lambda)}+f_{\lambda x}\phi_0, \\ &F_1^\mu=f_{xx}\phi_0w_1^{(\mu)}+f_{\mu x}\phi_0, \\ &F_1^\nu=f_{xx}\phi_0w_1^{(\nu)}+f_{\nu x}\phi_0. \end{aligned} \tag{3.19}$$

Similar procedure provides from (3.11)

$$u_2= c\phi_2 +c_0w_3^{(\lambda)}+c_1w_3^{(\mu)}+c_2w_3^{(\nu)} \tag{3.20}$$

where $\phi_2, w_3^{(\lambda)}, w_3^{(\mu)}, w_3^{(\nu)} \in V_0$ are uniquely decided by

$$\begin{aligned} &f_x\phi_2+P_3(\phi_0,\phi_1;f) = 0 \\ &f_xw_3^{(\lambda)}= -Q[F_2^\lambda], \\ &f_xw_3^{(\mu)}= -Q[F_2^\mu] \\ &f_xw_3^{(\nu)}= -Q[F_2^\nu] \end{aligned} \tag{3.21}$$

and c_i must satisfy

$$c_0\psi_0[F_2^\lambda]+c_1\psi_0[F_2^\mu]+c_2\psi_0[F_2^\nu] = 0 \tag{3.22}$$

where

$$\begin{aligned} &F_2^\lambda =f_{xxx}\phi_0\phi_0 w_1^{(\lambda)}+f_{xx}(\phi_1 w_1^{(\lambda)}+2\phi_0 w_2^{(\lambda)})+f_{xx\lambda}\phi_0\phi_0+f_{\lambda x}\phi_1, \\ &F_2^\mu =f_{xxx}\phi_0\phi_0 w_1^{(\mu)}+f_{xx}(\phi_1 w_1^{(\mu)}+2\phi_0w_2^{(\mu)})+f_{xx\mu}\phi_0\phi_0+f_{\mu x}\phi_1, \\ &F_2^\nu =f_{xxx}\phi_0\phi_0w_1^{(\nu)}+f_{xx}(\phi_1w_1^{(\nu)}+2\phi_0w_2^{(\nu)})+f_{xx\nu}\phi_0\phi_0+f_{\nu x}\phi_1. \end{aligned} \tag{3.23}$$

If the determinant

$$\begin{vmatrix} \psi_0[f_\lambda] & \psi_0[f_\mu] & \psi_0[f_\nu] \\ \psi_0[F_1^\lambda] & \psi_0[F_1^\mu] & \psi_0[F_1^\nu] \\ \psi_0[F_2^\lambda] & \psi_0[F_2^\mu] & \psi_0[F_2^\nu] \end{vmatrix} \neq 0 \tag{3.24}$$

we can know $c_0=c_1=c_2=0$ from (3.15), (3.18), (3.22) and $u_0= c\phi_1, u_1= c\phi_1, u_2=c\phi_2$ from (3.13), (3.16), (3.20). Substituting $u_k=c\phi_k (k=0,1,2)$ into (3.12a) and using (3.7) we obtain

$$cP_4(\phi_0,\phi_1,\phi_2;f)+f_x u_3 =0$$

$c=0$ and $u_k=0$ $(k=0,1,2)$ can be concluded because $\psi_0[P_4(\phi_0,\phi_1,\phi_2;f)]\neq 0$ at the folds of degree 4. Finally we get $u_3=0$ from $f_x u_3=0$ and $eu_3=0$. So (3.8) only has the solution $U=0$.

Next we are going to prove the surjectivity of Jacobian J under the condition (3.24). In other words, if (3.24) is satisfied the following equation

$$JU = T \tag{3.25}$$

has the solution $U=(u_0,u_1,u_2,u_3,c_0,c_1,c_2)^T$ for any $T=(t_0,t_1,t_2,t_3,b_0,b_1,b_2)^T$ where u_i, $t_i \in X$, $c_j, b_j \in \mathbb{R} (i=0,1,2,3;\ j=0,1,2)$.

Expanding (3.25) yields

$$f_x u_0+c_0 f_\lambda+c_1 f_\mu+c_2 f_\nu =t_0, \tag{3.26}$$

$$f_{xx}\phi_0 u_0+f_x u_1 +c_0 f_{\lambda x}\phi_0 +c_1 f_{\mu x}\phi_0+c_2 f_{\nu x}\phi_0 = t_1 , \tag{3.27a}$$

$$eu_1 = b_0, \tag{3.27b}$$

$$[g_2]_x u_0 +2f_{xx}\phi_0 u_1+f_x u_2 +c_0[g_2]_\lambda+c_1[g_2]_\mu+ c_2[g_2]_\nu= t_2 , \tag{3.28a}$$

$$eu_2 = b_1 , \tag{3.28b}$$

$$[g_3]_x u_0 +[g_3]_{\phi_0} u_1 +[g_3]_{\phi_1} u_2+ f_x u_3 +c_0[g_3]_\lambda+c_1[g_3]_\mu+ c_2[g_3]_\nu= t_3 , \tag{3.29a}$$

$$eu_3= b_2 . \tag{3.29b}$$

From (3.26) we get

$$u_0= c\phi_0 +c_0 w_1^{(\lambda)}+ c_1 w_1^{(\mu)}+c_2 w_1^{(\nu)}+s_0 \tag{3.30}$$

where $w_1^{(\lambda)}, w_1^{(\mu)}, w_1^{(\nu)} \in V_0$ are given in (3.14), $s_0 \in V_0$ is uniquely determined by

$$f_x s_0 = Q[t_0] \tag{3.31}$$

and c_i should satisfy

$$c_0\psi_0[f_\lambda]+c_1\psi_0[f_\mu]+c_2\psi_0[f_\nu] = \psi_0[t_0] . \tag{3.32}$$

The procedure of substituting and projecting in (3.27) gives us

$$u_1 = b_0\phi_0 + c\phi_1 + c_0 w_2^{(\lambda)} + c_1 w_2^{(\mu)} + c_2 w_2^{(\nu)} + s_1 \tag{3.33}$$

where $\phi_1, w_2^{(\lambda)}, w_2^{(\mu)}, w_2^{(\nu)} \in V_0$ are given in (3.17), $s \in V_0$ is uniquely determined by

$$f_x s_1 = Q[t_1 - f_{xx}\phi_0 s_0] \tag{3.34}$$

and c_i should satisfy

$$c_0\psi_0[F_1^\lambda] + c_1\psi_0[F_1^\mu] + c_2\psi_0[F_1^\nu] = \psi_0[t_1 - f_{xx}\phi_0 s_0] \tag{3.35}$$

where $F_1^\lambda, F_1^\mu, F_1^\nu$ are defined in (3.19).

Similarly from (3.28)

$$u_2 = b_1\phi_0 + c\phi_2 + c_0 w_3^{(\lambda)} + c_1 w_3^{(\nu)} + c_2 w_3^{(\nu)} + s_2 \tag{3.36}$$

where $\phi_2, w_3^{(\lambda)}, w_3^{(\mu)}, w_3^{(\nu)} \in V_0$ are given in (3.21) and $s_2 \in V_0$ is uniquely determined by

$$f_x s_2 = Q[t_2 - [g_2]_x s_0 - [g_2]_{\phi_0}(b_0\phi_0 + s_1)], \tag{3.37}$$

and c_i should satisfy

$$c_0\psi_0[F_2^\lambda] + c_1\psi_0[F_2^\mu] + c_2\psi_0[F_2^\nu] = \psi_0[t_2 - [g_2]_x s_0 - [g_2]_{\phi_0}(b_0\phi_0 + s_1)] \tag{3.38}$$

where F_2^λ, F_2^μ, F_2^ν are defined in (3.23).

Under the condition (3.24) we can solve (3.32), (3.35), (3.38)

$$c_i = c_i^* \qquad (i=0,1,2) \tag{3.39}$$

Substituting (3.39) into (3.30), (3.33), (3.36) we know that $u_k (k=0,1,2)$ are decided except the constant c. Plunging $u_k (k=0,1,2)$ into (3.29a) and noticing (3.29b), we obtain

$$u_3 = b_2\phi_0 + c\phi_3 + c_0^* w_4^{(\lambda)} + c_1^* w_4^{(\mu)} + c_2^* w_4^{(\nu)} + s_3 \tag{3.40}$$

where $\phi_3, w_4^{(\lambda)}, w_4^{(\mu)}, w_4^{(\nu)} \in V_0$ are uniquely determined by

$$\begin{aligned} & f_x\phi_3 + Q[P_4(\phi_0, \phi_1, \phi_2; f)] = 0, \\ & f_x w_4^{(\lambda)} = -Q[F_3^\lambda], \\ & f_x w_4^{(\mu)} = -Q[F_3^\mu], \\ & f_x w_4^{(\nu)} = -Q[F_3^\nu] \end{aligned} \tag{3.41}$$

and $s_3 \in V_0$ satisfies

$$f_x s_3 = Q[t_3 - [g_3]_x s_0 - [g_3]_{\phi_0} (b_0\phi_0 + s_1) - [g_3]_{\phi_1} (b_1\phi_0 + s_2)] \quad (3.42)$$

Here

$$F_3^\lambda = [g_3]_x w_1^{(\lambda)} + [g_3]_{\phi_0} w_2^{(\lambda)} + [g_3]_{\phi_1} w_3^{(\lambda)} + [g_3]_\lambda ,$$
$$F_3^\mu = [g_3]_x w_1^{(\mu)} + [g_3]_{\phi_0} w_2^{(\mu)} + [g_3]_{\phi_1} w_3^{(\mu)} + [g_3]_\mu , \quad (3.43)$$
$$F_3^\nu = [g_3]_x w_1^{(\nu)} + [g_3]_{\phi_0} w_2^{(\nu)} + [g_3]_{\phi_1} w_3^{(\nu)} + [g_3]_\nu .$$

Applying ψ_0 on (3.29a) and noticing (3.40), we have

$$c = c^* = \psi_0 [F_3^* - c_0^* F_3^\lambda - c_1^* F_3^\mu - c_2^* F_3^\nu] / \psi_0 [P_4(\phi_0, \phi_1, \phi_2; f)] \quad (3.44)$$

where $F_3^* = t_3 - [g_3]_x s_0 - [g_3]_{\phi_0} (b_0\phi_0 + s_1) - [g_3]_{\phi_1} (b_1\phi_0 + s_2)$ and $\psi_0 [P_4(\phi_0, \phi_1, \phi_2; f)] \neq 0$ at the folds of degree 4. Now $c = c^*$, $c_i = c_i^* (i=0,1,2)$ and $u_i (i=0,1,2,3)$ in (3.30), (3.33), (3.36), (3.40) are completely determined. In other words there is always the solution of (3.25) for any right side T. So far we have finished the proof of the following

Theorem 3.1 The extended system (3.1) of (1.1) for the folds of degree 4 is regular if and only if the condition (3.24) is satisfied at the folds.

Now we consider a perturbation $a = (\lambda, \mu, \nu, x)$ to the fold of degree 4 $a_0 = (\lambda_0, \mu_0, \nu_0, x_0)$ such that

$$\lambda = \lambda_0 + \hat{\lambda} ,$$
$$\mu = \mu_0 + \hat{\mu} . \quad (3.45)$$
$$\nu = \nu_0 + \hat{\nu} ,$$
$$x = x_0 + \hat{x}\phi_0 + w$$

where $\hat{\lambda}, \hat{\mu}, \hat{\nu}, \hat{x} \in R$, $w \in V_0$. Applying a Liapunov-Schmidt decomposition to f we can get the bifurcation equation of (1.1)

$$H(\hat{\lambda}, \hat{\mu}, \hat{\nu}, \hat{x}) = \psi_0 [f(\lambda_0 + \hat{\lambda}, \mu_0 + \hat{\mu}, \nu_0 + \hat{\nu}, x_0 + \hat{x}\phi_0 + w)] = 0$$

where $w \in V_0$ is uniquely decided by

$$Q[f(\lambda_0 + \hat{\lambda}, \mu_0 + \hat{\mu}, \nu_0 + \hat{\nu}, x_0 + \hat{x}\phi_0 + w)] = 0$$

It can be checked by straight calculation that

$$H(0,0,0,0) = 0,$$
$$H_{\hat{x}}(0,0,0,0) = \psi_0 [f_x^0 \phi_0] = 0,$$
$$H_{\hat{x}\hat{x}}(0,0,0,0) = \psi_0 [f_{xx}^0 \phi_0 \phi_0] = 0,$$

$$H_{\hat{x}\hat{x}\hat{x}}(0,0,0,0)=\psi_0[P_3(\phi_0,\phi_1;f)]=0$$

$$H_{\hat{x}\hat{x}\hat{x}\hat{x}}(0,0,0,0)=\psi_0[P_4(\phi_0,\phi_1,\phi_2;f)]\neq 0,$$

$$H_{\hat{\lambda}}(0,0,0,0)=\psi_0[f^0_\lambda],$$

$$H_{\hat{\mu}}(0,0,0,0)=\psi_0[f^0_\mu],$$

$$H_{\hat{\nu}}(0,0,0,0)=\psi_0[f^0_\nu],$$

$$H_{\hat{\lambda}\hat{x}}(0,0,0,0)=\psi_0[F^\lambda_1],$$

$$H_{\hat{\mu}\hat{x}}(0,0,0,0)=\psi_0[F^\mu_1],$$

$$H_{\hat{\nu}\hat{x}}(0,0,0,0)=\psi_0[F^\nu_1],$$

$$H_{\hat{\lambda}\hat{x}\hat{x}}(0,0,0,0)=\psi_0[F^\lambda_2],$$

$$H_{\hat{\mu}\hat{x}\hat{x}}(0,0,0,0)=\psi_0[F^\mu_2],\quad H_{\hat{\nu}\hat{x}\hat{x}}(0,0,0,0)=\psi_0[F^\nu_2]$$

where $\phi_1,\phi_2\in V_0$ are given in (3.17),(3.21)and $F^\lambda_1, F^\mu_1, F^\nu_1, F^\lambda_2, F^\mu_2, F^\nu_2$ are defined in (3.19), (3.23) valued at the folds of degree 4 $a_0=(\lambda_0,\mu_0,\nu_0,x_0)$. The catastrophe theory of Tom ([1],[4]) shows that under the condition (3.24) there exists smooth functions

$$\bar{\lambda}=\eta_1(\hat{\lambda},\hat{\mu},\hat{\nu}),\qquad \bar{\mu}=\eta_2(\hat{\lambda},\hat{\mu},\hat{\nu}),$$

$$\bar{\nu}=\eta_3(\hat{\lambda},\hat{\mu},\hat{\nu}),\qquad \bar{x}=\xi(\hat{\lambda},\hat{\mu},\hat{\nu},\hat{x})$$

such that H is contact equivalent to the normal form of the swallowtail catastrophe

$$\bar{H}(\bar{\lambda},\bar{\mu},\bar{\nu},\bar{x})=\bar{x}^4+\bar{\nu}\bar{x}^2+\bar{\mu}\bar{x}+\bar{\lambda}=0.$$

Thus we have

Proposition 3.2 Under the condition (3.24) the fold of degree 4 corresponds to the swallowtail catastrophe.

Because of the regularity of (3.1) we can use the Newton's method to solve the extended system. By using an algorithm which is similar with the algorithm for the folds of degree 3 in [6] we can reduce one step in the Newton's iteration for the extended system to a problem of solving 16 linear system of the original dimensions with the same matrix. The amount of work is decreased greatly.

4 Numerical procedure

The axial dispersion steady-state problem in a tubular nonadiabatic reaction with the first order exothermic can be characterized by two differential equations ([2])

$$\frac{1}{Pey}\frac{d^2y}{dz^2}-\frac{dy}{dz}+Da(1-y)\exp(\theta/(1+\theta/\gamma))=0 \qquad (4.1)$$

$$\frac{1}{Pe\theta}\frac{d^2\theta}{dz^2}-\frac{d\theta}{dz}+B\,Da(1-y)\exp(\theta/(1+\theta/\gamma))-\beta(\theta-\theta_c)=0$$

with the boundary conditions

$$z=0: \quad Pey^{*}y = \frac{dy}{dz}, \qquad Pe\theta^{*}\theta = \frac{d\theta}{dz}$$

$$z=1: \quad \frac{dy}{dz} = \frac{d\theta}{dz} = 0 . \tag{4.2}$$

Here y is dimensionless conversion, $y\varepsilon\ [0,1)$, θ is the dimensionless temperature, Da is the Damköhler number, γ is dimensionless activation energy, Pey is the Péclet number for axial mass transport, $Pe\theta$ is the Péclet number for axial heat transport, B is the dimensionless parameter of heat evolution, and θ_c is the dimensionless cooling temperature. All parameters are positive, θ_c can also be negative.

In order to discretize (4.1)(4.2) we use the central differences on the mesh points $z_j=jh(j=0,\dots,m)$, where $mh=1$. We take $m=40$. Fixing Pey = 10, $Pe\theta$ =5, B = 15, γ= 20 we have a nonlinear problem with 3 parameters Da,β ,θ_c

$$f(\ Da,\beta\ ,\theta_c,\ x) = 0 \tag{4.3}$$

where $x=(y_0,\theta_0,y_1,\theta_1,\dots,\ y_m,\theta_m)$, f has 2(m+1) components.

At first we take $\beta=2,\theta_c=0$ and vary Da. Starting from the trivial solution of (4.3) at Da=0, we can get the solution of (4.3) at Da=0.1 by continuation method as follows

$y(0)$	$y(\frac{1}{2})$	$y(1)$	$\theta(0)$	$\theta(\frac{1}{2})$	$\theta(1)$
0.015	0.110	0.222	0.272	0.955	1.453

(4.4)

Using the algorithm in [3] we can obtain the fold of degree 2 and its family along with β (take (4.4) as the initial guess for the fold of degree 2 in the case Da=0.1 and vary β)

β	Da	$y(0)$	$y(\frac{1}{2})$	$y(1)$	$\theta(0)$	$\theta(\frac{1}{2})$	$\theta(1)$
2.0	0.10573	0.017	1.40	0.327	0.321	1.274	2.271
2.5	0.12373	0.020	0.170	0.386	0.351	1.358	2.326
3.0	0.14375	0.025	0.209	0.460	0.392	1.493	2.428
3.5	0.16713	0.032	0.285	0.597	0.471	1.844	2.731

Taking the fold of degree 2 at β= 3.5 as the initial guess in our algorithm for the fold of degree 3(see [6] for detail) we get the fold of degree 3 and its family along with θ_c.

θ_c	β	Da	y(0)	$y(\frac{1}{2})$	y(1)	$\theta(0)$	$\theta(\frac{1}{2})$	$\theta(1)$
-0.1	3.59639	0.18408	0.037	0.352	0.688	0.520	2.156	2.831
0.0	3.54926	0.16982	0.034	0.316	0.655	0.501	2.045	2.927
3.0	2.81394	0.02366	0.011	0.168	0.599	1.076	3.713	6.544

The fold of degree 3 at $\theta_c = -0.1$ can be chosen as the initial guess for the fold of degree 4. Using the algorithm here we have computed the fold of degree 4 as follows

θ_c	β	Da	y(0)	$y(\frac{1}{2})$	y(1)	$\theta(0)$	$\theta(\frac{1}{2})$	$\theta(1)$
-0.10147	3.59766	0.18434	0.038	0.358	0.694	0.525	2.189	2.845

which corresponds to the swallowtail catastrophe of the discretized axial dispersion problem in a tubular non-adiabatic reaction.

References

[1] S.Chow, J.Hale and J.Mallet-Paret. Application of generic bifurcation I, Archs.Rat.Mech.Analysis 59(1975), pp159-188.

[2] M.Kubicek and M.Marek, Computational methods in bifurcation theory and dissipative structure, Springer Series in Compu.Physics, 1983.

[3] G.Moore and A.Spence, The calculation of turning points of nonlinear equations, SIAM J.Num.Analysis 17 (1980) pp.567-576.

[4] T.Poston and I.Stewart, Catastrophe theory and its application, Pitman Publishing Limited, 1978.

[5] A.Spence and B.Werner, Non-simple turning points and cusps, IMA J. of Numer. Anal.2(1982), pp.413-427.

[6] Z.H.Yang and H.B.Keller, A Direct Method for Computing Higher Order Folds, SIAM J. on Scientific and Statistical Computing Vol.7, No.2(1986), pp351-361.

CONVERGENCE STUDY FOR VISCOUS SPLITTING IN BOUNDED DOMAINS

Ying Lung-an
Department of Mathematics
Peking University
Beijing, China

The purpose of this paper is the analysis of an approximate methed for solving the following initial boundary value problem of the Navier--Stokes equation

$$\frac{\partial u}{\partial t} + (u\cdot\nabla)u + \frac{1}{\rho}\nabla p = \nu\Delta u + f \ , \tag{0.1}$$

$$\nabla\cdot u = 0 \ , \tag{0.2}$$

$$u\Big|_{x\in\partial\Omega} = 0 \ , \tag{0.3}$$

$$u\Big|_{t=0} = u_0(x) \ , \tag{0.4}$$

where Ω is a bounded domain of $\mathbb{R}^2$ or $\mathbb{R}^3$ with smooth boundaries, u is the velocity, p is the pressure, f is the body force, positive constants ρ, ν are the density and viscosity respectively, ∇ is the gradient operator, $\Delta = \nabla^2$, and the initial datum u_0 satisfies $\nabla\cdot u_0 = 0$ and $u_0\Big|_{x\in\partial\Omega} = 0$. The approximate solution is obtained by induction, and the equation (0.1) is splitted into an Euler equation and a nonstationary Stokes equation in each time step.

To calculate flow with high Reynold's number, Chorin applied this approach in his work [4], satisfactory results were obtained. Afterwards this method was extensively studied (see [5], [9], [12], [13]). Because vortex blobs was used for simulating the actual vortex field, this method was often named "the vortex method".

The mathematical foundation of this method was also the subject of some works. Beale and Majda considered the initial value problem and estimated the rate of convergence[3]. A product formula was given for the initial boundary value problem in [6], with no convergence proof. It was proved in [2] that if the solutions of the Euler equation were replaced by polynomials, then that scheme was convergent. The author considered two and three dimensional initial boundary value problems in [14]-[17], and indicated that Chorin's scheme would cause divergence, but if it was modified by adding a nonhomogeneous term, adequate convergence theorems could be proved.

In section one of the current paper we give a mathematical formulation of the vortex method and study it. In section two we consider the modified method and give a shorter proof of the convergence theorem for simply connected domains in $\mathbb{R}^2$. In section three we state the theorems proved in [14]-[17].

§1. A mathematical formulation of the vortex method

Suppose $\Omega \subset \mathbb{R}^2$ and Ω is simply connected. Let vorticity $\omega = -\nabla\wedge u$, where $\nabla\wedge = (\frac{\partial}{\partial x_2}, -\frac{\partial}{\partial x_1})$. We introduce stream function ψ, such that $u = (\nabla\wedge\psi)^T$, then (0.2) is satisfied automatically. ψ and ω satisfy

$$-\Delta\psi = \omega , \qquad \psi\Big|_{x \in \partial\Omega} = \frac{\partial\psi}{\partial n}\Big|_{x \in \partial\Omega} = 0 .$$

Applying $-\nabla\wedge$ to equation (0.1), we obtain

$$\frac{\partial\omega}{\partial t} + u\cdot\nabla\omega = \nu\Delta\omega + F , \tag{1.1}$$

where $F = -\nabla\wedge f$. The time interval $[0,T]$ is divided into equal subintervals with length k. For each interval $[ik, (i+1)k)$, $i=0, 1, \cdots$, the first step of Chorin's scheme is to solve the following initial boundary value problem of the Euler equation

$$\frac{\partial\tilde{\omega}_k}{\partial t} + \tilde{u}_k\cdot\nabla\tilde{\omega}_k = F , \tag{1.2}$$

$$-\Delta\tilde{\psi}_k = \tilde{\omega}_k , \tag{1.3}$$

$$\tilde{u}_k = (\nabla\wedge\tilde{\psi}_k)^T , \tag{1.4}$$

$$\tilde{\psi}_k \Big|_{x \in \partial\Omega} = 0 \ , \tag{1.5}$$

$$\tilde{\omega}_k(ik) = \omega_k(ik-0), \tag{1.6}$$

where $\omega_k(ik-0)$ is the solution of the last step for $i > 0$ and equal to $-\nabla\Lambda u_0$ for $i = 0$. Problem (1.2)-(1.6) is solved by means of the method of characteristics and $\tilde{\omega}_k$ is replaced by a sum of vortex blobs.

The second step is to create vortex blobs at the boundary to maintain the no-slip condition at the surface. Let the created vortex field be ω', then we may define an operator P as

$$P: \ \tilde{\omega}_k((i+1)k-0) \to \tilde{\omega}_k((i+1)k-0) + \omega' .$$

Let V be a subspace of $L^2(\Omega)$, such that $\theta \in V$ iff there is a $\phi \in H^2(\Omega)$ satisfying

$$-\Delta\phi = \theta \ , \tag{1.7}$$

$$\phi \Big|_{x \in \partial\Omega} = \frac{\partial\phi}{\partial n} \Big|_{x \in \partial\Omega} = 0 \ . \tag{1.8}$$

It is easy to see that $\tilde{\omega}_k((i+1)k-0) + \omega' \in V$, and $P^2 = P$, thus P is a projection from $L^2(\Omega)$ to V. The mathematical formulation of the second step is to find $P\tilde{\omega}_k((i+1)k-0)$.

The third step is to solve the following initial boundary value problem of the Stokes equation

$$\frac{\partial\omega_k}{\partial t} = \nu\Delta\omega_k \ , \tag{1.9}$$

$$-\Delta\psi_k = \omega_k \ , \tag{1.10}$$

$$\psi_k \Big|_{x \in \partial\Omega} = \frac{\partial\psi_k}{\partial n} \Big|_{x \in \partial\Omega} = 0 \ , \tag{1.11}$$

$$\omega_k(ik) = P\tilde{\omega}_k((i+1)k-0) \ . \tag{1.12}$$

It is solved by means of a random walk procedure.

However, scheme (1.2)-(1.6)(1.9)-(1.12) is divergent even for linear problems. We consider a very simple case. If equation (0.1) is replaced by a linear equation

$$\frac{\partial u}{\partial t} + \frac{1}{\rho} \nabla p = \nu \Delta u + f \ , \tag{1.13}$$

then (1.1) becomes

$$\frac{\partial \omega}{\partial t} = \nu \Delta \omega + F \tag{1.14}$$

and (1.2) becomes

$$\frac{\partial \tilde{\omega}_k}{\partial t} = F \ . \tag{1.15}$$

We prove that under some reasonable conditions the solution of (1.15), (1.3)-(1.6),(1.9)-(1.12) does not converge to the solution of equation (1.14), instead, it converges to the solution of equation

$$\frac{\partial \omega}{\partial t} = \nu \Delta \omega + PF \ . \tag{1.16}$$

For an arbitrary $u' \in (H^1(\Omega))^2$, let $\omega' = -\nabla \wedge u'$, set $\theta = P\omega'$, then $\theta \in V$. Determine function ϕ according to (1.7),(1.8), set $v = (\nabla \wedge \phi)^T$ and denote $v = \Theta u'$. Θ is a projection from $(H^1(\Omega))^2$ to $(H_0^1(\Omega))^2 \cap X$, where $X =$ closure in $(L^2(\Omega))^2$ of $\{u \in (C_0^\infty(\Omega))^2;\ \nabla \cdot u = 0\}$.[7] The equation associated with (1.16) is

$$\frac{\partial u}{\partial t} + \frac{1}{\rho} \nabla p = \nu \Delta u + \Theta f \ . \tag{1.17}$$

<u>Theorem 1.</u> If functions u_0 , f are sufficiently smooth, so is the solution u of problem (1.17)(0.2)-(0.4), and if

$$\| P\omega' \|_s \leq C \| \omega' \|_s \ , \qquad \forall \omega' \in H^s(\Omega),\ 0 \leq s < \frac{3}{2} \ , \tag{1.18}$$

where C is a constant depending on s, then

$$\| u(t) - u_k(t) \|_{s+1} \leq C_0 k \ , \qquad 0 \leq s < \frac{3}{2},\ 0 \leq t \leq T \ , \tag{1.19}$$

where u_k corresponds to the solution ω_k of problem (1.15),(1.3)-(1.6), (1.9)-(1.12), and C_0 is a constant.

<u>Proof</u> By C we always refer to a generic constant. By (1.18)

$$\| \Theta u' \|_{s+1} \leq C \| u' \|_{s+1} \ , \qquad 0 \leq s < \frac{3}{2} \ . \tag{1.20}$$

Let A be the Stokes operator[7]. The solution u of problem (1.17) (0.2)-(0.4) can be expressed in terms of

$$u(t) = e^{-\nu tA}u_0 + \int_0^t e^{-\nu(t-\tau)A}\,\Theta f(\tau)d\tau .$$

And the expression for solution u_k is

$$u_k(t) = e^{-\nu tA}u_0 + \sum_{i=0}^{[\frac{t}{k}]} e^{-\nu(t-ik)A}\int_{ik}^{(i+1)k}\Theta f(\tau)d\tau ,$$

where [ ] denotes the integral part of a number. We obtain

$$u(t)-u_k(t) = \sum_{i=0}^{[\frac{t}{k}]-1}\int_{ik}^{(i+1)k}(e^{-\nu(t-\tau)A}-e^{-\nu(t-ik)A})\Theta f(\tau)d\tau$$

$$+ \int_{[\frac{t}{k}]k}^{t}(e^{-\nu(t-\tau)A}-e^{-\nu(t-[\frac{t}{k}]k)A})\Theta f(\tau)d\tau$$

$$- \int_{t}^{([\frac{t}{k}]+1)k} e^{-\nu(t-[\frac{t}{k}]k)A}\,\Theta f(\tau)d\tau .$$

The first term is

$$I_1 = \sum_{i=0}^{[\frac{t}{k}]-1}\nu A\int_{ik}^{(i+1)k} e^{-\nu(t-\tau)A}\int_0^{\tau-ik} e^{-\nu\zeta A}d\zeta\,\Theta f(\tau)d\tau .$$

We consider the fractional powers of the Stokes operator with domain[7]

$$D(A^\alpha) = X \cap [(L^2(\Omega))^2,\ D(-\Delta)]_\alpha ,\quad 0 \leq \alpha \leq 1,$$

where $[\cdot,\cdot]_\alpha$ is the interpolation space[1]. It can be shown that $D(A^{\frac{s+1}{2}}) = D(A) \cap (H^{s+1}(\Omega))^2$ for $1 \leq s < \frac{3}{2}$, and

$$C^{-1}\|A^\alpha u'\|_0 \leq \|u'\|_{2\alpha} \leq C\|A^\alpha u'\|_0 ,\quad \forall u' \in D(A^\alpha) .$$

We take a constant s_1, such that $s < s_1 < \frac{3}{2}$, since $\Theta f(\tau) \in D(A)$,

$$\|I_1\|_{s+1} \leq C\sum_{i=0}^{[\frac{t}{k}]-1}\int_{ik}^{(i+1)k}\|A^{1+\frac{s-s_1}{2}} e^{-\nu(t-\tau)A}\int_0^{\tau-ik}e^{-\nu\zeta A}d\zeta\,A^{\frac{s_1+1}{2}}\Theta f(\tau)\|_0\,d\tau.$$

We use the following property of semigroup e^{-tA}:

$$\| A^{\alpha} e^{-tA} \| \leq Ct^{-\alpha} \quad , \quad \alpha \geq 0, \; t > 0 \; ,$$

and obtain

$$\| I_1 \|_{s+1} \leq C \sum_{i=0}^{[\frac{t}{k}]-1} \int_{ik}^{(i+1)k} (t-\tau)^{-1+\frac{s_1-s}{2}} \int_0^{\tau-ik} \| A^{\frac{s_1+1}{2}} \Theta f(\tau) \|_0 \, d\zeta d\tau$$

$$\leq C \sup_{0\leq\tau\leq T} \| \Theta f(\tau) \|_{s_1+1} \, k \int_0^t (t-\tau)^{-1+\frac{s_1-s}{2}} d\tau$$

$$\leq Ck \sup_{0\leq\tau\leq T} \| f(\tau) \|_{s_1+1} \quad .$$

The other terms can be estimated in a similar way. QED.

§2. A modified scheme

The modified viscous splitting scheme for solving (0.1)-(0.4) is

$$\frac{\partial \tilde{u}_k}{\partial t} + (\tilde{u}_k \cdot \nabla) \tilde{u}_k + \frac{1}{\rho} \nabla \tilde{p}_k = f \; , \tag{2.1}$$

$$\nabla \cdot \tilde{u}_k = 0 \; , \tag{2.2}$$

$$\tilde{u}_k \cdot n \Big|_{x \in \partial\Omega} = 0 \; , \tag{2.3}$$

$$\tilde{u}_k(ik) = u_k(ik-0) \; , \tag{2.4}$$

$$\frac{\partial u_k}{\partial t} + \frac{1}{\rho} \nabla p_k = \nu \Delta u_k + \frac{1}{k}(I-\Theta)\tilde{u}_k((i+1)k-0) \; , \tag{2.5}$$

$$\nabla \cdot u_k = 0 \; , \tag{2.6}$$

$$u_k \Big|_{x \in \partial\Omega} = 0 \; , \tag{2.7}$$

$$u_k(ik) = \Theta \tilde{u}_k((i+1)k-0) \; , \tag{2.8}$$

where $ik \leq t < (i+1)k$, $i=0,1,\cdots$, $u_k(-0) = u_0$. The only modification is to add a nonhomogeneous term in (2.5).

<u>Theorem 2</u> Let u be the solution of problem (0.1)-(0.4), $\tilde{u}_k$, u_k be the solution of problem (2.1)-(2.8). If u_0 is sufficiently smooth on $\overline{\Omega}$ and so are u, f on $\overline{\Omega}\times[0,T]$, $0 \leq s < \frac{3}{2}$, then

$$\max(\| u_k(t) \|_{s+1}, \| \tilde{u}_k(t) \|_{s+1}) \leq M, \quad 0 \leq t \leq T, \tag{2.9}$$

$$\max(\| u(t)-u_k(t) \|_0, \| u(t)-\tilde{u}_k(t) \|_0) \leq M' k, \quad 0 \leq t \leq T, \tag{2.10}$$

where the constants M, M' depend only on the domain Ω, constants ν, s, T, operator Θ, functions u_0, f and u.

We premise the proof of Theorem 2 with the following lemmas.

<u>Lemma 1</u> If $u_0 \in X \cap (H^3(\Omega))^2$, $f \in L^\infty(0,T;(H^3(\Omega))^2)$, $\| u_0 \|_3 \leq M_1$, $\sup_{0\leq t\leq T} \| f(t) \|_3 \leq M_2$, $-1 \leq s \leq 2$, u is the solution of problem

$$\frac{\partial u}{\partial t} + (u\cdot\nabla)u + \frac{1}{\rho}\nabla p = f,$$

$$\nabla\cdot u = 0,$$

$$u\cdot n\Big|_{x\in\partial\Omega} = 0,$$

$$u\Big|_{t=0} = u_0(x),$$

then there exists a constant $C_1 > 0$, such that if

$$k_0 = \frac{1}{C_1(M_1 + M_2 + 1)} \tag{2.11}$$

and $0 \leq t \leq k_0$, then u satisfies

$$\| u(t) \|_{s+1} \leq C_2(\| u_0 \|_{s+1} + 1), \tag{2.12}$$

where the constant C_2 depends only on the domain Ω and constants s, T, M_2.

<u>Proof</u> Using [11] we have

$$\| u(t) \|_3 \leq y(t),$$

where $y(t)$ is the solution of the following initial value problem of ordinary differential equation

$$y' = Cy^2 + C\|f(t)\|_3 ,$$

$$y(0) = \|u_0\|_3 .$$

We restrict $|y| \leq 3(M_1 + M_2)$, then clearly it will suffice that $t \leq k_0$ for a certain constant C_1 in (2.11).

We consider an auxiliary linear problem

$$\frac{\partial\tilde{u}}{\partial t} + (u\cdot\nabla)\tilde{u} + \nabla\tilde{\pi} = f ,$$

$$\nabla\cdot\tilde{u} = 0 ,$$

$$\tilde{u}\cdot n\Big|_{x\in\partial\Omega} = 0 ,$$

$$\tilde{u}\Big|_{t=0} = u_0(x) .$$

By uniqueness $\tilde{u}\equiv u$. In a similar way to [11], we can get

$$\|\nabla\tilde{\pi}(t)\|_3 \leq C(\|f(t)\|_3 + \|u(t)\|_3 \|\tilde{u}(t)\|_3),$$

then we can prove

$$\frac{1}{2}\frac{d}{dt}\|\tilde{u}(t)\|_3^2 \leq C(\|f(t)\|_3 + \|u(t)\|_3 \|\tilde{u}(t)\|_3)\|\tilde{u}(t)\|_3 .$$

But $\|u(t)\|_3 \leq 3(M_1 + M_2)$, by Gronwall inequality

$$\|\tilde{u}(t)\|_3 \leq e^{C(M_1+M_2)t}(\|u_0\|_3 + Ck_0 \sup_{0\leq\tau\leq t}\|f(\tau)\|_3) .$$

By (2.11)

$$\|\tilde{u}(t)\|_3 \leq C(\|u_0\|_3 + \sup_{0\leq\tau\leq t}\|f(\tau)\|_3) .$$

On the other hand

$$(\frac{\partial\tilde{u}}{\partial t}, \tilde{u}) = (f,\tilde{u}) ,$$

hence

$$\|\tilde{u}(t)\|_0 \leq \|u_0\|_0 + \int_0^t \|f(\tau)\|_0 d\tau .$$

Since mapping $(u_0, f) \to \tilde{u}$ is linear, (2.12) is obtained by interpolation[10]. QED.

Now we estimate the solution of (2.1)-(2.8).

<u>Lemma 2</u> If $1 < s < \frac{3}{2}$, then there is a constant C_3 which depends only on the domain Ω, constants s, ν, T and $\sup_{0 \leq t \leq T} \|f(t)\|_1$, $\|u_0\|_{s+1}$, such that

$$\|u_k(jk-0)\|_{s+1} \leq C_3 + C \sup_{0 \leq \tau < jk} \|\tilde{u}_k(\tau)\|_1^{2-\frac{1}{s}} \|\tilde{u}_k(\tau)\|_{s+1}^{\frac{1}{s}} ,$$

$$j=0,1,\cdots . \tag{2.13}$$

<u>Proof</u> The solution u_k can be expressed in terms of

$$u_k(t) = e^{-\nu tA} u_0 + \sum_{i=0}^{[\frac{t}{k}]} e^{-\nu(t-ik)A} \int_{ik}^{(i+1)k} \Theta f_1(\tau) d\tau$$

$$+ \sum_{i=0}^{[\frac{t}{k}]-1} \int_{ik}^{(i+1)k} e^{-\nu(t-\tau)A} \frac{1}{k} \int_{ik}^{(i+1)k} P'(I-\Theta) f_1(\zeta) d\zeta d\tau$$

$$+ \int_{[\frac{t}{k}]k}^{t} e^{-\nu(t-\tau)A} \frac{1}{k} \int_{[\frac{t}{k}]k}^{([\frac{t}{k}]+1)k} P'(I-\Theta) f_1(\zeta) d\zeta d\tau , \tag{2.14}$$

where P' is the orthogonal projection from $(L^2(\Omega))^2$ to X, $f_1(\tau) = f(\tau) - (\tilde{u}_k(\tau) \cdot \nabla) \tilde{u}_k(\tau)$. Using the properties of the Stokes operator A and semigroup $e^{-\nu tA}$, we get an estimate

$$\|u_k(jk-0)\|_{s+1} \leq C(\|u_0\|_{s+1} + \sup_{0 \leq \tau < jk} \|f_1(\tau)\|_1) ,$$

then (2.13) is obtained by using the interpolation inequality[10]. QED.

<u>Lemma 3</u> If $1 < s < \frac{3}{2}$, and there is a constant M_0, such that

$\| \tilde{u}_k(t) \|_1 \leq M_0$ for $0 \leq t \leq T$, and there are constants C_2, k_0, such that

$$\| \tilde{u}_k(t) \|_{s+1} \leq C_2 (\| \tilde{u}_k(ik) \|_{s+1} + 1) \tag{2.15}$$

for $ik \leq t < (i+1)k$, $i=0,1,\cdots$, and $0 < k \leq k_0$, then

$$\sup_{0 \leq t \leq T} \| \tilde{u}_k(t) \|_{s+1} \leq M_3 , \tag{2.16}$$

where the constant M_3 depends only on the domain Ω, constants s, ν, T, C_2, M_0, and $\sup_{0 \leq t \leq T} \| f(t) \|_1$, $\| u_0 \|_{s+1}$.

Proof By Lemma 2, initial condition (2.4) and (2.15),

$$\sup_{0 \leq t < ik} \| \tilde{u}_k(t) \|_{s+1} \leq C_2 (C_3 + C M_0^{2-\frac{1}{s}} \sup_{0 \leq t < ik} \| \tilde{u}_k(t) \|_{s+1}^{\frac{1}{s}} + 1) ,$$

which yields (2.16). QED.

Lemma 4 If $\| \tilde{u}_k(t) \|_{s+1} \leq M_3$ for $ik \leq t < (i+1)k$, $1 \leq s < \frac{3}{2}$, then

$$\frac{1}{k} \| (I-\Theta) \tilde{u}_k((i+1)k-0) \|_1 \leq C_4 , \tag{2.17}$$

where the constant C_4 depends only on the domain Ω, operator Θ, constants M_3, s, and function f.

Proof We denote by C_4 a generic constant which satisfies the above condition. Because $u_k(ik-0) \in D(A)$, $(I-\Theta) u_k(ik-0)=0$. And $I-\Theta$ is a bounded operator, hence

$$\begin{aligned} \|(I-\Theta) \tilde{u}_k((i+1)k-0) \|_1 &= \|(I-\Theta) (\tilde{u}_k((i+1)k-0) - u_k(ik-0)) \|_1 \\ &\leq C \| \tilde{u}_k((i+1)k-0) - u_k(ik-0) \|_1 . \end{aligned}$$

By initial condition (2.4),

$$\|(I-\Theta) \tilde{u}_k((i+1)k-0) \|_1 \leq C \| \tilde{u}_k((i+1)k-0) - \tilde{u}_k(ik) \|_1 . \tag{2.18}$$

Let $\tilde{\omega}_k = -\nabla \wedge \tilde{u}_k$, then by (2.1) $\tilde{\omega}_k$ is the solution of

$$\frac{\partial \tilde{\omega}_k}{\partial t} + \tilde{u}_k \cdot \nabla \tilde{\omega}_k = F .$$

By integrating we obtain

$$\|\tilde{\omega}_k((i+1)k-0)-\tilde{\omega}_k(ik)\|_0 \leq \int_{ik}^{(i+1)k} (\|\tilde{u}_k(\tau)\|_{0,\infty}\|\tilde{\omega}_k(\tau)\|_1+\|F(\tau)\|_0)d\tau$$

$$\leq C_4 k . \tag{2.19}$$

Since $\tilde{u}_k = (\nabla\wedge\tilde{\psi}_k)^T$ and $\tilde{\psi}_k$ is the solution of

$$-\Delta\tilde{\psi}_k = \tilde{\omega}_k \ , \quad \tilde{\psi}_k\big|_{x\in\partial\Omega} = 0 \ ,$$

we have

$$\|\tilde{u}_k((i+1)k-0)-\tilde{u}_k(ik)\|_1 \leq C\|\tilde{\omega}_k((i+1)k-0)-\tilde{\omega}_k(ik)\|_0 \ .$$

Together with (2.18)(2.19) it yields (2.17). QED.

Lemma 5 If $1 \leq s < \frac{3}{2}$, $k \leq 1$, $\|\tilde{u}_k(t)\|_{s+1} \leq M_3$ for $ik \leq t < (i+1)k$, then

$$\|u_k(t)\|_3 \leq C_5(t-ik)^{\frac{s}{2}-1} \tag{2.20}$$

on the same interval, where the constant C_5 depends only on the domain Ω, operator Θ, constants M_3, ν, s and function f.

Proof Let $w = \frac{\partial u_k}{\partial t}$, $\pi = \frac{\partial p_k}{\partial t}$, differentiating equations (2.5)-(2.7) formally with respect to t, we obtain

$$\frac{\partial w}{\partial t} + \frac{1}{\rho}\nabla\pi = \nu\Delta w \ ,$$

$$\nabla\cdot w = 0 \ ,$$

$$w\big|_{x\in\partial\Omega} = 0 \ ,$$

$$w(ik) = \frac{\partial u_k}{\partial t}\Big|_{t=ik} = -\nu A\Theta\tilde{u}_k((i+1)k-0) + \frac{1}{k}P'(I-\Theta)\tilde{u}_k((i+1)k-0) \ .$$

By the results in [8], $\frac{\partial u_k}{\partial t}$ is the weak solution of it. But the above problem possesses a strong solution

$$w(t) = e^{-\nu(t-ik)A}w(ik) \ ,$$

therefore

$$\frac{\partial u_k}{\partial t} = e^{-\nu(t-ik)A} w(ik) \ .$$

By Lemma 4 we get estimate

$$\left\| \frac{\partial u_k}{\partial t} \right\|_1 \leq C(t-ik)^{\frac{s}{2}-1} (M_3 + C_4) \ .$$

Equation (2.5) can be rewritten as

$$-\nu\Delta u_k + \frac{1}{\rho} \nabla p_k = \frac{1}{k}(I-\Theta)\tilde{u}_k((i+1)k-0) - \frac{\partial u_k}{\partial t} \ .$$

Using the estimate of the solution of the Stokes problem, we get (2.20). QED.

Lemma 6 If $1 < s < \frac{3}{2}$, $k \leq 1$, $\| \tilde{u}_k(t) \|_{s+1} \leq M_4$ for $0 \leq t \leq T$, then

$$\max(\| u(t)-u_k(t) \|_0 \ , \ \| u(t)-\tilde{u}_k(t) \|_0) \leq C_6 k \ , \quad 0 \leq t \leq T,$$

where the constant C_6 depends only on the domain Ω, operator Θ, constants T, ν, s, M_4, functions u_0, f, and the solution u of problem (0.1)-(0.4).

Proof We denote by C_6 a generic constant which satisfies the above condition. If we regard the solution u of (0.1)-(0.4) as a known function, and let $f_1 = f-(u\cdot\nabla)u$, then (0.1) becomes

$$\frac{\partial u}{\partial t} + \frac{1}{\rho}\nabla p = \nu\Delta u + f_1 \ . \tag{2.21}$$

We apply scheme (2.1)-(2.8) to problem (2.21),(0.2)-(0.4), the solution of which is denoted by $\tilde{u}^*$, $\tilde{p}^*$, u^*, p^*, then

$$\max(\| u(t)-u^*(t) \|_{s+1} \ , \ \| u(t)-\tilde{u}^*(t) \|_{s+1}) \leq C_6 k, \ 0 \leq t \leq T \ . \tag{2.22}$$

The proof of (2.22) is analogous to that of Theorem 1.

By (2.14),

$$u^*(t)-u_k(t) = \sum_{i=0}^{[\frac{t}{k}]} e^{-\nu(t-ik)A} \int_{ik}^{(i+1)k} \Theta f_2(\tau)\,d\tau$$

$$+\sum_{i=0}^{[\frac{t}{k}]-1}\int_{ik}^{(i+1)k} e^{-\nu(t-\tau)A}\frac{1}{k}\int_{ik}^{(i+1)k} P'(I-\Theta)f_2(\zeta)d\zeta d\tau$$

$$+\int_{[\frac{t}{k}]k}^{t} e^{-\nu(t-\tau)A}\frac{1}{k}\int_{[\frac{t}{k}]k}^{([\frac{t}{k}]+1)k} P'(I-\Theta)f_2(\zeta)d\zeta d\tau,$$

where $f_2 = ((u-\tilde{u}_k)\cdot\nabla)u + (\tilde{u}_k\cdot\nabla)(u-\tilde{u}_k)$. Recomposing the terms, we have

$$u^*(t)-u_k(t) = \sum_{i=0}^{[\frac{t}{k}]-1}\int_{ik}^{(i+1)k} e^{-\nu(t-\tau)A}\frac{1}{k}\int_{ik}^{(i+1)k} P'f_2(\zeta)d\zeta d\tau$$

$$+\sum_{i=0}^{[\frac{t}{k}]-1}\int_{ik}^{(i+1)k}(e^{-\nu(t-ik)A}-e^{-\nu(t-\tau)A})\frac{1}{k}\int_{ik}^{(i+1)k}\Theta f_2(\zeta)d\zeta d\tau$$

$$+e^{-\nu(t-[\frac{t}{k}]k)A}\int_{[\frac{t}{k}]k}^{([\frac{t}{k}]+1)k}\Theta f_2(\tau)d\tau$$

$$+\int_{[\frac{t}{k}]k}^{t} e^{-\nu(t-\tau)A}\frac{1}{k}\int_{[\frac{t}{k}]k}^{([\frac{t}{k}]+1)k} P'(I-\Theta)f_2(\zeta)d\zeta d\tau . \quad (2.23)$$

We estimate the first term

$$I_1 = \sum_{i=0}^{[\frac{t}{k}]-1}\int_{ik}^{(i+1)k} e^{-\nu(t-\tau)A}\frac{1}{k}\int_{ik}^{(i+1)k} P'f_2(\zeta)d\zeta d\tau .$$

We can prove

$$\| e^{-\nu(t-\tau)A} P'f_2(\zeta)\|_0 \le C(t-\tau)^{-1+\frac{1}{q}}(\|u\|_{0,r}\|u-\tilde{u}_k\|_0 + \|\tilde{u}_k\|_{0,r}\|u-\tilde{u}_k\|_0),$$

where $q>0$, $r>0$, $\frac{1}{q}+\frac{1}{r}=\frac{1}{2}$, which is similar to the Lemma 1.1 of [2]. Therefore

$$\| I_1 \|_0 \leq C \sum_{i=0}^{[\frac{t}{k}]-1} \int_{ik}^{(i+1)k} (t-\tau)^{-1+\frac{1}{q}} \frac{1}{k} \int_{ik}^{(i+1)k} (\|u\|_{0,r} + \|\tilde{u}_k\|_{0,r}) \|u-\tilde{u}_k\|_0$$

$$d\zeta d\tau \; . \tag{2.24}$$

By equation (2.1),

$$\frac{\partial(\tilde{u}^* - \tilde{u}_k)}{\partial t} + ((u-\tilde{u}_k)\cdot\nabla)u + (\tilde{u}_k\cdot\nabla)(u-\tilde{u}_k) + \frac{1}{\rho}\nabla(\tilde{p}^* - \tilde{p}_k) = 0 \; .$$

Taking the inner product of the left hand side with $\tilde{u}^* - \tilde{u}_k$, observing $\tilde{u}^* - \tilde{u}_k \in X$, and

$$((\tilde{u}_k\cdot\nabla)(\tilde{u}^* - \tilde{u}_k), \tilde{u}^* - \tilde{u}_k) = 0 \; ,$$

we have

$$\frac{1}{2}\frac{d}{dt}\|\tilde{u}^* - \tilde{u}_k\|_0^2 + (((u-\tilde{u}_k)\cdot\nabla)u, \tilde{u}^* - \tilde{u}_k) + ((\tilde{u}_k\cdot\nabla)(u-\tilde{u}^*), \tilde{u}^* - \tilde{u}_k) = 0 \; .$$

Then using (2.22) we obtain

$$\frac{1}{2}\frac{d}{dt}\|\tilde{u}^* - \tilde{u}_k\|_0^2 \leq C_6(\|\tilde{u}^* - \tilde{u}_k\|_0 + k)\|\tilde{u}^* - \tilde{u}_k\|_0 \; .$$

By Gronwall inequality and initial condition (2.4),

$$\|\tilde{u}^*(t) - \tilde{u}_k(t)\|_0 \leq e^{C_6(t-ik)} (\|u^*(ik-0) - u_k(ik-0)\|_0 + C_6 k^2) \; . \tag{2.25}$$

Substituting (2.25),(2.22) into (2.24), we get

$$\| I_1 \|_0 \leq C_6 \sum_{i=0}^{[\frac{t}{k}]-1} \int_{ik}^{(i+1)k} (t-\tau)^{-1+\frac{1}{q}} (\|u^*(ik-0) - u_k(ik-0)\|_0 + k)d\tau \; .$$

Set $\psi(t) = \sup_{0\leq\tau<t} \|u^*(\tau) - u_k(\tau)\|_0$, then

$$\| I_1 \|_0 \leq C_6 \int_0^t (t-\tau)^{-1+\frac{1}{q}} (\psi(\tau)+k)d\tau \; .$$

The L^2 norm of other terms of (2.23) are bounded by $C_6 k$, thus

$$\|u^*(t) - u_k(t)\|_0 \leq C_6 \int_0^t (t-\tau)^{-1+\frac{1}{q}} \psi(\tau)d\tau + C_6 k \; .$$

Taking the supremum with respect to t, we obtain

$$\psi(t) \leq C_6 \int_0^t (t-\tau)^{-1+\frac{1}{q}} \psi(\tau)d\tau + C_6 k .$$

The corresponding Volterra integral equation is

$$y(t) = C_6 \int_0^t (t-\tau)^{-1+\frac{1}{q}} y(\tau)d\tau + C_6 k .$$

It can be checked that $\psi(t) \leq y(t)$ and $y(t) \leq C_6 k$.

Using (2.22) again, we get the estimate for $u-u_k$. Inequalities (2.22), (2.25) give the estimate for $u-\tilde{u}_k$. QED.

<u>Lemma 7</u> If $1 < s < \frac{3}{2}$, $k \leq 1$, $\| \tilde{u}_k(t) \|_{s+1} \leq M_4$ for $ik \leq t < (i+1)k$, then $\| u_k(t) \|_{s+1} \leq M_5$ on the same interval, where the constant M_5 depends only on the domain Ω, constants ν, s, M_4, operator Θ, and function f.

<u>Proof</u> The solutions u_k of problem (2.5)-(2.8) can be expressed in terms of

$$u_k(t) = e^{-\nu(t-ik)A} \Theta \tilde{u}_k((i+1)k-0)$$
$$+ \int_{ik}^t e^{-\nu(t-\tau)A} P' \frac{1}{k}(I-\Theta)\tilde{u}_k((i+1)k-0)d\tau .$$

We have estimate

$$\| u_k(t) \|_{s+1} \leq C(\|\Theta\tilde{u}_k((i+1)k-0)\|_{s+1} + \|\frac{1}{k}(I-\Theta)\tilde{u}_k((i+1)k-0)\|_1)$$

Then Lemma 4 yields the conclusion. QED.

<u>Proof of Theorem 2</u> We assume that $1 < s < \frac{3}{2}$. Let $M_0 = 2 \max_{0 \leq t \leq T} \| u(t) \|_1$. We determine constant C_2 in Lemma 1, then determine constant M_3 according to Lemma 3, and constant C_5 according to Lemma 5. By (2.11), we take $k_0 \in (0,1]$, which satisfies

$$k_0 \leq \frac{1}{C_1(C_5 k_0^{\frac{s}{2}-1} + M_2 + 1)} , \tag{2.26}$$

that is

$$C_1(C_5 k_0^{\frac{s}{2}} + M_2 k_0 + k_0) \leq 1 ,$$

which always holds if k_0 is small enough. By Lemma 2, set

$$M_6 = C_3 + CM_0^{2-\frac{1}{s}} M_3^{\frac{1}{s}} . \tag{2.27}$$

By Lemma 1, set

$$M_4 = \max(C_2(M_6 + 1), M_3). \tag{2.28}$$

We determine constant C_6 according to Lemma 6, constant M_5 according to Lemma 7, and reduce k_0 , if necessary, such that

$$\| u_0 \|_3 \leq C_5 k_0^{\frac{s}{2}-1} , \tag{2.29}$$

$$C(C_6 k_0)^{\frac{s}{s+1}} \left(\max_{0 \leq t \leq T} \| u(t) \|_{s+1} + \max(M_4, M_5)\right)^{\frac{1}{s+1}} \leq \frac{M_0}{2} , \tag{2.30}$$

where C is a fixed constant to be determined in inequality (2.32) below.

With the determined constants, we prove by induction that if $0<k\leq k_0$, then

$$\| \tilde{u}_k(t) \|_1 \leq M_0 , \quad \| u_k(t) \|_1 \leq M_0 ,$$

$$\| \tilde{u}_k(t) \|_{s+1} \leq M_3 ,$$

$$\| u(t)-u_k(t) \|_0 \leq C_6 k , \quad \| u(t)-\tilde{u}_k(t) \|_0 \leq C_6 k .$$

Two cases are considered simultaneously: (i) $j=0$, (ii) $j>0$ and the above assertion is valid for $0 \leq t < jk$. If $j > 0$, then by Lemma 2 and (2.27)

$$\| u_k(jk-0) \|_{s+1} \leq M_6 \tag{2.31}$$

((2.31) also holds for $j=0$). If $j>0$, then by Lemma 5,

$$\| u_k(jk-0) \|_3 \leq C_5 k^{\frac{s}{2}-1} .$$

By (2.29) it also holds for $j=0$. By Lemma 1 and (2.26)(2.28)

$$\| \tilde{u}_k(t) \|_{s+1} \leq M_4$$

for $jk \leq t < (j+1)k$. By Lemma 6,

$$\| u(t)-u_k(t) \|_0 \ , \ \| u(t)-\tilde{u}_k(t) \|_0 \leq C_6 k$$

always holds for $0 \leq t < (j+1)k$. By Lemma 7,

$$\| u_0(t) \|_{s+1} \leq M_5$$

for $jk \leq t < (j+1)k$. By the interpolation inequality,

$$\| u-\tilde{u}_k \|_1 \leq C \| u-\tilde{u}_k \|_0^{\frac{s}{s+1}} \| u-\tilde{u}_k \|_{s+1}^{\frac{1}{s+1}} \tag{2.32}$$

for $jk \leq t < (j+1)k$, similarly

$$\| u-u_k \|_1 \leq C \| u-u_k \|_0^{\frac{s}{s+1}} \| u-u_k \|_{s+1}^{\frac{1}{s+1}} .$$

By (2.30),

$$\| u-\tilde{u}_k \|_1 \ , \ \| u-u_k \|_1 \leq \frac{M_0}{2} .$$

Hence

$$\| \tilde{u}_k \|_1 \ , \ \| u_k \|_1 \leq M_0$$

for $jk \leq t < (j+1)k$. By Lemma 1 and Lemma 3, $\| \tilde{u}_k(t) \|_{s+1} \leq M_3$ for $jk \leq t < (j+1)k$. The induction is complete.

Using Lemma 7, we can establish the estimate for $\| u_k(t) \|_{s+1}$.

If $k > k_0$, then there are at most $\frac{T}{k_0}$ steps and (2.9), (2.10) can be easy verified. QED.

§3. Remarks

The following conclusions were proved in [14]-[17].

1. Theorem 2 is also true for three dimensional domain Ω, if Ω is homeomorphic to a ball, and there is a smooth solution u on $\bar{\Omega} \times [0,T]$, and $0 < k \leqq k_0$ for a small k_0. The difference is that here only the existence of local classical solutions has been proved, so T depends on the data u_0, f, and k can not be "large".

2. If operator P is an orthogonal projection, then the inequality (2.10) can be improved. The following inequality is true:

$$\max(\| u(t)-u_k(t) \|_1 , \quad \| u(t)-\tilde{u}_k(t) \|_1) \leqq M'k , \quad 0 \leqq t \leqq T . \tag{3.1}$$

(2.9),(3.1) are also true for multi-connected domain, the boundary $\partial\Omega$ of which consists of N+1 simple closed curves $\Gamma_0, \Gamma_1, \cdots, \Gamma_N$, $N \geqq 0$, where $\Gamma_j (j=1,\cdots, N)$ are inside of Γ_0 and outside of one another.

REFERENCES

[1] Adams, R.A., Sobolev Spaces, Academic Press, New York (1975).
[2] Alessandrini, G., Douglis, A. and Fabes, E., An approximate layering method for the Navier-Stokes equations in bounded cylinders, Annali di Matematica, 135, (1983), 329-347.
[3] Beale, J.T. and Majda, A., Rate of convergence for viscous splitting of the Navier-Stokes equations, Math. Comp., 31, (1981), 243-259.
[4] Chorin, A.J., Numerical study of slightly viscous flow, J. Fluid Mech., 59, (1973), 785-796.
[5] Chorin, A.J., Vortex sheet approximation of boundary layers, J. Comput. Phys., 27, (1978), 428-442.
[6] Chorin, A.J., Hughes, T.J.R., McCracken, M.F. and Marsden, J.E., Product formulas and numerical algorithms, Comm. Pure Appl. Math., 31, (1978), 205-256.
[7] Fujita, H. and Morimoto, H., On fractional powers of the Stokes operator, Proc. Japan Acad., 46, (1970), 1141-1143.
[8] Ladyzhenskaya, O.A., The Mathematical Theory of Viscous Incompressible Flow, Gordon and Breach, New York (1969).
[9] Leonard, A., Vortex methods for flow simulation, J. Comput. Phys., 37, (1980), 289-335.
[10] Lions, J.L. and Magenes, E., Nonhomogeneous Boundary Value Problems and Applications, Springer-Verlag, Berlin (1972).
[11] Temam, R., On the Euler equations of incompressible perfect fluids, J. Functional Analysis, 20, (1975), 32-43.
[12] Teng, Z-H., Elliptic-vortex method for incompressible flow at high Reynolds number, J. Comput. Phys., 46, (1982), 54-68.
[13] Teng, Z.-H., Variable-elliptic-vortex method for incompressible flow simulation, J. Comput. Math., 4(1986), 255-262.
[14] Ying, L.-a., The viscosity splitting method in bounded domains, (to appear).
[15] Ying, L.-a., The viscosity splitting method for the Navier-Stokes

equations in bounded domains, (to appear).
[16] Ying, L.-a., On the viscosity splitting method of initial boundary value problems of the Navier-Stokes equations, (to appear).
[17] Ying, L.-a., Viscosity splitting method for three dimensional Navier-Stokes equations, (to appear).

AN EXPLICIT SCHEME FOR AN INVERSE SCATTERING PROBLEM AND ITS STABILITY ANALYSIS

Zhu Ben-ren and Jin Mao-yuan
Mathematics Department, Shandong University
Jinan , Shandong, P.R.China

Abstract

In this paper an explicit scheme for solving numerically an inverse scattering problem is proposed. An estimate of the difference solution has been obtained by using discrete Gelfand-Levitan equations and two derived equations, and two criteria for stability have been found. Several numerical experiments give strong support to our investigation.

1. INTRODUCTION

In applied sciences, such as theoretical physics, geophysics , quantum mechanics, optics, acoustics, oceanography etc, a class of inverse boundary value problems of wave equations are encountered very often. Through proper Liouville transformation those inverse problems always are reduced to the following inverse BVP, the so-called inverse scattering problem :

$$\begin{aligned} &\frac{\partial^2 u}{\partial t^2} = \frac{\partial^2 u}{\partial x^2} - q(x)u \quad , \quad t,x>0 \\ &u(x,0) = \delta(x), \qquad \frac{\partial}{\partial t}u(x,0) = 0, \qquad\qquad (1) \\ &\left(\frac{\partial}{\partial x}u + \alpha u\right)(0,t) = 0, \quad u(0,t) = \tilde{f}(t). \end{aligned}$$

Here 'stimulative' function $\delta(x)$ is the Dirac function, and $\tilde{f}(t)$ is so-called response to $\delta(x)$, that is recorded data. The question is to find α and $q(x)$ from given response $\tilde{f}(t)$.

In this work an explicit difference scheme for problem (1) is proposed and stability analysis is carried out together with some estimates of the solutions and criteria of stability. Because of nonlinearity of the problem, our investigation of stability would be entirely different from those of classical initial-boundary value problems. Here the discrete Gelfand-Levitan theory is needed.

2. AN EXPLICIT SCHEME FOR INVERSE SCATTERING PROBLEM

In this section we construct the difference scheme of problem (1). First the square mesh and mesh function are defined as follows:

$$x_\ell = (\ell-\tfrac{1}{2})h, \quad t_k = (k-\tfrac{1}{2})h, \quad u_{\ell k} = u(x_\ell, t_k)$$
$$\ell = 0,1,2,\dots, \quad k = 0,\pm1,\pm2,\dots .$$

Corresponding to (1) we propose the following scheme:

$$\begin{aligned} &u_{\ell k+1} + u_{\ell k-1} = u_{\ell+1 k} + u_{\ell-1 k} - q_\ell h^2 u_{\ell k}, \quad \ell=1,2,\dots,\ k=0,\pm1,\pm2,\dots, \\ &u_{11} = u_{10} = 1/h,\ u_{01} = u_{00} = (1+\alpha h)/h, \quad u_{\ell 0} = u_{\ell 1} = 0, \text{ for } \ell > 1, \\ &u_{0k} = (1+\alpha h)u_{1k} . \end{aligned} \tag{2}$$

It is easy to verify the following relations:

$$\begin{aligned} &u_{\ell k} = 0, \text{ for } 0 \leq k < \ell, \quad u_{11} = u_{22} = \dots = u_{kk} = 1/h, \\ &q_\ell = -(u_{\ell \ell+1} - u_{\ell-1 \ell})/h, \ \ell = 2,3,\dots, \quad q_1 = -(u_{02} - \alpha)/h + \alpha u_{02}/(1+\alpha h), \\ &u_{kk+1} = -\sum_{\ell=1}^{k-1} q_\ell h + \alpha = -\sum_{\ell=1}^{k-1} q_\ell h + \tilde{u}_{01}, \quad \tilde{u}_{01} = \alpha . \end{aligned} \tag{3}$$

The discrete inverse problem is to find $\alpha, q_\ell, \ell=1,2,\dots$ from given $\tilde{u}_{01}$ and $u_{0,2\ell}, \ell=1,2,\dots$. One can solve this problem in the following way:

(0) take $\tilde{u}_{01}$ as the extrapolation value of u_{0k}, then $\alpha = \tilde{u}_{01}$;

(1) compute u_{12} and q_1 from given u_{02} and boundary condition in (2): $u_{12} = u_{02}/(1+\alpha h)$, $q_1 = -(u_{12} - \tilde{u}_{01})/h$;

(2) for $\ell \geq 1$ compute $u_{\ell \ell+1}, u_{\ell-1 \ell+3}, \dots, u_{1 2\ell+1}$ and $u_{0 2\ell+1}$;

(3) from $u_{0 2\ell+2}$, compute $u_{1 2\ell+2}, u_{2 2\ell+1}, \dots, u_{\ell+1 \ell+2}$ and $q_{\ell+1} = -(u_{\ell+1 \ell+2} - u_{\ell \ell+1})/h$.

The whole procedure can be seen in figure 1.. We close this section by presenting a useful relation

$$q_\ell + \sum_{j=1}^{\ell-1} q_j (u_{j 2\ell-j} + u_{j 2\ell-j-1})h + (u_{0 2\ell} - u_{0 2\ell-2}) \times h^{-1} - \alpha(1+\alpha h)^{-1}(u_{0 2\ell-2} - u_{0 2\ell-1}) = 0, \ \ell = 2,3,\dots \tag{4}$$

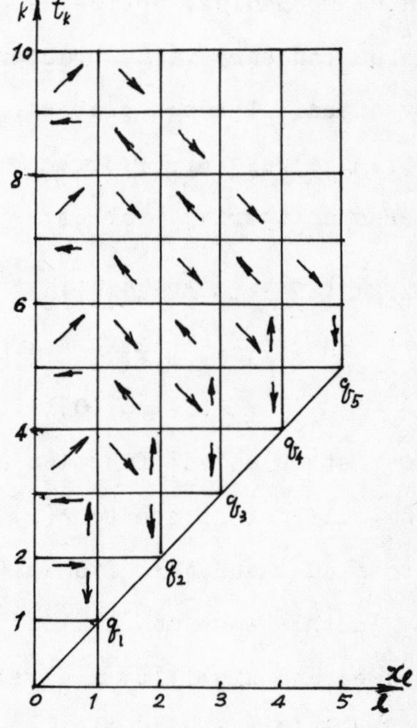

Figure 1.

3. DISCRETE GELFAND-LEVITAN THEORY

In this section we point out that the problem (2) is closely related to the discrete Gelfand-Levitan theory. First let us consider the discrete Sturm-Liouville problem

$$\begin{aligned} &-\phi_{\ell-1}+(2+q_\ell h^2)\phi_\ell-\phi_{\ell+1}=\lambda h^2\phi_\ell, \quad -1,2,\dots \\ &\phi_0=(1+\alpha h)\phi_1, \quad \phi_1=1 \end{aligned} \tag{5}$$

For any real λ the solution $\phi_\ell(\lambda)$ is called eigenfunction, and the spectral function $\rho(\lambda)$ is nondecreasing which satisfies (1) $\rho(-\infty)=0$, $\rho(+\infty)=1/h$; (2) $\int_{-\infty}^{+\infty}\phi_k(\lambda)\phi_\ell(\lambda)d\rho(\lambda)=\delta_{k\ell}/h$, here $\delta_{k\ell}=0$, for $k\neq\ell$ and $\delta_{kk}=1$. Specially when $\alpha=q_\ell\equiv0$ the corresponding eigenfunction and spectral function are denoted by $\phi_\ell^\circ(\lambda)$ and $\rho^\circ(\lambda)$.

We note that the equation in (2) is separable, and the eigenfunctions of left and right sides are $\phi_k^\circ(\lambda)$, $\phi_\ell(\lambda)$, so any linear combination of $\phi_k^\circ(\lambda)\phi_\ell(\lambda)$ satisfies the equation and boundary condition . Namely we are interested in two of them:

$$u_{\ell k}=\int_{-\infty}^{+\infty}\phi_k^\circ(\lambda)\phi_\ell(\lambda)d\rho(\lambda), \quad v_{\ell k}=\int_{-\infty}^{+\infty}\phi_k^\circ(\lambda)\phi_\ell(\lambda)d\rho^\circ(\lambda).$$

It is easy matter to verify that $u_{\ell k}$ is the only solution of problem (2), and $v_{\ell k}$ is the only solution of the following problem:

$$\begin{aligned} &v_{\ell\,k+1}+v_{\ell\,k-1}=v_{\ell+1\,k}+v_{\ell-1\,k}-q_\ell h^2 v_{\ell k}, \quad \begin{array}{l}\ell=0,1,2,\dots,\\ k=0,\pm1,\pm2,\dots,\end{array} \\ &v_{01}=v_{00}=(1+\alpha h)/h, \ v_{\ell 0}=v_{\ell 1}, \text{ for } \ell=0,1,2,\dots \\ &v_{11}=v_{10}=1/h, \ v_{0k}=v_{1k}=0, \text{ for } k>1. \end{aligned} \tag{6}$$

Also it is easy to see that

$$u_{\ell k}=0=v_{k\ell}, \text{for } k<\ell, \quad u_{kk}=v_{kk}=1/h, \quad k=1,2,\dots,$$

and there are relationships between $\{\phi_k^\circ\}$ and $\{\phi_k\}$:

$$\phi_k^\circ=\phi_k+\sum_{\ell=1}^{k-1}u_{\ell k}\phi_\ell h, \quad \phi_k=\phi_k^\circ+\sum_{\ell=1}^{k-1}v_{k\ell}\phi_\ell^\circ h \tag{7}$$

We denote $\sigma(\lambda)=\rho(\lambda)-\rho^\circ(\lambda)$ and define function $f_{\ell k}$ as follows

$$f_{\ell k}=f_{k\ell}=\int_{-\infty}^{+\infty}\phi_k^\circ(\lambda)\phi_\ell^\circ(\lambda)d\sigma(\lambda).$$

Using (7) and orthogonality of $\phi_k(\phi_k^\circ)$ with respect to spectral function $\rho(\rho^\circ)$, we obtain the following three equations:

$$f_{\ell k}=u_{\ell k}+\sum_{j=1}^{\ell-1}u_{j\ell}u_{jk}h, \quad 1\leq\ell<k \tag{8-1}$$

the Gelfand-Levitan nonlinear 'integral' equation,

$$f_{k\ell} + v_{k\ell} + \sum_{j=1}^{k-1} f_{\ell j} v_{kj} h = 0 , \quad 1 \leq \ell < k \tag{8-2}$$

the G-L linear 'integral' equation and an equation relating $u_{\ell k}$ and $v_{\ell k}$ linearly

$$u_{\ell k} = f_{\ell k} + \sum_{j=1}^{\ell-1} v_{\ell j} f_{jk} h , \quad 1 \leqslant \ell < k. \tag{8-3}$$

Besides, function $f_{\ell k}$ satisfies the following relations:

(1) $f_{k\ell} + f_{k\ell+1} = f_{1k+\ell} + f_{1k-\ell}$, $1 \leq \ell < k$;

(2) $f_{2rk} = \frac{1}{2}(f_{1k+2r-1} + f_{1k-2r+1}) + \frac{1}{2}\sum_{\ell=-r+1}^{r-1} \Delta^2 f_{1k+2\ell}$ $k > 2r$, $f_{2r+1k} = \frac{1}{2}(f_{1k+2r} + f_{1k-2r}) + \frac{1}{2}\sum_{\ell=-r+1}^{r-1} \Delta^2 f_{1k+2\ell+1}$, $k > 2r+1$

where $\Delta^2 f_{\ell k} = f_{\ell k-1} - 2f_{k\ell} - f_{\ell k+1}$ and $f_{1k} = u_{1k} = (1+\alpha h)^{-1} u_{ok}$, e.i. $f_{\ell k}$ can be represented in terms of $\{u_{ok}\}$, the boundary values of $u_{\ell k}$.

Finally we introduce the vector-matrix notations:

$v^k = (v_{k1}, v_{k2}, \ldots, v_{kk-1})^T$, $\quad u^k = (u_{1k}, u_{2k}, \ldots, u_{k-1k})^T$,

$f^k = (f_{1k}, f_{2k}, \ldots, f_{k-1k})^T$, $\quad f_k^\ell = (f_{1k}, f_{2k}, \ldots, f_{\ell-1k})^T$,

$F_k = \{f_{ij}\}_1^K$ ----- a real symmetric matrix.

Thus the equation (8-2) can be rewritten in matrix form:

$$f^k + (I + F_{k-1} h) v^k = 0. \tag{8-2'}$$

Denote the least eigenvalue of matrix $(I + F_k h)$ by m_k:

$$m_k = \min_{x \neq 0} \, (x, (I + F_k h)x)/(x, x) .$$

Obviously $m_1 = 1 \geqslant m_2 \geqslant m_3 \geqslant \ldots \geqslant m_k \geqslant \ldots$ and $m_k > 0$, for all k.

<u>Definition</u> If there exist a constant $\eta > 0$ independent of k,h such that $m_k \geqslant \eta > 0$, then G-L equation (8-2) is well-conditioned.

If G-L equation (8-2) is well-conditioned, then we have

$$\|v^k\| \leqslant m_{k-1}^{-1} \|f^k\| \leq \eta^{-1} \|f^k\| , \quad k = 1, 2, \ldots \tag{9}$$

Condition (9) is the key for our stability analysis.

4. THE ESTIMATES OF THE SOLUTIONS AND THE CRITERIA FOR STABILITY

By means of (4), (8-2), (8-3) and (9), it is not hard to get the estimate of the difference solution of inverse scattering problem (2):

$$\begin{aligned} \max_{1 \leq \ell \leq k} |q_\ell| \leqslant (h^{-1}|u_{02} - \tilde{u}_{01}| + \tilde{u}_{01}^2)(1+\alpha h)^{-1} + 2(\max_{2 \leq \ell < k} |g_\ell|) \times \\ \times \exp\{2x_k(\|f^{2k}\|_0 + \eta^{-1} \|f^{2k}\| \, |||f^{2k}|||_0) , \end{aligned} \tag{10}$$

where $g_\ell = (1+\alpha h)^{-1}\{(u_{02\ell} - u_{02\ell-2})(2h)^{-1} - \frac{1}{2}\alpha(u_{02\ell-1} - u_{02\ell-2})\}$

$\|f^k\|_0 = \max_{1 \leq \ell < k} |f_{\ell k}|$, $\|f^k\| = (\sum_{\ell=1}^{k-1} f_{\ell k}^2 h)^{1/2}$, $|||f^k|||_0 = \max_{1 \leq \ell < k} \|f^\ell\|$.

Similarly the difference quotients of q_ℓ of high order can be estimated (the results are omitted). Under the condition of boundedness of various norms of $f_{\ell k}$ the estimate (10) shows that the boundedness of the solution might be violated if $m_k \to 0$. One can see this fact from Example 3 in Section 5.

The condition (9) also guarantees continuous dependence of q_ℓ on $u_{0k}(\tilde{f}(t_k))$, i.e., stability of the algorithm.

Now let us turn to the criteria under which (9) is guaranteed. In practice $\tilde{f}(t)$ is recorded data, that involve some error. In order that the algorithm is stable, the principles of handling these data so that (9) holds are mostly important. The following criteria are useful.

CRITERION 1. If $f_{ij} = \sum_{\ell=0}^{L} \hat{f}_\ell \cos((j-\frac{1}{2})\omega_\ell)/\cos\frac{1}{2}\omega_\ell$, and for any $k \neq \ell$, $\omega_k \pm \omega_\ell \neq 2r\pi$, r—integer, then

⟨1⟩ if all $\hat{f}_\ell \geqslant 0$, then (9) holds, and $\eta=1$;

⟨2⟩ if there is at least one $\hat{f}_\ell < 0$, then (9) is violated for x_k large enough.

CRITERION 2. If $f_{ij} = \frac{2}{\pi}\int_0^{\Omega_h} \hat{f}(\omega)\cos(\omega x_j)\,d\omega$, $\Omega_h = \pi/h$, then

⟨1⟩ if $\hat{f}(\omega) \geqslant 0$, then (9) holds, and $\eta=1$;

⟨2⟩ if $\min_\omega \hat{f}(\omega) < -4$, $\hat{f}(\omega)$ is continuous in the neiborhood of the minimum, then (9) is violated for x_k large enough.

REMARK Whether (9) is valid for $-4 \leqslant \min \hat{f}(\omega) < 0$, is still an open question. But for continuous case for G-L equation (8-2) being well-conditioned the necessary and sufficient condition is

$$\min_\omega \hat{f}(\omega) > -2/\pi .$$

5. NUMERICAL RESULTS

EXAMPLE 1. For $\alpha=0$ and $q(x)=\sin(16\pi x)$, we choose $h=1/960$ and solve the direct scattering problem to get $\tilde{f}(t)(u^h_{0\ell}, \ell=1,2,\ldots,960)$. Then using coarse mesh $H=4h$ and $u^H_{0\ell} = u^h_{04\ell}$, $\ell=1,2,\ldots,240$, we solve the inverse problem to get $\tilde{q}(kH)$, $k=1,2,\ldots,120$. The curves depicted in figure 2. are $\tilde{q}(kH)$ and $q(kH)$, $k=31\sim46$ for comparison.

EXAMPLE 2. For $\alpha=0$ and

$$q_\ell = \begin{cases} 0, & \text{for } \ell \in (0,80) \text{ or } (200,300) \text{ or } (420,960), \\ \text{random number}, & \text{for } \ell \in (81,200) \text{ or } (301,419), \end{cases}$$

we choose h=1/960 and solve the direct problem to get $\tilde{f}(t)$, then using coarse mesh H=4h and $u^H_{0\ell} = u^h_{04\ell}$, $\ell = 1,2,\ldots,240$, solve the inverse problem to get $\tilde{q}(kH)$, k=1,2,...,120. The curves depicted in figures 3.a and 3.b are $\tilde{q}(kH)$ and $q(kH)$.

EXAMPLE 3. For h=1/240, $\tilde{f}(t_k) = \gamma(\cos((k-\frac{1}{2})h\pi)-1)$, γ=1,2,4, according to Criterion 1. the inverse problem is unstable. Numerical results give apparent evidences. The larger γ, the faster computation becomes overflow:

$$\tilde{q}_{110} = 1.16 \times 10^{17}, \text{ for } \gamma = 1;$$

$$\tilde{q}_{85} = 6.79 \times 10^{13}, \text{ for } \gamma = 2;$$

$$\tilde{q}_{67} = -6.69 \times 10^{14}, \text{ for } \gamma = 4.$$

The curves depicted in figure 4. are $\tilde{q}_\ell$'s.

REFERENCES

(1) I.M.Gelfand & B.M.Levitan
On the determination of differential equation from its spectral function
Izv. Akad. Nauk SSSR Ser. Mat. 15 (1951) p309-

(2) W.Symes
Numerical stability in an inverse scattering problem
SIAM J. Numer. Anal. vol.17 No 5 (1980) p707-

(3) K.P.Bube & R.Burridge
The one-dimensional inverse problem of reflection seismology
SIAM Rev. vol 25 No 4 (1983) p497-

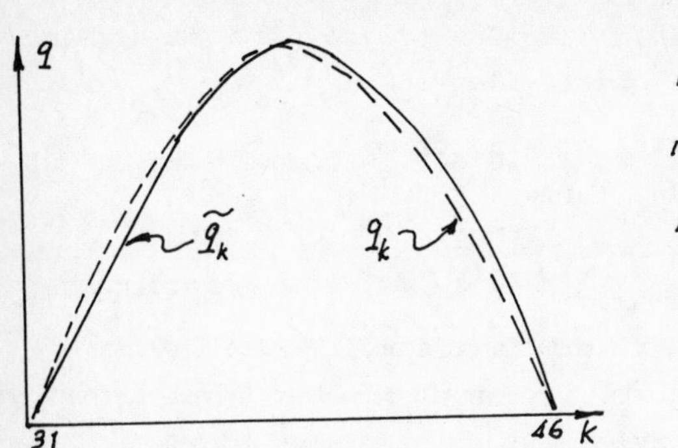

Figure 2.

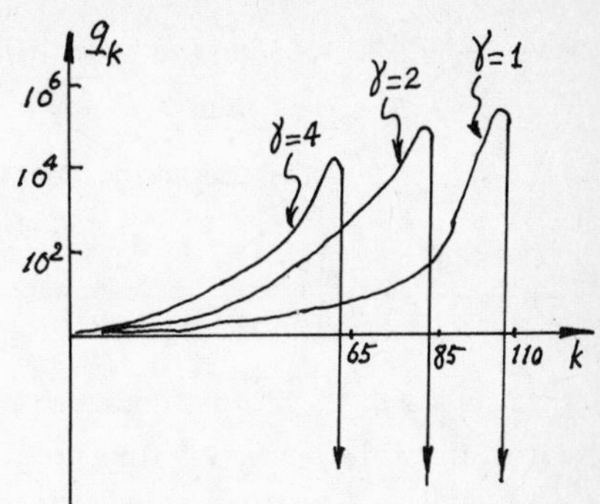

Figure 4.

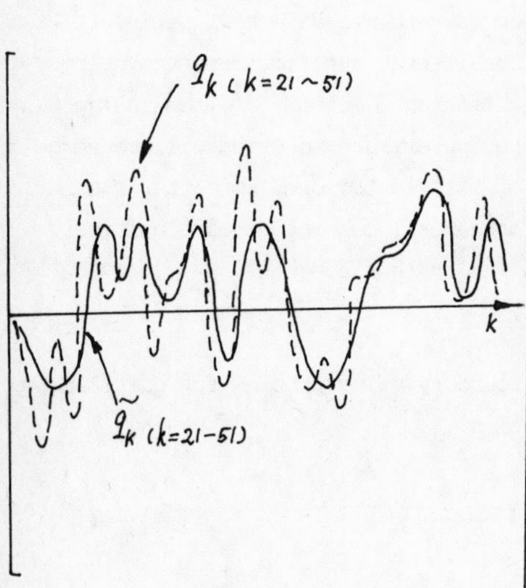

Figure 3.a.

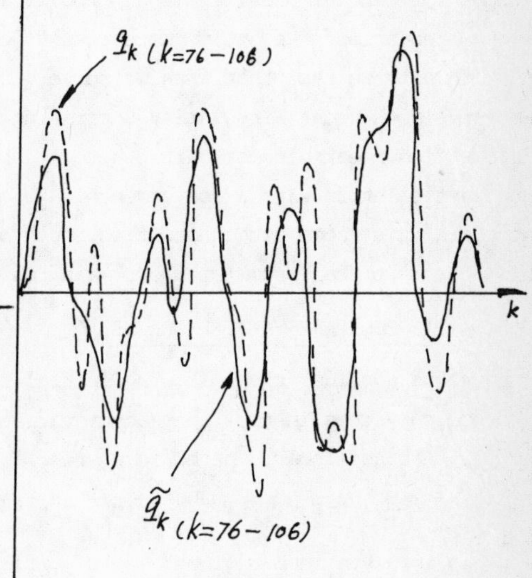

Figure 3.b.

ON STABILITY AND CONVERGENCE OF DIFFERENCE SCHEMES
FOR QUASILINEAR HYPERBOLIC INITIAL-BOUNDARY-VALUE PROBLEMS

Zhu You-lan*,** and Yong Wen-an**

The Computing Center, Academia Sinica
Beijing, China

Abstract

In the seventies[2,3] we obtained some results on stability of difference schemes for initial-boundary-value problems of linear diagonalized hyperbolic systems in two independent variables. Later[4] these results were extended to general linear hyperbolic systems with "moving boundaries" and some convergence theorems were established. In [1], we completed some proofs of 'global' convergence of difference schemes for general quasilinear hyperbolic initial-boundary-value problems with moving boundaries. Recently, more results on convergence have been derived. From these results we know that when we solve a quasilinear hyperbolic system using certain second order Singularity-Separating difference methods[3] (separating discontinuities, weak discontinuities etc.), the approximate solution will converge to the exact solution with a convergence rate of Δt^2 in L_2 norm, no matter whether or not there exist some discontinuities, such as shocks, contact discontinuities. In this paper we shall summarize our main results on this subject.

1. INITIAL-BOUNDARY-VALUE PROBLEMS

Let us consider the following initial-boundary-value problem for quasilinear hyperbolic systems in two independent variables.

1. A quasilinear hyperbolic system

$$\frac{\partial \overline{U}}{\partial t} + \overline{A}(\overline{U}, x, t) \frac{\partial \overline{U}}{\partial x} = \overline{F}(\overline{U}, x, t) \tag{1.1}$$

is given in L regions: $x_{\ell-1}(t) \leq x \leq x_\ell(t),\ 0 \leq t \leq T,$ $\ell = 1,2,\cdots,L.$

2. On external boundaries and internal boundaries $x = x_\ell(t)$, $\ell = 0,1,\cdots,L$, a number of nonlinear boundary conditions are prescribed:

$$\begin{cases} B_0(\overline{U}_0^+, x_0, z_0, t) = 0, \\ B_\ell(\overline{U}_\ell^-, \overline{U}_\ell^+, x_\ell, z_\ell, t) = 0, \quad \ell = 1,2,\cdots, L-1, \\ B_L(\overline{U}_L^-, x_L, z_L, t) = 0, \end{cases} \tag{1.2}$$

* This work was supported in part by the Institute for Mathematics and its Applications with funds provided by the National Science Foundation of U.S.A.

** This work was supported in part by the National Science Foundation of China.

where

$$z_\ell = \frac{dx_\ell(t)}{dt}, \qquad \ell = 0,1,\cdots,L. \tag{1.3}$$

3. At t = 0 initial values are specified:

$$\begin{cases} \bar{U}(x,0) = \bar{D}_\ell(x), \text{ for } x_{\ell-1}(0) \leq x \leq x_\ell(0), & \ell = 1,2,\cdots,L, \\ x_\ell(0) = c_{0,\ell},\ z_\ell(0) = c_{1,\ell}, & \ell = 0,1,\cdots,L. \end{cases} \tag{1.4}$$

It is required to determine $\bar{U}$ in the L regions and $x_\ell(t)$, $z_\ell(t)$ for $0 \leq t \leq T$, $\ell = 0,1,\cdots,L$. Here $\bar{U},\bar{F},\bar{D}_\ell$ are N_1 - dimensional vectors, $\bar{A}$— an $N_1 \times N_1$ matrix, B_ℓ— ν_ℓ - dimensional vectors and $\bar{U}_\ell^{\pm} \equiv \bar{U}_\ell^{\pm}(t) = \lim_{x \to x_\ell(t) \pm 0} \bar{U}(x,t)$. Without loss of generality, we assume that the initial values (1.4) satisfy the boundary conditions (1.2). In fact, if initial values are discontinuous at some point, we should solve a Riemann problem to determine how many and what types of internal boundaries need to be introduced; if initial values do not satisfy some external boundary conditions, a similar procedure should be done. Therefore the initial values always match the boundary conditions. The hyperbolic type means that $\bar{A}$ has the following expression

$$\bar{A} = \bar{G}^{-1}\ \bar{\Lambda}\ \bar{G}, \tag{1.5}$$

where

$$\bar{G} = \begin{bmatrix} \bar{G}_1^* \\ \bar{G}_2^* \\ \vdots \\ \bar{G}_{N_1}^* \end{bmatrix} \quad \text{and} \quad \bar{\Lambda} = \begin{bmatrix} \bar{\lambda}_1 & & & 0 \\ & \bar{\lambda}_2 & & \\ & & \ddots & \\ 0 & & & \bar{\lambda}_{N_1} \end{bmatrix}$$

are a nonsingular real matrix and a real diagonal matrix respectively, $\bar{G}_n^*$ being the transpose of a column vector $\bar{G}_n$. In the following we assume $\bar{\lambda}_1 \geq \bar{\lambda}_2 \geq \cdots \geq \bar{\lambda}_{N_1}$ and define $p_\ell \equiv$ the number of $\bar{\lambda}_{n,\ell}^{+}$ which are less than or equal to $\frac{dx_\ell}{dt}$, and $q_\ell \equiv$ the number of $\bar{\lambda}_{n,\ell}^{-}$ which are greater than or equal to $\frac{dx_\ell}{dt}$, where $\bar{\lambda}_{n,\ell}^{\pm} = \lim_{x = x_\ell(t) \pm 0} \bar{\lambda}_n(x,t)$. We require the boundary conditions (1.2) are compatible with the equation (1.1). That is, we suppose that ν_ℓ satisfy

$$\begin{cases} \nu_0 + p_0 = N_1 + 1, \\ \nu_\ell + p_\ell + q_\ell = 2N_1 + 1, \quad \ell = 1,2,\cdots,L-1, \\ \nu_L + q_L = N_1 + 1, \end{cases} \tag{1.6}$$

and furthermore, the systems

$$\begin{cases} B_0(\bar{U}_0^+, x_0, z_0, t) = 0, \\ \bar{G}_{(0)}^+\ \bar{U}_0^+ = F_0^+, \end{cases}$$

$$\begin{cases} \bar{G}^{-}_{(\ell)}\ \bar{U}^{-}_{\ell} = F^{-}_{\ell}\ , \\ B_{\ell}(\bar{U}^{-}_{\ell},\ \bar{U}^{+}_{\ell},\ x_{\ell},\ z_{\ell},\ t) = 0, \qquad \ell = 1,2,\cdots, L-1, \\ \bar{G}^{+}_{(\ell)}\ \bar{U}^{+}_{\ell} = F^{+}\ , \end{cases} \tag{1.7}$$

$$\begin{cases} \bar{G}^{--}_{(L)}\ \bar{U}^{-}_{L} = F^{-}_{L}\ , \\ B_{L}(\bar{U}^{-}_{L},\ x_{L},\ z_{L},\ t) = 0 \end{cases}$$

always have "entropy-satisfying" solutions $\bar{U}^{+}_{0}$, z_0; $\bar{U}^{-}_{\ell}$, $\bar{U}^{+}_{\ell}$, z_{ℓ}, $\ell=1,2,\cdots,L-1$; $\bar{U}^{-}_{L}$, z_L, respectively if reasonable x_{ℓ}, F^{+}_{ℓ}, F^{-}_{ℓ} and t are given. Here F^{+}_{ℓ}, F^{-}_{ℓ} are p_{ℓ}-, q_{ℓ} - dimensional vectors and

$$\bar{G}^{+}_{(\ell)} = \left.\begin{bmatrix} \bar{G}^{*}_{N_1-p_{\ell}+1} \\ \vdots \\ \bar{G}^{*}_{N_1} \end{bmatrix}\right|_{x=x_{\ell}(t)+0}, \qquad \bar{G}^{-}_{(\ell)} = \left.\begin{bmatrix} \bar{G}^{*}_{1} \\ \vdots \\ \bar{G}^{*}_{q_{\ell}} \end{bmatrix}\right|_{x=x_{\ell}(t)-0}.$$

In order to make numerical computation easy, we introduce the following coordinate transformation

$$\begin{cases} \xi = \dfrac{x - x_{\ell-1}(t)}{x_{\ell}(t)-x_{\ell-1}(t)} + \ell - 1, \qquad \text{if } x_{\ell-1}(t) \le x \le x_{\ell}(t), \\ \qquad\qquad\qquad\qquad\qquad\qquad \ell = 1,2,\cdots, L, \\ t = t. \end{cases}$$

Clearly, the boundary $x = x_{\ell}(t)$ in the x - t coordinate system corresponds to the straight line $\xi = \ell$ in the ξ - t coordinate system. For $\bar{U}$

$$\frac{\partial}{\partial t} = \frac{\partial}{\partial t} + \frac{\partial \xi}{\partial t}\ \frac{\partial}{\partial \xi}\ ,$$

$$\frac{\partial}{\partial x} = \frac{\partial \xi}{\partial x}\ \frac{\partial}{\partial \xi},$$

where

$$\begin{cases} \dfrac{\partial \xi}{\partial t} = \dfrac{(-z_{\ell-1})(x_\ell - x_{\ell-1}) - (x - x_{\ell-1})(z_\ell - z_{\ell-1})}{(x_\ell - x_{\ell-1})^2}, \\ \dfrac{\partial \xi}{\partial x} = \dfrac{1}{x_\ell - x_{\ell-1}}, \quad \text{for} \quad x_{\ell-1} \le x \le x_\ell, \ \ell = 1, 2, \cdots, L. \end{cases}$$

Therefore (1.1) can be transformed into

$$\frac{\partial \overline{U}}{\partial t} + \overline{A}_1 \frac{\partial \overline{U}}{\partial \xi} = \overline{F}, \tag{1.8}$$

where

$$\overline{A}_1(\overline{U}, X, Z, \xi, t) = \frac{\partial \xi}{\partial t} I_{N_1} + \frac{\partial \xi}{\partial x} \overline{A},$$

I_{N_1} being an $N_1 \times N_1$ unit matrix, $X = (x_0, x_1, \cdots, x_L)^T$, and $Z = (z_0, z_1, \cdots, z_L)^T$.

For convenience to theoretical proof, in what follows, the ordinary differential relations (1.3) are understood as some hyperbolic partial differential equations. Therefore, defining

$$U = \begin{bmatrix} \overline{U} \\ X \end{bmatrix}, \qquad A = \begin{bmatrix} \overline{A}_1 & 0 \\ 0 & 0 \end{bmatrix}, \qquad F = \begin{bmatrix} \overline{F} \\ Z \end{bmatrix},$$

we can combine (1.8) and (1.3) to

$$\frac{\partial U}{\partial t} + A(U, Z, \xi, t) \frac{\partial U}{\partial \xi} = F(U, Z, \xi, t).$$

Thus, (1.1)-(1.4) can be rewritten as:

1. A quasilinear hyperbolic system

$$\frac{\partial U}{\partial t} + A(U, Z, \xi, t) \frac{\partial U}{\partial \xi} = F(U, Z, \xi, t) \tag{1.9}$$

is given in L regions: $\ell - 1 \le \xi \le \ell$, $0 \le t \le T$, $\ell = 1, 2, \cdots, L$.

2. On the straight boundaries $\xi = \ell$, $\ell = 0, 1, \cdots, L$, a number of nonlinear boundary conditions are prescribed:

$$\begin{cases} B_0(U_0^+, z_0, t) = 0, \\ B_\ell(U_\ell^-, U_\ell^+, z_\ell, t) = 0, \quad \ell = 1, 2, \cdots, L-1, \\ B_L(U_L^-, z_L, t) = 0. \end{cases} \tag{1.10}$$

3. At $t = 0$, initial values are specified:

$$\begin{cases} U(\xi, 0) = D_\ell(\xi) \quad \text{for} \quad \ell - 1 \le \xi \le \ell, \quad \ell = 1, 2, \cdots, L, \\ Z(0) = C_1, \end{cases} \tag{1.11}$$

where D_ℓ and C_1 are $(N_1 + L + 1)$- and $(L + 1)$-dimensional vectors respectively. We need to determine U in the L regions and Z for $0 \le t \le T$. Let

$$G = \begin{bmatrix} \overline{G} & 0 \\ 0 & I_{L+1} \end{bmatrix}, \qquad \Lambda = \begin{bmatrix} \frac{\partial \xi}{\partial t} I_{N_1} + \frac{\partial \xi}{\partial x} \overline{\Lambda} & 0 \\ 0 & 0 \end{bmatrix},$$

we have

$$A = G^{-1} \Lambda G.$$

Therefore, (1.9) can be written as

$$G \frac{\partial U}{\partial t} + \Lambda G \frac{\partial U}{\partial \xi} = GF. \tag{1.12}$$

2. DIFFERENCE SCHEMES

The system (1.12) can further be rewritten in the form

$$G_n^* \frac{\partial U}{\partial t} + \lambda_n G_n^* \frac{\partial U}{\partial \xi} = f_n, \quad n = 1,2,\cdots,N, \tag{2.1}$$

where G_n^* is the n-th row of G, λ_n is the n-th diagonal element of Λ, f_n is the n-th component of GF and $N = N_1 + L + 1$. In what follows, we will discuss the discretization of (2.1).

In each subregion $\ell-1 \leq \xi \leq \ell$, $0 \leq t \leq T$, we make a rectangular mesh with mesh sizes $\Delta\xi = 1/M_\ell$ and Δt, M_ℓ being an integer. For convenience, in what follows, we assume $M = M_\ell$ for all ℓ, $f^k_{\ell,m}$ denotes the value of f at $\xi = \ell-1+ m\Delta\xi$, $t = k\Delta t$ and we define $\sigma = \lambda\Delta t/\Delta\xi$. Moreover, we assume $\frac{\Delta t}{\Delta \xi}$ is bounded. In the ℓ-th subregion for each $\lambda^k_{n,\ell}$ we define a set $\underline{g}(\lambda^k_{n,\ell})$ as follows:

$$\underline{g}(\lambda^k_{n,\ell}) = \begin{cases} \{1,2,\cdots,M\}, & \text{if } \lambda^k_{n,\ell,0} > 0 \text{ and } \lambda^k_{n,\ell,M} \geq 0, \\ \{0,1,\cdots,M-1\}, & \text{if } \lambda^k_{n,\ell,0} \leq 0 \text{ and } \lambda^k_{n,\ell,M} < 0, \\ \{0,1,\cdots,M\}, & \text{if } \lambda^k_{n,\ell,0} \leq 0 \text{ and } \lambda^k_{n,\ell,M} \geq 0, \\ \{1,2,\cdots,M-1\}, & \text{if } \lambda^k_{n,\ell,0} > 0 \text{ and } \lambda^k_{n,\ell,M} < 0. \end{cases} \tag{2.2}$$

We suppose that when n, ℓ is fixed, $\underline{g}(\lambda^k_{n,\ell})$ is the same for all k. Therefore, from now on, we use the symbol $\underline{g}(\lambda_{n,\ell})$ instead of $\underline{g}(\lambda^k_{n,\ell})$. For each $\lambda_{n,\ell}$ (2.1) is approximated by a second order accurate system of difference equations of the following form

$$\sum_{h=h_1(m)}^{h_2(m)} R^{k+\frac{1}{2}}_{h,n,\ell,m} U^{k+1}_{\ell,m+h} = \sum_{h=h_1(m)}^{h_2(m)} S^{k+\frac{1}{2}}_{h,n,\ell,m} U^k_{\ell,m+h} + \Delta t f^{k+\frac{1}{2}}_{n,\ell,m}, \tag{2.3}$$

$$m \in \underline{g}(\lambda_{n,\ell}),$$

where for any m hold the relations

$$0 \leq m+h_1(m) \leq m + h_2(m) \leq M, \quad \max_m \{|h_1(m)|, \ |h_2(m)|\} \leq H,$$

H being a positive integer,

and $R^{k+\frac{1}{2}}_{h,n,\ell,m}$, $S^{k+\frac{1}{2}}_{h,n,\ell,m}$ and $f^{k+\frac{1}{2}}_{n,\ell,m}$ depend on $U^{k+\frac{1}{2}}_{\ell,m+j}$, $Z^{k+\frac{1}{2}}$, $U^{k}_{\ell,m+j}$, Z^{k} besides ξ and t, j satisfying $h_1(m) \le j \le h_2(m)$. According to the consistency, among $R^{k+\frac{1}{2}}_{h,n,\ell,m}$, $S^{k+\frac{1}{2}}_{h,n,\ell,m}$ exists the relation

$$\sum_{h=h_1(m)}^{h_2(m)} R^{k+\frac{1}{2}}_{h,n,\ell,m} = \sum_{h=h_1(m)}^{h_2(m)} S^{k+\frac{1}{2}}_{h,n,\ell,m} = 0(1). \tag{2.4}$$

If $\lambda_{n,\ell,m} > 0$ (or < 0) for all m, using the implicit second order scheme for (2.1)

$$\frac{1}{2}\mu G^{*k+\frac{1}{2}}_{m\mp\frac{1}{2}}(U^{k+1}_{m} + U^{k+1}_{m\mp1}) + \frac{1}{2}\mu\sigma^{k+\frac{1}{2}}_{m\mp\frac{1}{2}}\mu G^{*k+\frac{1}{2}}_{m\mp\frac{1}{2}}\Delta_{\mp}U^{k+1}_{m}$$

$$= \frac{1}{2}\mu G^{*k+\frac{1}{2}}_{m\mp\frac{1}{2}}(U^{k}_{m} + U^{k}_{m\mp1}) - \frac{1}{2}\mu\sigma^{k+\frac{1}{2}}_{m\mp\frac{1}{2}}\mu G^{*k+\frac{1}{2}}_{m\mp\frac{1}{2}}\Delta_{\mp}U^{k}_{m} + \Delta t\,\mu f^{k+\frac{1}{2}}_{m\mp\frac{1}{2}} \tag{2.5}$$

we can have a system of difference equations in the form (2.3). In (2.5) the subscripts n and ℓ are omitted, and we use the notation

$$\Delta_{+}U_m = U_{m+1} - U_m,\quad \Delta_{-}U_m = U_m - U_{m-1},$$

$$\mu f_{m\mp\frac{1}{2}} = \frac{1}{2}(f_m + f_{m\mp1}).$$

And the minus sign of $\mp$ should be adopted if $\lambda > 0$ and the plus sign if $\lambda < 0$.

If $\lambda_{n,\ell,0} > 0$ and $\lambda_{n,\ell,M} < 0$, using the explicit second order scheme for (2.1)

$$G^{*k+\frac{1}{2}}_{m'}U^{k+1}_{m} = G^{*k+\frac{1}{2}}_{m'}\Big(\frac{1}{2}(1+\sigma^{k+\frac{1}{2}}_{m'})\sigma^{k+\frac{1}{2}}_{m'}U^{k}_{m-1} + (1+\sigma^{k+\frac{1}{2}}_{m'})(1-\sigma^{k+\frac{1}{2}}_{m'})U^{k}_{m}$$

$$-\frac{1}{2}(1-\sigma^{k+\frac{1}{2}}_{m'})\sigma^{k+\frac{1}{2}}_{m'}U^{k}_{m+1}\Big) + \Delta t\, f^{k+\frac{1}{2}}_{m'}, \tag{2.6}$$

we can obtain a system of difference equations in form (2.3); if $\lambda_{n,\ell,0} \le 0$ and $\lambda_{n,\ell,M} \ge 0$, using the one-sided second order explicit scheme

$$G^{*k+\frac{1}{2}}_{m'}U^{k+1}_{m} = G^{*k+\frac{1}{2}}_{m'}\Big(-\frac{1}{2}(1-|\sigma^{k+\frac{1}{2}}_{m'}|)|\sigma^{k+\frac{1}{2}}_{m'}|U^{k}_{m\mp2}$$

$$+|\sigma^{k+\frac{1}{2}}_{m'}|(2-|\sigma^{k+\frac{1}{2}}_{m'}|)U^{k}_{m\mp1} + \frac{1}{2}(2-|\sigma^{k+\frac{1}{2}}_{m'}|)(1-|\sigma^{k+\frac{1}{2}}_{m'}|)U^{k}_{m}\Big)$$

$$+ \Delta t f^{k+\frac{1}{2}}_{m'}, \tag{2.7}$$

a system in form (2.3) can also be obtained. In (2.6) and (2.7) the subscripts n and ℓ are also omitted and in (2.7) the minus sign of $\mp$ should be chosen if $\lambda \ge 0$ and the plus sign if $\lambda \le 0$. (Noticing that (2.7) becomes $G^{*k+\frac{1}{2}}_{m'}U^{k+1}_{m} = G^{*k+\frac{1}{2}}_{m'}U^{k}_{m} + \Delta t\, f^{k+\frac{1}{2}}_{m'}$ if $\lambda = 0$, no matter which sign you choose, we know that it is no problem for the sign of equality to appear in both cases.) In order to guarantee a second order accuracy, $G^{*k+\frac{1}{2}}_{m'}$ and $\sigma^{k+\frac{1}{2}}_{m'}$ are computed by the linear interpolation formula

$$f_{m'} = (1 \mp (m - m'))f_m \mp (m' - m)f_{m\mp1},$$

where

$$m' = m - \frac{\lambda_m^{k+\frac{1}{2}} \frac{\Delta t}{2\Delta\xi}}{1 \mp (\lambda_{m\mp 1}^{k+\frac{1}{2}} - \lambda_m^{k+\frac{1}{2}}) \frac{\Delta t}{2\Delta\xi}} ,$$

and the minus sign of $\mp$ should be chosen if $\lambda \geq 0$ and the plus sign if $\lambda \leq 0$.

Consequently, second order accurate systems in form (2.3) are existent. We have given some schemes, which are in form (2.3) and can be applied to various cases, in [2] and [3]. For more details, please see [2] and [3].

In order to use scheme (2.3), we have to know $U^0_{\ell,m}$, Z^0 , $U^{\frac{1}{2}}_{\ell,m}$ and $Z^{\frac{1}{2}}$ at the beginning. We use a first order scheme to get $U^{\frac{1}{2}}_{\ell,m}$ and $Z^{\frac{1}{2}}$ from $U^0_{\ell,m}$ and Z^0. If the superscripts of G and σ in (2.5)-(2.7) are changed to k from $k+\frac{1}{2}$, (2.5)-(2.7) can be applied to this purpose.

Therefore, the numerical procedure can be described as follows. At first, in order to get $U^{\frac{1}{2}}_{\ell,m}$ and $Z^{\frac{1}{2}}$ from $U^0_{\ell,m}$ and Z^0, the system

$$\begin{cases} \sum\limits_{h=h_1(m)}^{h_2(m)} R^0_{h,n,\ell,m} U^{\frac{1}{2}}_{\ell,m+h} = \sum\limits_{h=h_1(m)}^{h_2(m)} S^0_{h,n,\ell,m} U^0_{\ell,m+h} + \frac{1}{2}\Delta t\, f^0_{n,\ell,m} \\ \qquad \ell = 1,2,\cdots,L, \quad n=1,2,\cdots,N, \quad m \in \underline{g}(\lambda_{n,\ell}), \\ B_0(U^{\frac{1}{2}}_{1,0}, Z^{\frac{1}{2}}_0, t^{\frac{1}{2}}) = 0, \\ B_\ell(U^{\frac{1}{2}}_{\ell,M}, U^{\frac{1}{2}}_{\ell+1,0}, Z^{\frac{1}{2}}_\ell, t^{\frac{1}{2}}) = 0, \quad \ell = 1,2,\cdots, L-1, \\ B_L(U^{\frac{1}{2}}_{L,M}, Z^{\frac{1}{2}}_L, t^{\frac{1}{2}}) = 0 \end{cases} \tag{2.8}$$

should be solved. Here $R^0_{h,n,\ell,m}$, $S^0_{h,n,\ell,m}$, $f^0_{n,\ell,m}$ depend only on $U^0_{\ell,m}$ and Z^0. Then for $k = 0, \frac{1}{2}, 1,\cdots, T/\Delta t - 1$, we solve

$$\begin{cases} \sum\limits_{h=h_1(m)}^{h_2(m)} R^{k+\frac{1}{2}}_{h,n,\ell,m} U^{k+1}_{\ell,m+h} = \sum\limits_{h=h_1(m)}^{h_2(m)} S^{k+\frac{1}{2}}_{h,n,\ell,m} U^k_{\ell,m+h} + \Delta t\, f^{k+\frac{1}{2}}_{n,\ell,m}, \\ \qquad \ell = 1,2,\cdots,L, \quad n = 1,2,\cdots,N, \ m \in \underline{g}(\lambda_{n,\ell}), \\ B_0(U^{k+1}_{1,0}, Z^{k+1}_0, t^{k+1}) = 0, \\ B_\ell(U^{k+1}_{\ell,M}, U^{k+1}_{\ell+1,0}, Z^{k+1}_\ell, t^{k+1}) = 0, \quad \ell = 1,2,\cdots,L-1, \\ B_L(U^{k+1}_{L,M}, Z^{k+1}_L, t^{k+1}) = 0. \end{cases} \tag{2.9}$$

Let $\underline{U}$ be a vector whose components are $U_{1,0}, U_{1,1},\cdots,U_{1,M},\cdots,U_{L,0}, U_{L,1},\cdots,U_{L,M}$ from the top to the bottom and U_b be a vector whose components are $U_{1,0}, U_{1,M}, U_{2,0}, U_{2,M}, U_{3,0}, U_{3,M},\cdots,U_{L,0}, U_{L,M}$ from the top to the bottom. $\underline{R}$ denotes a matrix whose every row consists of the coefficients on the left hand side of a difference equation in (2.8) or (2.9), i.e., consists of $R_{h,n,\ell,m}$, $h = h_1(m),\cdots,h_2(m)$, and a great number of zeros. The rows of matrix $\underline{S}$ and the components of vector $\underline{F}$ may be defined equivalently. Then (2.8) and (2.9) can be rewritten as

$$\begin{cases} \underline{R}^0\ \underline{U}^{\frac{1}{2}} = \underline{S}^0\underline{U}^0 + \frac{1}{2}\,\Delta t\ \underline{F}^0, \\ B(U_b^{\frac{1}{2}},\ Z^{\frac{1}{2}},\ t^{\frac{1}{2}}) = 0, \end{cases} \tag{2.10}$$

$$\begin{cases} \underline{R}^{k+\frac{1}{2}}\ \underline{U}^{k+1} = \underline{S}^{k+\frac{1}{2}}\ \underline{U}^k + \Delta t\ \underline{F}^{k+\frac{1}{2}}, \\ B(U_b^{k+1},\ Z^{k+1},\ t^{k+1}) = 0, \\ \qquad k = 0,\ \frac{1}{2}, \cdots, T/\Delta t - 1. \end{cases} \tag{2.11}$$

Here $B(U_b, Z, t) = 0$ are nonlinear equations which correspond to these nonlinear boundary conditions in (2.8) or (2.9). According to their definitions, $\underline{R}^0$, $\underline{S}^0$, $\underline{F}^0$ depend on $\underline{U}^0$, Z^0, t^0, and $\underline{R}^{k+\frac{1}{2}}$, $\underline{S}^{k+\frac{1}{2}}$, $\underline{F}^{k+\frac{1}{2}}$ depend on $\underline{U}^{k+\frac{1}{2}}$, $Z^{k+\frac{1}{2}}$, $t^{k+\frac{1}{2}}$, $\underline{U}^k, Z^k, t^k$, $k=0, \frac{1}{2}, 1, \cdots, T/\Delta t - 1$, i.e., $\underline{R}^0 \equiv \underline{R}(\underline{U}^0, Z^0, t), \cdots$, and $\underline{R}^{k+\frac{1}{2}} \equiv \underline{R}(\underline{U}^{k+\frac{1}{2}}, Z^{k+\frac{1}{2}}, t^{k+\frac{1}{2}}, \underline{U}^k, Z^k, t^k), \cdots$, $k = 0, \frac{1}{2}, \cdots, T/\Delta t - 1$.

Noticing (2.4), we can rewrite (2.3) in the form

$$\sum_{h=h_1(m)}^{h_2(m)} R^{k+\frac{1}{2}}_{h,n,\ell,m}\ \delta U^k_{\ell,m+h} = \sum_{h=h_1(m)}^{h_2(m)-1} Q^{k+\frac{1}{2}}_{h,n,\ell,m}\ \Delta_+ U^k_{\ell,m+h} + \Delta t\ f^{k+\frac{1}{2}}_{n,\ell,m}, \tag{2.12}$$

where $\delta U^k_{\ell,m+h} = U^{k+1}_{\ell,m+h} - U^k_{\ell,m+h}$, $\Delta_+ U^k_{\ell,m+h} = U^k_{\ell,m+h+1} - U^k_{\ell,m+h}$, and

$$Q^{k+\frac{1}{2}}_{h,n,\ell,m} = \sum_{j=h_1(m)}^{h} (R^{k+\frac{1}{2}}_{j,n,\ell,m} - S^{k+\frac{1}{2}}_{j,n,\ell,m}).$$

Therefore (2.10) and (2.11) can also be written in the following form

$$\begin{cases} \underline{R}^0\ \delta_{\frac{1}{2}}\underline{U}^0 = \underline{Q}^0\ \Delta_+\ \underline{U}^0 + \frac{1}{2}\,\Delta t\ \underline{F}^0, \\ B(U_b^{\frac{1}{2}},\ Z^{\frac{1}{2}},\ t^{\frac{1}{2}}) = 0, \end{cases} \tag{2.13}$$

and

$$\begin{cases} \underline{R}^{k+\frac{1}{2}}\ \delta\underline{U}^k = \underline{Q}^{k+\frac{1}{2}}\ \Delta_+\ \underline{U}^k + \Delta t\ \underline{F}^{k+\frac{1}{2}}, \\ B(U_b^{k+1},\ Z^{k+1},\ t^{k+1}) = 0, \\ \qquad k=0, \frac{1}{2}, \cdots, T/\Delta t - 1. \end{cases} \tag{2.14}$$

Here $\delta_{\frac{1}{2}}\ \underline{U}^0 = \underline{U}^{\frac{1}{2}} - \underline{U}^0$, $\underline{Q}$ may be defined in the same way as $\underline{S}$ is done, and $\underline{Q}$, $\underline{R}$, $\underline{S}$ satisfy the relation

$$\underline{SU} = \underline{RU} + \underline{Q}\ \Delta_+\ \underline{U}. \tag{2.15}$$

It is clear that we can also construct a second order scheme in the following form

$$\begin{cases} \underline{R}_1^{k+\frac{1}{2}}\ \underline{U}^{k+1} = \underline{S}_1^{k+\frac{1}{2}}\ \underline{U}^k + \Delta t\ \underline{F}_1^{k+\frac{1}{2}}, \\ B(U_b^{k+1},\ Z^{k+1},\ t^{k+1}) = 0, \end{cases} \tag{2.16}$$

where $\underline{R}_1^{k+\frac{1}{2}}$, $\underline{S}_1^{k+\frac{1}{2}}$ and $\underline{F}_1^{k+\frac{1}{2}}$ depend on $\underline{U}^{k+1}$, Z^{k+1}, t^{k+1}, $\underline{U}^k$, Z^k, t^k.

3. BASIC ASSUMPTIONS

Consider the scalar equation

$$\frac{\partial u}{\partial t} + \lambda(u,\xi,t)\frac{\partial u}{\partial \xi} = 0, \quad 0 \le \xi \le 1, \qquad 0 \le t \le T, \tag{3.1}$$

and suppose that it is approximated by difference equations of the following form

$$\sum_{h=h_1(m)}^{h_2(m)} \gamma_{h,m}^{k+\frac{1}{2}} u_{m+h}^{k+1} = \sum_{h=h_1(m)}^{h_2(m)} s_{h,m}^{k+\frac{1}{2}} u_{m+h}^{k}, \tag{3.2}$$

$$m \in \underline{g}(\lambda)$$

$$0 \le m+h_1(m) \le m + h_2(m) \le M, \qquad \max_m\{|h_1(m)|, \quad |h_2(m)|\} \le H,$$

$$k=0,1,\cdots .$$

For scheme (3.2), besides the consistency condition, some stability condition and "well-conditioned" condition should be required. We suppose that the following "most weak" stability condition and "well-conditioned" condition are satisfied:

(1) The von Neumann condition

$$(\gamma_m^{k+\frac{1}{2}}(\theta))^* \gamma_m^{k+\frac{1}{2}}(\theta) - (s_m^{k+\frac{1}{2}}(\theta))^* s_m^{k+\frac{1}{2}}(\theta) \ge 0 \tag{3.3}$$

holds for any $m \in \underline{g}(\lambda)$, where

$$\gamma_m^{k+\frac{1}{2}}(\theta) = \sum_{h=h_1(m)}^{h_2(m)} \gamma_{h,m}^{k+\frac{1}{2}}(0,0)e^{ih\theta},$$

$$s_m^{k+\frac{1}{2}}(\theta) = \sum_{h=h_1(m)}^{h_2(m)} s_{h,m}^{k+\frac{1}{2}}(0,0)e^{ih\theta}$$

and $(\gamma_m^{k+\frac{1}{2}}(\theta))^*$, $(s_m^{k+\frac{1}{2}}(\theta))^*$ are conjugate complex numbers of $\gamma_m^{k+\frac{1}{2}}(\theta)$, $s_m^{k+\frac{1}{2}}(\theta)$, $\gamma_{h,m}^{k+\frac{1}{2}}(0,0)$, $s_{h,m}^{k+\frac{1}{2}}(0,0)$ being the linear main parts of $\gamma_{h,m}^{k+\frac{1}{2}} \equiv \gamma_{h,m}^{k+\frac{1}{2}}(\Delta\xi, \Delta t) =$ $\gamma_h(m\Delta\xi, (k+\frac{1}{2})\Delta t, \Delta\xi, \Delta t)$, $s_{h,m}^{k+\frac{1}{2}} \equiv s_{h,m}^{k+\frac{1}{2}}(\Delta\xi, \Delta t) = s_h(m\Delta\xi, (k+\frac{1}{2})\Delta t, \Delta\xi, \Delta t)$.

(2) The 'well-conditioned' condition

$$(\gamma_m^{k+\frac{1}{2}}(\theta))^* \gamma_m^{k+\frac{1}{2}}(\theta) \ge c_1 > 0 \tag{3.4}$$

holds for any $m \in \underline{g}(\lambda)$, c_1 being a positive constant. For explicit schemes, (3.4) always holds.

(3) If u is Lipschitz continuous with respect to ξ and t, then $\gamma_{h,m}^{k+\frac{1}{2}}$, $s_{h,m}^{k+\frac{1}{2}}$ are Lipschitz continuous with respect to ξ and t, i.e.,

$$|\gamma_{h,m}^{k+\frac{1}{2}} - \gamma_{h,m-1}^{k+\frac{1}{2}}| \le c\Delta\xi, \qquad |s_{h,m}^{k+\frac{1}{2}} - s_{h,m-1}^{k+\frac{1}{2}}| \le c\Delta\xi \tag{3.5}$$

$$|\gamma^{k+\frac{1}{2}}_{h,m} - \gamma^{k-\frac{1}{2}}_{h,m}| \leq c\Delta t, \qquad |s^{k+\frac{1}{2}}_{h,m} - s^{k-\frac{1}{2}}_{h,m}| \leq c\Delta t$$

and the differences between $\gamma^{k+\frac{1}{2}}_{h,m}(0,0)$; $s^{k+\frac{1}{2}}_{h,m}(0,0)$ and $\gamma^{k+\frac{1}{2}}_{h,m}$, $s^{k+\frac{1}{2}}_{h,m}$ satisfy

$$\begin{aligned} |\gamma^{k+\frac{1}{2}}_{h,m}(0,0) - \gamma^{k+\frac{1}{2}}_{h,m}| &\leq c(\Delta t + \Delta\xi), \\ |s^{k+\frac{1}{2}}_{h,m}(0,0) - s^{k+\frac{1}{2}}_{h,m}| &\leq c(\Delta t + \Delta\xi), \end{aligned} \tag{3.6}$$

where c is a constant. From now on, c_i denotes a constant, and we use c to express different constants if it is not necessary to give a specified subscript i.

We define

$$T^k = \sum_{m\in \underline{g}(\lambda)} (\sum_h \gamma^{k+\frac{1}{2}}_{h,m} u^k_{m+h})^2 \Delta\xi, \tag{3.7}$$

$$||u^k||^2 = \sum_{m=0}^{M} |u^k_m|^2 \Delta\xi, \tag{3.8}$$

i.e., $||u^k||$ is the L_2 norm of u^k. (In this paper, $||f||$ always means the L_2-norm of f .)

We say that scheme (3.2) possesses Property A if from (3.3)-(3.6) we can derive the two inequalities:

$$\begin{aligned} \text{(i)} \quad T^k + c\Delta\xi||u^k||^2 \geq &- c|u^k_0|^2 \Delta\xi\delta_0(\lambda^k_0) - \\ &- c|u^k_M|^2 \Delta\xi\delta_1(\lambda^k_M) + c_2 \sum_{m\in \underline{g}(\lambda^k)} |u^k_m|^2 \Delta\xi; \end{aligned} \tag{3.9}$$

$$\begin{aligned} \text{(ii)} \quad T^{k+1} - T^k \leq &[c\delta_0(\lambda^k_0) - c_2|\sigma^k_0|(1 - \delta_0(\lambda^k_0))]|u^k_0|^2 \Delta\xi \\ &+ [c\delta_1(\lambda^k_M) - c_2|\sigma^k_M|(1 - \delta_1(\lambda^k_M))]|u^k_M|^2\Delta\xi + c\Delta\xi||u^k||^2, \end{aligned} \tag{3.10}$$

where

$$\delta_0(\lambda^k_0) \equiv \begin{cases} 1, & \lambda^k_0 > 0 \\ 0 & \lambda^k_0 \leq 0 \end{cases}, \qquad \delta_1(\lambda^k_M) \equiv \begin{cases} 1, & \lambda^k_M < 0 \\ 0, & \lambda^k_M \geq 0 \end{cases},$$

c is a constant and c_2 is a positive constant. Actually, using $\delta_0(\lambda^k_0)$ and $\delta_1(\lambda^k_M)$, $\underline{g}(\lambda^k)$ can be expressed as follows

$$\underline{g}(\lambda^k) = \{\delta_0(\lambda^k_0),\ \delta_0(\lambda^k_0) + 1, \cdots, M - \delta_1(\lambda^k_M)\} .$$

Therefore if $\underline{g}(\lambda^k)$ is the same for all k, then $\delta_0(\lambda^k_0)$, $\delta_1(\lambda^k_M)$ are also the same for all k. We have used $\underline{g}(\lambda)$ instead of $\underline{g}(\lambda^k)$. Consequently, in what follows, we use $\delta_0(\lambda_0)$, $\delta_1(\lambda_M)$ instead of $\delta_0(\lambda^k_0)$, $\delta_1(\lambda^k_M)$.

Suppose $\lambda =$ constant > 0. Using scheme (2.5), we can obtain a system of difference equations of form (3.2) approximating (3.1) as follows:

$$\frac{u_m^{k+1} + u_{m-1}^{k+1}}{2} + \sigma\frac{u_m^{k+1} - u_{m-1}^{k+1}}{2} = \frac{u_m^{k} + u_{m-1}^{k}}{2} - \sigma\frac{u_m^{k} - u_{m-1}^{k}}{2}, \quad m=1,2,\cdots,M. \tag{3.11}$$

Therefore

$$T^k = [\sum_{m=1}^{M} \frac{1}{4}((1+\sigma)u_m + (1-\sigma)u_{m-1})^2 \quad]^k \Delta\xi$$

$$= [\frac{1}{4}(1-\sigma)^2 u_0^2 + \frac{1+\sigma^2}{2}\sum_{m=1}^{M-1} u_m^2 + \frac{(1+\sigma)^2}{4} u_M^2 + \frac{1}{2}(1-\sigma^2)\sum_{m=1}^{M} u_{m-1}u_m]^k \Delta\xi$$

$$= \begin{cases} [\frac{1}{4}((1-\sigma)^2-(1-\sigma^2))u_0^2 + \sigma^2\sum_{m=1}^{M-1} u_m^2 + \frac{1}{4}((1+\sigma)^2 - (1-\sigma^2))u_M^2 \\ \quad + \frac{1}{2}(1-\sigma^2)(\frac{1}{2}u_0^2 + \sum_{m=1}^{M-1} u_m^2 + \frac{1}{2}u_M^2 + \sum_{m=1}^{M} u_{m-1}u_m)]^k\Delta\xi \\ \quad \geq [\frac{1}{2}\sigma(\sigma-1)u_0^2 + \sigma^2\sum_{m=1}^{M} u_m^2]^k \Delta\xi, \text{ if } 0 \leq \sigma \leq 1, \\ [\frac{1}{4}((1-\sigma)^2-(\sigma^2-1))u_0^2 + \sum_{m=1}^{M-1} u_m^2 + \frac{1}{4}((1+\sigma)^2 - (\sigma^2-1))u_M^2 \\ \quad + \frac{1}{2}(\sigma^2-1)(\frac{1}{2}u_0^2 + \sum_{m=1}^{M-1} u_m^2 + \frac{1}{2}u_M^2 - \sum_{m=1}^{M} u_{m-1}u_m)]^k\Delta\xi \\ \quad \geq [\frac{1}{2}(1-\sigma)u_0^2 + \sum_{m=1}^{M} u_m^2]^k \Delta\xi, \text{ if } 1 \leq \sigma \end{cases} \tag{3.12}$$

and

$$T^{k+1} - T^k = [\sum_{m=1}^{M} \frac{1}{4}((1-\sigma)u_m + (1+\sigma)u_{m-1})^2 - \sum_{m=1}^{M} \frac{1}{4}((1+\sigma)u_m - (1-\sigma)u_{m-1})^2]^k \Delta\xi$$

$$= [\sum_{m=1}^{M} \frac{1}{4}((1-\sigma)^2 - (1+\sigma)^2)u_m^2 + \frac{1}{4}\sum_{m=0}^{M-1}((1+\sigma)^2 - (1-\sigma)^2)u_m^2]^k \Delta\xi$$

$$= [\sigma u_0^2 - \sigma u_M^2]^k \Delta\xi, \tag{3.13}$$

where $[\ldots.]^k$ means that every quantity in $[\cdots\cdot]$ has a superscript k. Let c = max $\{\frac{1}{2}\sigma(1-\sigma), \frac{1}{2}(\sigma-1), \sigma\}$, $c_2 = \min\{\sigma^2, 1\}$, noticing $\delta_0(\lambda_0) = 1$ and $\delta_1(\lambda_M) = 0$ in the present case, we can write (3.12) and (3.13) in the form of (3.9) and (3.10) respectively if $0 < \sigma$. For scheme (2.5), if $\lambda > 0$, (3.3) always holds and (3.4) is equivalent to $0 < \sigma$. Therefore, scheme (2.5) has Property A if λ = constant > 0.

The following scheme approximating (3.1) with $\lambda = \text{constant} \geq 0$,

$$u_m^{k+1} = u_m^k - \sigma(u_m^k - u_{m-1}^k), \qquad m = 1,2,\cdots, M \tag{3.14}$$

is also in the form (3.2). In this case

$$T^k = \sum_{m=1}^{M} |u_m^k|^2 \Delta\xi$$

and if $0 \leq \sigma \leq 1$, we have

$$T^{k+1} - T^k = [\sum_{m=1}^{M} ((1-\sigma)u_m + \sigma u_{m-1})^2 - \sum_{m=1}^{M} u_m^2]^k \Delta\xi$$

$$=[(1-\sigma)^2 \sum_{m=1}^{M} u_m^2 + \sigma^2 \sum_{m=0}^{M-1} u_m^2 + 2(1-\sigma)\sigma \sum_{m=1}^{M} u_m u_{m-1} - \sum_{m=1}^{M} u_m^2]^k \Delta\xi$$

$$\leq [(1-\sigma)^2 \sum_{m=1}^{M} u_m^2 + \sigma^2 \sum_{m=0}^{M-1} u_m^2 + (1-\sigma)\sigma(u_0^2 + 2\sum_{m=1}^{M-1} u_m^2 + u_M^2) - \sum_{m=1}^{M} u_m^2]^k \Delta\xi$$

$$= [\sigma u_0^2 - \sigma u_M^2]^k \Delta\xi.$$

Let $c = \sigma$, $c_2 = 1$, they can be written in forms (3.9) and (3.10). For (3.14) the condition (3.3) is equivalent to $0 \leq \sigma \leq 1$ and (3.4) always holds. That is, (3.14) also possesses Property A if $\lambda = \text{constant} \geq 0$. Actually, many schemes are of Property A even for the case $\lambda = \lambda(\xi, t)$. In fact, for several schemes with variable coefficients, including schemes (2.5), (2.6), (2.7) and some combinations of them, we have proved[3] that they possess Property A. Some of results are given in Section 6.

As we have done in Section 2, the boundary conditions can be written in the following form

$$B(U_b, Z, t) = 0. \tag{3.15}$$

Its first variation equation is

$$\frac{\partial B}{\partial (U_b, Z)} \begin{bmatrix} \delta U_b \\ \delta Z \end{bmatrix} = 0.$$

We further rewrite it as

$$B_g \begin{bmatrix} G_b \ \delta U_b \\ \delta Z \end{bmatrix} = 0, \tag{3.16}$$

where

$$G_b = \begin{bmatrix} G_{1,0} & & & & \\ & G_{1,M} & & & \\ & & \ddots & & \\ & & & G_{L,0} & \\ & & & & G_{L,M} \end{bmatrix} \tag{3.17}$$

and

$$B_g = \frac{\partial B}{\partial(U_b, Z)} \begin{bmatrix} G_b^{-1} & \\ & I_{L+1} \end{bmatrix}, \tag{3.18}$$

I_{L+1} being an $(L + 1) \times (L + 1)$ unit matrix. Clearly $G_b \delta U_b$ is a 2LN-vector whose components are $(G_n^* \delta U)_{\ell,0}$ and $(G_n^* \delta U)_{\ell,M}$, $n=1,2,\cdots,N$, $\ell = 1,2,\cdots,L$. $(G_b \delta U_b)_i$ denotes the i-th component of $G_b \delta U_b$. Λ_b is defined as follows: its i-th component λ_{bi} is $\lambda_{n,\ell,0}$ if $(G_b \delta U_b)_i = (G_n^* \delta U)_{\ell,0}$ or $\lambda_{n,\ell,M}$ if $(G_b \delta U_b)_i = (G_n^* \delta U)_{\ell,M}$. The set $I = \{ 1,2,\cdots,2LN \}$ is divided into I_0 and I_1 in the following way: in the case $\lambda_{bi} = \lambda_{n,\ell,0}$ for some i, we say that i belongs to I_0 if $\delta_0(\lambda_{bi}) = \delta_0(\lambda_{n,\ell,0}) = 0$ and to I_1 if $\delta_0(\lambda_{bi}) = \delta_0(\lambda_{n,\ell,0}) = 1$; in the case $\lambda_{bi} = \lambda_{n,\ell,M}$ for some i, we say that i belongs to I_0 if $\delta_1(\lambda_{bi}) = \delta_1(\lambda_{n,\ell,M}) = 0$, and to I_1 if $\delta_1(\lambda_{bi}) = \delta_1(\lambda_{n,\ell,M}) = 1$.

We say the boundary condition (3.15) possesses Property B if for any V_b, Y satisfying

$$B_g \begin{pmatrix} V_b \\ Y \end{pmatrix} = E, \tag{3.19}$$

the following inequality holds:

$$|Y|^2 + \sum_{i \in I_1} |v_{bi}|^2 \leq c\left(\sum_{i \in I_0} |\lambda_{bi}| |v_{bi}|^2 + |E|^2 \right), \tag{3.20}$$

where c is a constant and v_{bi} is the i-th component of V_b.

In practical problems we usually have such an inequality. From the definitions of $\delta_0(\lambda_{n,\ell,0})$ and $\delta_1(\lambda_{n,\ell,M})$ we know that if $i \in I_0$, the characteristics line corresponding to $\lambda_{bi} = \lambda_{n,\ell,0}$ or $\lambda_{n,\ell,M}$ arrives at a boundary. This means that if $i \in I_0$, the value of v_{bi} should be obtained from partial differential equations. Therefore, when all such v_{bi} are given, the boundary conditions should be able to give all values at boundaries. That is, the system

$$\begin{cases} B_g \begin{pmatrix} V_b \\ Y \end{pmatrix} = E \\ V_b^{(0)} = E^{(0)} \end{cases} \tag{3.21}$$

should have a solution, where $V_b^{(0)}$ denotes the vector consisting of all the v_{bi}, $i \in I_0$ and $E^{(0)}$ is a given vector. If (3.21) has a unique solution then we have

$$|Y|^2 + \sum_{i \in I_1} |v_{bi}|^2 \leq c\left(\sum_{i \in I_0} |v_{bi}|^2 + |E|^2 \right). \tag{3.22}$$

Moreover, if all $|\lambda_{bi}| \geq$ a positive constant, then (3.22) can be changed to (3.20) immediately. (However, generally speaking, (3.22) is "weaker" than (3.20)). If

$\lambda_{bi^*} = 0$ for certain i^* and the i^*-th column of B_g is a zero-vector, which happens in several practical problems, then we will have

$$|Y|^2 + \sum_{i \in I_i} |v_{bi}|^2 \leq c\,[\sum_{i \in I_0 \text{ but } i \neq i^*} |v_{bi}|^2 + |E|^2]. \tag{3.23}$$

(3.23) can also be changed to (3.20). Therefore, (3.20) holds for many practical problems.

For the coefficients and the nonhomogeneous terms of differential equations and difference equations and boundary conditions, we assume that they are Lipschitz continuous with respect to their arguments. For the exact solution, besides the Lipschitz continuity in every subregion, we further assume that the truncation error will really be $0(\Delta t^3)$ for second order schemes.

4. EXISTENCE OF SOLUTIONS OF DIFFERENCE EQUATIONS

In this section, we shall prove that if

(i) (3.9) and (3.22) hold;

(ii) the errors at $t = k\Delta t$, $(k + 1/2)\Delta t$ are $0(\Delta t^2)$;

(iii) the coefficients in difference equations and the functions in boundary conditions are Lipschitz continuous with respect to their arguments,

then (2.11) has a solution $\{\underline{U}^{k+1}, Z^{k+1}\}$, and the difference between the solution and the exact solution is $0(\Delta t^2)$.

Suppose $\underline{\tilde{U}}, \tilde{Z}$ be the exact solution and $\tilde{U}_b$ is the equivalent of U_b in the case of exact solution. Since (2.11) is a second order scheme, we have

$$\begin{cases} \underline{\tilde{R}}^{k+\frac{1}{2}}\,\underline{\tilde{U}}^{k+1} = \underline{\tilde{S}}^{k+\frac{1}{2}}\,\underline{\tilde{U}}^{k} + \Delta t\,\underline{\tilde{F}}^{k+\frac{1}{2}} + 0(\Delta t^3) \\ B(\tilde{U}_b^{k+1}, \tilde{Z}^{k+1}, t^{k+1}) = 0, \quad k = 0, 1/2, \cdots, T/\Delta t - 1, \end{cases} \tag{4.1}$$

where $\underline{\tilde{R}}$, $\underline{\tilde{S}}$, $\underline{\tilde{F}}$ are almost the same as $\underline{R}$, $\underline{S}$, $\underline{F}$, but arguments $\underline{U}$, Z are substituted by $\underline{\tilde{U}}$, $\tilde{Z}$, and $0\,(\Delta t^3)$ denotes a vector whose L_2 norm is of order Δt^3. Noticing (2.15), we can rewrite the first equation of (4.1) as

$$\begin{aligned} \underline{R}^{k+\frac{1}{2}}\,\underline{\tilde{U}}^{k+1} &= (\underline{R}^{k+\frac{1}{2}} - \underline{\tilde{R}}^{k+\frac{1}{2}})\underline{\tilde{U}}^{k+1} + \underline{\tilde{S}}^{k+\frac{1}{2}}\underline{\tilde{U}}^{k} + \Delta t\,\underline{\tilde{F}}^{k+\frac{1}{2}} + 0(\Delta t^3) \\ &= \underline{S}^{k+\frac{1}{2}}\underline{U}^{k} + \Delta t\,\underline{F}^{k+\frac{1}{2}} + (\underline{R}^{k+\frac{1}{2}} - \underline{\tilde{R}}^{k+\frac{1}{2}})\,\underline{\tilde{U}}^{k+1} \\ &\quad - \underline{S}^{k+\frac{1}{2}}(\underline{U}^{k} - \underline{\tilde{U}}^{k}) - (\underline{S}^{k+\frac{1}{2}} - \underline{\tilde{S}}^{k+\frac{1}{2}})\,\underline{\tilde{U}}^{k} - \Delta t(\underline{F}^{k+\frac{1}{2}} - \underline{\tilde{F}}^{k+\frac{1}{2}}) + 0(\Delta t^3) \\ &= \underline{S}^{k+\frac{1}{2}}\underline{U}^{k} + \Delta t\,\underline{F}^{k+\frac{1}{2}} + (\underline{R}^{k+\frac{1}{2}} - \underline{\tilde{R}}^{k+\frac{1}{2}})\,\delta\underline{\tilde{U}}^{k} \\ &\quad - (\underline{Q}^{k+\frac{1}{2}} - \underline{\tilde{Q}}^{k+\frac{1}{2}})\Delta_+\underline{\tilde{U}}^{k} - \underline{S}^{k+\frac{1}{2}}(\underline{U}^{k} - \underline{\tilde{U}}^{k}) - \Delta t(\underline{F}^{k+\frac{1}{2}} - \underline{\tilde{F}}^{k+\frac{1}{2}}) + 0(\Delta t^3). \end{aligned}$$

According to the given condition, the L_2 norms of the errors $\begin{bmatrix} \underline{U}^k - \underline{\tilde{U}}^k \\ Z^k - \tilde{Z}^k \end{bmatrix}$ and $\begin{bmatrix} \underline{U}^{k+\frac{1}{2}} - \underline{\tilde{U}}^{k+\frac{1}{2}} \\ Z^{k+\frac{1}{2}} - \tilde{Z}^{k+\frac{1}{2}} \end{bmatrix}$ are $0(\Delta t^2)$, so their L_∞ norms are $0(\Delta t^{3/2})$. Moreover, in each row of $\underline{R}$ or $\underline{Q}$ there are only several non-zero components and each element of $\underline{R}$, $\underline{Q}$ or $\underline{F}$

has only several arguments. Therefore ,

$$||(\underline{R}^{k+\frac{1}{2}}-\underline{\tilde{R}}^{k+\frac{1}{2}})\delta\underline{\tilde{U}}^{k} - (\underline{Q}^{k+\frac{1}{2}} - \underline{\tilde{Q}}^{k+\frac{1}{2}})\Delta_{+}\underline{\tilde{U}}^{k}-\underline{S}^{k+\frac{1}{2}}(\underline{U}^{k}-\underline{\tilde{U}}^{k})-$$

$$-\Delta t(\underline{F}^{k+\frac{1}{2}} - \underline{\tilde{F}}^{k+\frac{1}{2}}) + O(\Delta t^3)|| \quad \text{is } O(\Delta t^2).$$

Consequently, (4.1) can be written in the form

$$\begin{cases} \underline{R}^{k+\frac{1}{2}}\,\underline{\tilde{U}}^{k+1} = \underline{S}^{k+\frac{1}{2}}\,\underline{U}^{k} + \Delta t\,\underline{F}^{k+\frac{1}{2}} + O(\Delta t^2), \\ B(\tilde{U}_b^{k+1}, \tilde{Z}^{k+1}, t^{k+1}) = 0. \end{cases} \tag{4.2}$$

Consider the following system

$$\underline{A}(\underline{X}) = \underline{F} \tag{4.3}$$

where

$$\underline{X} = \begin{bmatrix} U \\ Z \end{bmatrix}, \quad \underline{A}(\underline{X}) = \begin{bmatrix} \underline{R}^{k+\frac{1}{2}}\,U \\ B(U_b, Z, t^{k+1}) \end{bmatrix}.$$

Here the numbers of components of B, U_b, Z are fixed, but the number of components of $\underline{U}$ is $O(\frac{1}{\Delta t})$. Also, we suppose every element of $\frac{\partial B}{\partial(U_b,Z)}$ is Lipschitz contin - uous with respect to its arguments. From (4.2) we know that if

$$\underline{F} = \underline{F}^* \equiv \begin{bmatrix} \underline{S}^{k+\frac{1}{2}}\,\underline{U}^{k} + \Delta t\,\underline{F}^{k+\frac{1}{2}} + O(\Delta t^2) \\ 0 \end{bmatrix},$$

(4.3) has a solution $\underline{X}^* = \begin{bmatrix} \tilde{U}^{k+1} \\ \tilde{Z}^{k+1} \end{bmatrix}$. Now we shall prove that if $||\underline{F}-\underline{F}^*|| \le c\Delta t^{\frac{1}{2}+\delta}$, $\delta > 0$ and Δt is small enough, then (4.3) has a solution which satisfies

$$||\underline{X} - \underline{X}^*|| \le 2||\underline{A}'^{-1}(\underline{X}^*)||\;||\underline{F} - \underline{F}^*|| \tag{4.4}$$

and the solution of (4.3) satisfying (4.4) is unique, where $\underline{A}'$ is the Jacobian of $\underline{A}$ with respect to $\underline{X}$.

First, we would like to point out that because the numbers of components of B, U_b, Z are fixed and every element of $\frac{\partial B}{\partial(U_b,Z)}$ is Lipschitz continuous with respect to its arguments, we have

$$||\underline{A}'(\underline{X}_1) - \underline{A}'(\underline{X}_2)|| \le c||\underline{X}_1 - \underline{X}_2||_{L_\infty}. \tag{4.5}$$

Consider the simplified Newton iteration with $\underline{X}^0 = \underline{X}^*$:

$$\underline{X}^{k+1} = \underline{X}^{k} + \underline{A}'^{-1}(\underline{X}^*)(\underline{F} - \underline{A}(\underline{X}^{k})), \quad k = 0,1,\cdots. \tag{4.6}$$

From (4.6) we have

$$\begin{aligned} \underline{X}^{k+1}-\underline{X}^* &= \underline{X}^{k}-\underline{X}^*+\underline{A}'^{-1}(\underline{X}^*)(\underline{A}(\underline{X}^*) - \underline{A}(\underline{X}^{k}))+\underline{A}'^{-1}(\underline{X}^*)(\underline{F} - \underline{F}^*) \\ &= \underline{A}'^{-1}(\underline{X}^*)(\underline{A}'(\underline{X}^*) - \underline{A}'(\underline{X}^* + \xi(\underline{X}^{k} - \underline{X}^*)))(\underline{X}^{k}- \underline{X}^*) \\ &\quad + \underline{A}'^{-1}(\underline{X}^*)(\underline{F} - \underline{F}^*), \quad 0 \le \xi \le 1. \end{aligned} \tag{4.7}$$

Therefore, if (4.4) holds for $\underline{X}^k$ and Δt is small enough, then (4.4) holds also for $\underline{X}^{k+1}$. In fact, when (4.4) holds for $\underline{X}^k$, we have

$$||\underline{X}^k - \underline{X}^*||_{L_\infty} \leq c \, \Delta t^\delta .$$

Therefore, letting Δt be small enough, from (4.5) we obtain

$$||\underline{A}'(\underline{X}^*) - \underline{A}'(\underline{X}^* + \xi(\underline{X}^k - \underline{X}^*))|| \leq \frac{1}{2||\underline{A}'^{-1}(\underline{X}^*)||} \tag{4.8}$$

and

$$||\underline{X}^{k+1} - \underline{X}^*|| \leq 2||\underline{A}'^{-1}(\underline{X}^*)|| \; ||\underline{F} - \underline{F}^*||$$

can be derived from (4.7). Noting

$$||\underline{X}^1 - \underline{X}^*|| \leq ||\underline{A}'^{-1}(\underline{X}^*)|| \; ||\underline{F} - \underline{F}^*|| ,$$

and using the inductive method, we know (4.4) holds for all X^k.

From (4.6) we have

$$\begin{aligned}\underline{X}^{k+2} - \underline{X}^{k+1} &= \underline{X}^{k+1} - \underline{X}^k + \underline{A}'^{-1}(\underline{X}^*)(\underline{A}(\underline{X}^k) - \underline{A}(\underline{X}^{k+1}))\\ &= \underline{A}'^{-1}(\underline{X}^*)(\underline{A}'(\underline{X}^*) - \underline{A}'(\underline{X}^k + \xi(\underline{X}^{k+1} - \underline{X}^k)))(\underline{X}^{k+1} - \underline{X}^k),\\ &\qquad 0 \leq \xi \leq 1.\end{aligned}$$

Since both $\underline{X}^{k+1}$ and $\underline{X}^k$ satisfy (4.4), we can choose a small Δt such that

$$||\underline{A}'(\underline{X}^*) - \underline{A}'(\underline{X}^k + \xi(\underline{X}^{k+1} - \underline{X}^k))|| \leq \frac{1}{2||\underline{A}'^{-1}(\underline{X}^*)||} . \tag{4.9}$$

Therefore we have

$$||\underline{X}^{k+2} - \underline{X}^{k+1}|| \leq \frac{1}{2}||\underline{X}^{k+1} - \underline{X}^k||,$$

and the convergence of the iteration (4.6) can be obtained immediately.

Suppose both $\underline{X}_1$ and $\underline{X}_2$ are solutions of (4.3) and satisfy (4.4). Noting that in this case

$$||\underline{A}'(\underline{X}^*) - \underline{A}'(\underline{X}_1 + \xi(\underline{X}_2 - \underline{X}_1))|| \leq \frac{1}{2||\underline{A}'^{-1}(\underline{X}^*)||} ,$$

we can have

$$\begin{aligned}||\underline{A}'^{-1}(\underline{X}_1 + \xi(\underline{X}_2 - \underline{X}_1))|| &\leq ||(I - \underline{A}'^{-1}(\underline{X}^*)(\underline{A}'(\underline{X}^*) - \underline{A}'(\underline{X}_1 + \xi(\underline{X}_1 - \underline{X}_2)))^{-1}|| \; ||\underline{A}'^{-1}(\underline{X}^*)||\\ &\leq \frac{||\underline{A}'^{-1}(\underline{X}^*)||}{1 - ||\underline{A}'^{-1}(\underline{X}^*)|| \; ||\underline{A}'(\underline{X}^*) - \underline{A}'(\underline{X}_1 + \xi(\underline{X}_1 - \underline{X}_2))||}\\ &\leq 2||\underline{A}'^{-1}(\underline{X}^*)|| .\end{aligned}$$

Moreover, there is the inequality

$$||\underline{X}_1 - \underline{X}_2|| \leq ||\underline{A}'^{-1}(\underline{X}_1 + \xi(\underline{X}_2 - \underline{X}_1))|| \; ||\underline{A}(\underline{X}_1) - \underline{A}(\underline{X}_2)|| .$$

Therefore, from $\underline{A}(\underline{X}_1) = \underline{A}(\underline{X}_2) = \underline{F}$ we can obtain $\underline{X}_1 = \underline{X}_2$, i.e., (4.3) have only one solution which satisfies (4.4).

Since both (4.2) and (2.11) are in the form (4.3) and the difference between two $\underline{F}$'s is $0(\Delta t^2)$, from what we have proved we know that if the boundedness of the L_2 norm of the matrix

$$\underline{A}'^{-1} \equiv \left(\frac{\partial}{\partial(\underline{U},Z)} \begin{bmatrix} \underline{R}^{k+\frac{1}{2}} \underline{U} \\ B(U_b, Z, t^{k+1}) \end{bmatrix} \right)^{-1} \Bigg|_{\underline{U}=\underline{\tilde{U}}^{k+1},\ Z=\tilde{Z}^{k+1}} \tag{4.10}$$

can be shown, our conclusion is proved.

Let us consider the system

$$\begin{cases} \underline{R}^{k+\frac{1}{2}} \underline{U}^{k+1} = \underline{S}^{k+\frac{1}{2}} \underline{U}^{k}, \\ \dfrac{\partial \tilde{B}^{k+1}}{\partial(U_b, Z)} \begin{pmatrix} U_b^{k+1} \\ Z^{k+1} \end{pmatrix} = E, \end{cases} \tag{4.11}$$

where

$$\frac{\partial \tilde{B}^{k+1}}{\partial(U_b, Z)} = \frac{\partial B(U_b, Z, t)}{\partial(U_b, Z)} \Bigg|_{U_b = \tilde{U}_b^{k+1},\ Z = \tilde{Z}^{k+1},\ t = t^{k+1}}.$$

It is easy to know that if when a scheme is used to

$$\frac{\partial u}{\partial t} + \lambda \frac{\partial u}{\partial x} = 0, \quad 0 \le x \le 1, \tag{4.12}$$

we get a system in the form

$$\sum_h \gamma_{h,m}^{k+\frac{1}{2}} u_{m+h}^{k+1} = \sum_h s_{h,m}^{k+\frac{1}{2}} u_{m+h}^{k},$$

$$m \in \underline{g}(\lambda),$$

then when the same scheme is used to

$$G_n^* \frac{\partial U}{\partial t} + \lambda_n G_n^* \frac{\partial U}{\partial \xi} = 0, \quad n = 1,2,\cdots,N,$$

for each $\lambda_{n,\ell}$ the system of difference equations

$$\sum_h R_{h,n,\ell,m}^{k+\frac{1}{2}} U_{\ell,m+h}^{k+1} = \sum_h S_{h,n,\ell,m}^{k+\frac{1}{2}} U_{\ell,m+h}^{k},$$

$$m \in \underline{g}(\lambda_{n,\ell}) \tag{4.13}$$

can be rewritten in the form

$$\sum_h (\gamma_{h,n,\ell,m}^{k+\frac{1}{2}} u_{n,\ell,m+h}^{k+1} + O(\Delta t)\, U_{\ell,m+h}^{k+1}) =$$

$$= \sum_h (s_{h,n,\ell,m}^{k+\frac{1}{2}} u_{n,\ell,m+h}^{k} + O(\Delta t) U_{\ell,m+h}^{k}), \tag{4.14}$$

$$m \in \underline{g}(\lambda_{n,\ell}),$$

where u_n is the n-th component of $\tilde{G}U$, each $O(\Delta t)$ is an $N \times N$ matrix and its components are quantites of order Δt.

Define

$$\begin{cases} c_3^{m/M}, & \text{if } \delta_0(\lambda_{n,\ell}) = 1 \text{ and } \delta_1(\lambda_{n,\ell}) = 0, \\ c_3^{1-m/M}, & \text{if } \delta_0(\lambda_{n,\ell}) = 0 \text{ and } \delta_1(\lambda_{n,\ell}) = 1, \end{cases}$$

$$b_{n,\ell,m} = \begin{cases} c_3, & \text{if} \quad \delta_0(\lambda_{n,\ell}) = 0 \quad \text{and} \quad \delta_1(\lambda_{n,\ell}) = 0, \\ 1, & \text{if} \quad \delta_0(\lambda_{n,\ell}) = 1 \quad \text{and} \quad \delta_1(\lambda_{n,\ell}) = 1, \end{cases} \tag{4.15}$$

where c_3 is a constant greater than 1. Clearly, on boundaries, if $i \in I_0$, the corresponding $b_{n,\ell,m} = c_3$ and if $i \in I_1$, $b_{n,\ell,m} = 1$.

Multiplying (4.14) by $b_{n,\ell,m}$, defining

$\hat{u}_{n,\ell,m} = b_{n,\ell,m}\, u_{n,\ell,m}$ and noticing $b_{n,\ell,m}/b_{n,\ell,m+h} = 1 + O(\Delta t)$,

we obtain

$$\sum_h \left(\gamma^{k+\frac{1}{2}}_{h,n,\ell,m}\, \hat{u}^{k+1}_{n,\ell,m+h} + O(\Delta t)\, \hat{U}^{k+1}_{\ell,m+h}\right) =$$

$$\sum_h \left(s^{k+\frac{1}{2}}_{h,n,\ell,m}\, \hat{u}^{k}_{n,\ell,m+h} + O(\Delta t)\hat{U}^{k}_{\ell,m+h}\right), \tag{4.16}$$

$$m \in \underline{g}(\lambda_{n,\ell}),$$

where $\hat{U}$ denotes the vector $(\hat{u}_1, \hat{u}_2, \cdots, \hat{u}_N)^T$.

Through straightforward derivation, we have

$$T^{k+1}_{n,\ell} = \sum_{m \in g(\lambda_{n,\ell})} \left(\sum_h \gamma^{k+\frac{1}{2}}_{h,n,\ell,m}\, \hat{u}^{k+1}_{n,\ell,m+h} + O(\Delta t)\hat{U}^{k+1}_{\ell,m+h}\right)^2 \Delta\xi$$

$$\geq \sum_{m \in \underline{g}(\lambda_{n,\ell})} \left(\sum_h \gamma^{k+\frac{1}{2}}_{h,n,\ell,m}\, \hat{u}^{k+1}_{n,\ell,m+h}\right)^2 \Delta\xi - c\,\Delta t\, \|\hat{\underline{U}}^{k+1}_{\ell}\|^2 ,$$

where $\|\hat{\underline{U}}_\ell\|^2 = \sum_{m=0}^{M} |\hat{U}_{\ell,m}|^2 \Delta\xi$ and c is a constant, $|\hat{U}_{\ell,m}|$ being the Euclidean norm of $\hat{U}_{\ell,m}$. Therefore, when (3.9) holds, we have

$$T^{k+1}_{n,\ell} \geq -c|\hat{u}^{k+1}_{n,\ell,0}|^2 \Delta\xi\, \delta_0(\lambda_{n,\ell,0}) - c|\hat{u}^{k+1}_{n,\ell,M}|^2 \Delta\xi\, \delta_1(\lambda_{n,\ell,M})$$

$$- c\,\Delta t\, \|\hat{\underline{U}}^{k+1}_{\ell}\|^2 + c_2 \sum_{m \in \underline{g}(\lambda_{n,\ell})} |\hat{u}^{k+1}_{n,\ell,m}|^2 \Delta\xi .$$

Consequently,

$$T^{k+1}_{a\ell\ell} \equiv \sum_{n=1}^{N} \sum_{\ell=1}^{L} T^{k+1}_{n,\ell} + c_3^2 \left| \frac{\partial \tilde{B}^{k+1}}{\partial (U_b, Z)} \begin{pmatrix} U_b^{k+1} \\ Z^{k+1} \end{pmatrix} \right|^2 \Delta\xi$$

$$\geq -c \sum_{i \in I_1} |\hat{u}^{k+1}_{bi}|^2 \Delta\xi + c_2 \sum_{n=1}^{N} \sum_{\ell=1}^{L} \sum_{m \in \underline{g}(\lambda_{n,\ell})} |\hat{u}^{k+1}_{n,\ell,m}|^2 \Delta\xi$$

$$+ c_3^2 |E|^2 \Delta\xi - c\,\Delta t\, \|\hat{\underline{U}}^{k+1}\|^2$$

$$\geq \frac{c_2}{2} \left(\|\hat{\underline{U}}^{k+1}\|^2 + |Z^{k+1}|^2 \Delta\xi \right)$$

$$-(c + \frac{c_2}{2})(\sum_{i \in I_1} |u_{bi}^{k+1}|^2 + |Z^{k+1}|^2)\Delta\xi$$

$$+ \frac{c_2 c_3^2}{2} \sum_{i \in I_0} |u_{bi}^{k+1}|^2 \Delta\xi + c_3^2 |E|^2 \Delta\xi - c\Delta t \, ||\hat{\underline{U}}^{k+1}||^2 .$$

Here the following relations are used:

$$\begin{cases} ||\hat{\underline{U}}^{k+1}||^2 \equiv \sum_{n=1}^{N} \sum_{\ell=1}^{L} \sum_{m=0}^{M} |\hat{u}_{n,\ell,m}^{k+1}|^2 \Delta\xi \\ \quad = \sum_{n=1}^{N} \sum_{\ell=1}^{L} \sum_{m \in g'(\lambda_{n,\ell})} |\hat{u}_{n,\ell,m}^{k+1}|^2 \Delta\xi \\ \quad + \sum_{i \in I_1} |\hat{u}_{bi}^{k+1}|^2 \Delta\xi , \\ \sum_{n=1}^{N} \sum_{\ell=1}^{L} \sum_{m \in \underline{g}(\lambda_{n,\ell})} |\hat{u}_{n,\ell,m}^{k+1}|^2 \Delta\xi \geq \sum_{i \in I_0} |\hat{u}_{bi}^{k+1}|^2 \Delta\xi, \\ \sum_{i \in I_0} |\hat{u}_{bi}^{k+1}|^2 = c_3^2 \sum_{i \in I_0} |u_{bi}^{k+1}|^2 , \\ \sum_{i \in I_1} |\hat{u}_{bi}^{k+1}|^2 = \sum_{i \in I_1} |u_{bi}^{k+1}|^2 . \end{cases} \tag{4.17}$$

The second equation of (4.11) can be transformed into

$$\frac{\partial \tilde{B}^{k+1}}{\partial (U_b, Z)} \begin{pmatrix} \tilde{G}_b^{k+1} & \\ & I_{L+1} \end{pmatrix}^{-1} \begin{pmatrix} (\tilde{G}_b U_b)^{k+1} \\ Z^{k+1} \end{pmatrix} = E,$$

where $\tilde{G}_b$ is a matrix similar to G_b defined by (3.17), but U_b should be replaced by $\tilde{U}_b$. Therefore it has the form of (3.19) and the following inequality in the form (3.22) holds:

$$|Z|^2 + \sum_{i \in I_1} |u_{bi}|^2 \leq c(\sum_{i \in I_0} |u_{bi}|^2 + |E|^2) . \tag{4.18}$$

Consequently, noticing $c_2 > 0$, we can choose such a c_3 that

$$T_{a\ell\ell}^{k+1} \geq \frac{c_2}{2} (||\hat{\underline{U}}^{k+1}||^2 + |Z^{k+1}|^2 \Delta\xi) - c \, \Delta t \, ||\hat{\underline{U}}^{k+1}||^2$$

$$\geq \frac{c_2}{4} (||\hat{\underline{U}}^{k+1}||^2 + |Z^{k+1}|^2 \Delta\xi). \tag{4.19}$$

Here suppose Δt be small enough. It is clear that

$$T_{a\ell\ell}^{k+1} = || \underline{B} \begin{bmatrix} \underline{R}^{k+\frac{1}{2}} \; \underline{U}^{k+1} \\ \frac{\partial \tilde{B}^{k+1}}{\partial (U_b, Z)} \begin{bmatrix} U_b^{k+1} \\ Z^{k+1} \end{bmatrix} \end{bmatrix} ||^2 ,$$

where $\underline{B}$ is a diagonal matrix whose components are $b_{n,\ell,m}$, $n=1,2,\cdots,N$, $\ell=1,2,\cdots,L$, $m \in \underline{g}(\lambda_{n,\ell})$ and a certain number of c_3. From (4.19) we can have

$$\|\underline{A}'^{-1}\|^2 = \sup \frac{\left\| \begin{bmatrix} \underline{U} \\ Z \end{bmatrix} \right\|^2}{\left\| \underline{A}' \begin{bmatrix} \underline{U} \\ Z \end{bmatrix} \right\|^2} \leq \sup \frac{\|\hat{\underline{U}}^{k+1}\|^2 + |Z^{k+1}|^2 \Delta\xi}{\|\underline{B}\|^{-2} T^{k+1}_{a\ell\ell}}$$

$$\leq \frac{4}{c_2 \|\underline{B}\|^{-2}} = \frac{4c_3^2}{c_2}. \tag{4.20}$$

This means we complete our proof.

It is not difficult to show the following results.

(1) If the conditions described at the beginning of this section hold, the errors of approximate solution at $t=(k+1+\delta)\Delta t$, $\delta=0, 1/2,\cdots,m/2$, m being a finite integer, will be $0(\Delta t^2)$.

(2) If the term $0(\Delta t^3)$ in (4.1) is changed to $0(\Delta t^2)$, the conclusion is still correct. Therefore there exists a solution $\begin{pmatrix} \underline{U}^{\frac{1}{2}} \\ Z^{\frac{1}{2}} \end{pmatrix}$ of (2.10) satisfying the inequality

$$\left\| \begin{pmatrix} \underline{U}^{\frac{1}{2}} - \tilde{\underline{U}}^{\frac{1}{2}} \\ Z^{\frac{1}{2}} - \tilde{Z}^{\frac{1}{2}} \end{pmatrix} \right\| \leq c\Delta t^2, \tag{4.21}$$

where c is a certain constant.

(3) For first order schemes, a similar conclusion is correct. The only difference is that the errors at the 'given' level(s) and at the 'unknown' level are $0(\Delta t)$.

(4) For scheme (2.16), if conditions (i) and (iii) are fulfilled and

(ii') the error at $t = k\Delta t$ is $0(\Delta t^2)$,

then the system (2.16) has a soluton $\{\underline{U}^{k+1}, Z^{k+1}\}$, and the difference between $\{\underline{U}^{k+1}, Z^{k+1}\}$ and the exact solution $\{\tilde{\underline{U}}^{k+1}, \tilde{Z}^{k+1}\}$ is $0(\Delta t^{3/2})$.

5. PROOF OF CONVERGENCE

As in Section 4, $\{\tilde{\underline{U}}, \tilde{Z}\}$ denotes the exact solution and suppose in each subregion $\{\tilde{\underline{U}}, \tilde{Z}\}$ is quite smooth so that $\{\tilde{\underline{U}}, \tilde{Z}\}$ satisfy (4.1). By means of (2.15), (4.1) can be rewritten as

$$\begin{cases} \tilde{\underline{R}}^{k+\frac{1}{2}} \delta \tilde{\underline{U}}^k = \tilde{\underline{Q}}^{k+\frac{1}{2}} \Delta_+ \tilde{\underline{U}}^k + \Delta t\, \tilde{\underline{F}}^{k+\frac{1}{2}} + O_1^{k+\frac{1}{2}} (\Delta t^3), \\ B(\tilde{U}_b^{k+1}, \tilde{Z}^{k+1}, t^{k+1}) = 0, \quad k = 0, 1/2, 1, \cdots, T/\Delta t-1, \end{cases} \tag{5.1}$$

where $\tilde{\underline{Q}}$ is equivalent to $\underline{Q}$, i.e., substituting $\tilde{\underline{U}}, \tilde{Z}$ for $\underline{U}$, Z in $\underline{Q}$, we can obtain $\tilde{\underline{Q}}$. Subtracting (5.1) from (2.14), we obtain the error equations:

$$\begin{cases} \underline{R}^{k+\frac{1}{2}}\delta(\underline{U}^k - \underline{\tilde{U}}^k) + (\underline{R}^{k+\frac{1}{2}} - \underline{\tilde{R}}^{k+\frac{1}{2}})\,\delta\,\underline{\tilde{U}}^k \\ = \underline{Q}^{k+\frac{1}{2}}\,\Delta_+(\underline{U}^k - \underline{\tilde{U}}^k) + (\underline{Q}^{k+\frac{1}{2}} - \underline{\tilde{Q}}^{k+\frac{1}{2}})\,\Delta_+\,\underline{\tilde{U}}^k \\ \quad + \Delta t(\underline{F}^{k+\frac{1}{2}} - \underline{\tilde{F}}^{k+\frac{1}{2}}) + O_1^{k+\frac{1}{2}}(\Delta t^3), \\ \dfrac{\partial B^{k+1}}{\partial(U_b, Z)}\begin{bmatrix} U_b^{k+1} - \tilde{U}_b^{k+1} \\ Z^{k+1} - \tilde{Z}^{k+1} \end{bmatrix} = 0, \quad k = 0, \tfrac{1}{2}, 1, \cdots, \end{cases} \tag{5.2}$$

where $U_b = U_b^{k+1} + \xi(U_b^{k+1} - \tilde{U}_b^{k+1})$ in $\dfrac{\partial B^{k+1}}{\partial(U_b, Z)}$, $0 \le \xi \le 1$. Since each $U_{\ell,m}^{k+\frac{1}{2}}$ or $U_{\ell,m}^k$ appears only in several rows of $\underline{R}^{k+\frac{1}{2}}$ and each row of $\underline{R}^{k+\frac{1}{2}}$ has only several nonzero components, $(\underline{R}^{k+\frac{1}{2}} - \underline{\tilde{R}}^{k+\frac{1}{2}})\,\delta\,\underline{\tilde{U}}^k$ has the following form

$$0(\Delta t)(\underline{U}^{k+\frac{1}{2}} - \underline{\tilde{U}}^{k+\frac{1}{2}}) + 0^*(\Delta t)(Z^{k+\frac{1}{2}} - \tilde{Z}^{k+\frac{1}{2}})$$
$$+ 0(\Delta t)(\underline{U}^k - \underline{\tilde{U}}^k) + 0^*(\Delta t)(Z^k - \tilde{Z}^k).$$

Here $0(\Delta t)$ denotes a matrix in each row or each column of which there are only several non-zero components of order Δt. Therefore, the L_2 norm of $0(\Delta t)$ is less than $c\Delta t$, c being a constant. $0^*(\Delta t)$ denotes a matrix whose number of columns is $L + 1$ and whose each component is a quantity of order Δt. Consequently, the L_2 norm of $0^*(\Delta t)$ is less than $c\Delta t^{\frac{1}{2}}$, c being a constant. Obviously, $(\underline{Q}^{k+\frac{1}{2}} - \underline{\tilde{Q}}^{k+\frac{1}{2}})\Delta_+\underline{\tilde{U}}^k$ and $\Delta t(\underline{F}^{k+\frac{1}{2}} - \underline{\tilde{F}}^{k+\frac{1}{2}})$ have the same form. Therefore, the first part of (5.2) can be written in the form

$$\underline{R}^{k+\frac{1}{2}}(\underline{U}^{k+1} - \underline{\tilde{U}}^{k+1}) = 0(\Delta t)(\underline{U}^{k+\frac{1}{2}} - \underline{\tilde{U}}^{k+\frac{1}{2}}) + 0^*(\Delta t)(Z^{k+\frac{1}{2}} - \tilde{Z}^{k+\frac{1}{2}})$$
$$+ R^{k+\frac{1}{2}}(\underline{U}^k - \underline{\tilde{U}}^k) + \underline{Q}^{k+\frac{1}{2}}\Delta_+(\underline{U}^k - \underline{\tilde{U}}^k)$$
$$+ 0(\Delta t)(\underline{U}^k - \underline{\tilde{U}}^k) + 0^*(\Delta t)(Z^k - \tilde{Z}^k) + 0_1^{k+\frac{1}{2}}(\Delta t^3).$$

Let $\underline{V}$ be a vector whose components are $\tilde{G}_{1,0}(U_{1,0} - \tilde{U}_{1,0}), \cdots, \tilde{G}_{1,M}(U_{1,M} - \tilde{U}_{1,M}), \cdots, \tilde{G}_{L,0}(U_{L,0} - \tilde{U}_{L,0}), \cdots, \tilde{G}_{L,M}(U_{L,M} - \tilde{U}_{L,M})$ from the top to the bottom, V_b be a vector whose components are $\tilde{G}_{1,0}(U_{1,0} - \tilde{U}_{1,0}), \tilde{G}_{1,M}(U_{1,M} - \tilde{U}_{1,M}), \cdots, \tilde{G}_{L,0}(U_{L,0} - \tilde{U}_{L,0})$ and $\tilde{G}_{L,M}(U_{L,M} - \tilde{U}_{L,M})$, and $Y = Z - \tilde{Z}$. Noticing (2.15), the error system can further be rewritten as

$$\begin{cases} \underline{R}_g^{k+\frac{1}{2}}\ \underline{V}^{k+1} = 0(\Delta t)\underline{V}^{k+\frac{1}{2}} + 0^*(\Delta t)\,Y^{k+\frac{1}{2}} \\ \qquad + (\underline{S}_g^{k+\frac{1}{2}} + 0(\Delta t))\underline{V}^k + 0^*(\Delta t)\,Y^k + 0_1^{k+\frac{1}{2}}(\Delta t^3), \\ \underline{B}_g^{k+1}\begin{pmatrix} \underline{V}^{k+1} \\ Y^{k+1} \end{pmatrix} = 0, \end{cases} \tag{5.3}$$

where $\underline{R}_g^{k+\frac{1}{2}} = \underline{R}^{k+\frac{1}{2}}(\underline{\tilde{G}}^{k+1})^{-1}$, $\underline{S}_g^{k+\frac{1}{2}} = \underline{S}^{k+\frac{1}{2}}(\underline{\tilde{G}}^{k})^{-1}$,

$$\underline{\tilde{G}} \equiv \begin{bmatrix} \tilde{G}_{1,0} & & & & & 0 \\ & \ddots & & & & \\ & & \tilde{G}_{1,M} & & & \\ & & & \ddots & & \\ & & & & \tilde{G}_{L,0} & \\ & & & & & \ddots \\ 0 & & & & & \tilde{G}_{L,M} \end{bmatrix} \tag{5.4}$$

and $\underline{B}_g^{k+1}$ is defined by

$$\underline{B}_g^{k+1} \begin{pmatrix} \underline{V}^{k+1} \\ \underline{Y}^{k+1} \end{pmatrix} \equiv B_g^{k+1} \begin{pmatrix} V_b^{k+1} \\ \underline{Y}^{k+1} \end{pmatrix}, \tag{5.5}$$

B_g^{k+1} denoting $\dfrac{\partial B^{k+1}}{\partial(U_b,Z)} \begin{bmatrix} \tilde{G}_b^{k+1} & \\ & I_{L+1} \end{bmatrix}^{-1}$, $\tilde{G}_b \equiv \begin{bmatrix} \tilde{G}_{1,0} & & & & 0 \\ & \tilde{G}_{1,M} & & & \\ & & \ddots & & \\ & & & \tilde{G}_{L,0} & \\ 0 & & & & \tilde{G}_{L,M} \end{bmatrix}$. Rewriting

the first equation of (5.3) as

$$\underline{R}_g^{k+\frac{1}{2}}\ \underline{V}^{k+1} + 0(\Delta t)\underline{V}^{k+\frac{1}{2}} + 0^*(\Delta t)\underline{Y}^{k+\frac{1}{2}}$$

$$= (\underline{S}_g^{k+\frac{1}{2}} + 0(\Delta t))V^k + 0^*(\Delta t)\underline{Y}^k + 0_1^{k+\frac{1}{2}}(\Delta t^3) \tag{5.6}$$

for $k = \frac{1}{2}, \frac{3}{2}, \cdots$, and combining (5.3) with $k = j$ and (5.6) with $k = j + \frac{1}{2}$, j being an integer, then we have the final form of the error system

$$\underline{\hat{R}}^k \underline{W}^{k+1} = \underline{\hat{S}}^k \underline{W}^k + 0^k(\Delta t^3), \tag{5.7}$$
$$k = 0,1,\cdots, T/\Delta t - 1,$$

where

$$\begin{cases} \underline{\hat{R}}^k = \begin{bmatrix} \underline{R}_g^{k+1} & 0 & 0(\Delta t) & 0^*(\Delta t) \\ & \underline{B}_g^{k+3/2} & 0 & 0 \\ 0 & 0 & \underline{R}_g^{k+\frac{1}{2}} & 0 \\ 0 & 0 & & \underline{B}_g^{k+1} \end{bmatrix} \\ \underline{\hat{S}}^k = \begin{bmatrix} \underline{S}_g^{k+1} + 0(\Delta t) & 0^*(\Delta t) & 0 & 0 \\ 0 & 0 & 0 & 0 \\ 0(\Delta t) & 0^*(\Delta t) & \underline{S}_g^{k+\frac{1}{2}} + 0(\Delta t) & 0^*(\Delta t) \\ 0 & 0 & 0 & 0 \end{bmatrix}, \end{cases} \tag{5.8}$$

$$\underline{W}^k = \begin{bmatrix} \underline{V}^{k+\frac{1}{2}} \\ Y^{k+\frac{1}{2}} \\ \underline{V}^k \\ Y^k \end{bmatrix}$$

$$0^k(\Delta t^3) = \begin{bmatrix} 0_1^{k+1}(\Delta t^3) \\ 0_1^{k+\frac{1}{2}}(\Delta t^3) \end{bmatrix} .$$

Therefore

$$\begin{aligned} \underline{W}^{k+1} &= (\hat{\underline{R}}^k)^{-1} \hat{\underline{S}}^k \underline{W}^k + (\hat{\underline{R}}^k)^{-1} 0^k(\Delta t^3) \\ &= (\hat{\underline{R}}^k)^{-1} \hat{\underline{S}}^k (\hat{\underline{R}}^{k-1})^{-1} \hat{\underline{S}}^{k-1} \underline{W}^{k-1} + (\hat{\underline{R}}^k)^{-1} \hat{\underline{S}}^k (\hat{\underline{R}}^{k-1})^{-1} 0^{k-1}(\Delta t^3) \\ &\quad + (\hat{\underline{R}}^k)^{-1} 0^k(\Delta t^3) \\ &= \prod_{j=0}^{k} (\hat{\underline{R}}^j)^{-1} \hat{\underline{S}}^j \underline{W}^0 + \sum_{i=0}^{k} \prod_{j=i+1}^{k} (\hat{\underline{R}}^j)^{-1} \hat{\underline{S}}^j (\hat{\underline{R}}^i)^{-1} 0^i(\Delta t^3). \end{aligned}$$

Consequently, if

$$\| (\hat{\underline{R}}^i)^{-1} \| \le c, \tag{5.9}$$

$$\| \prod_{j=j_1}^{j_2} (\hat{\underline{R}}^j)^{-1} \hat{\underline{S}}^j \| \le c, \quad \text{for } 0 \le j_1 \Delta t \le j_2 \Delta t \le T, \tag{5.10}$$

then

$$\| \underline{W}^{k+1} \| \le c_4 (\Delta t^2 + \| \underline{W}^0 \|), \text{ for any } (k+1)\Delta t \le T,$$

where c_4 is a certain constant. In Section 4 we pointed out $\| \begin{pmatrix} \underline{U}^{\frac{1}{2}} - \tilde{\underline{U}}^{\frac{1}{2}} \\ Z^{\frac{1}{2}} - \tilde{Z}^{\frac{1}{2}} \end{pmatrix} \| = 0(\Delta t^2)$, so $\| W^0 \| = 0(\Delta t^2)$. Therefore, noting the boundedness of $\| \tilde{\underline{G}}^{-1} \|$, we can obtain immediately

$$\| \begin{pmatrix} \underline{U} - \tilde{\underline{U}} \\ Z - \tilde{Z} \end{pmatrix}^{k+1} \| \le c\Delta t^2, \; k=0, \tfrac{1}{2}, \cdots, T/\Delta t - 1, \tag{5.11}$$

where c is a constant.

That is, if the systems for $\underline{W}^{k+1}$

$$\hat{\underline{R}}^k W^{k+1} = \hat{\underline{S}}^k \underline{W}^k, \qquad k = 0,1,\cdots, \tag{5.12}$$

are "well-conditioned" and the procedure (5.12) is stable, which means that (5.9) and (5.10) hold respectively, the approximate solution $\{ \underline{U}, Z \}$ obtained from (2.10)-(2.11) converges to the exact solution $\{ \tilde{\underline{U}}, \tilde{Z} \}$.

In what follows, we shall discuss the stability of the procedure (5.12).

If we can find an invertible matrix $\underline{B}$ such that for the procedure (5.12)

$$\| \underline{B}\hat{\underline{R}}^k \underline{W}^{k+1} \|^2 - \| \underline{B}\hat{\underline{R}}^{k-1} \underline{W}^k \|^2 \le c\Delta t \; \| W^k \|^2 \tag{5.13}$$

and

$$\max\{ ||\underline{B}\hat{\underline{R}}^k||^2, \quad ||(\underline{B}\hat{\underline{R}}^k)^{-1}||^2\} \leq c, \tag{5.14}$$

then we obtain

$$\begin{aligned} || \underline{B}\hat{\underline{R}}^k \underline{W}^{k+1}||^2 &\leq ||\underline{B}\hat{\underline{R}}^{k-1}\underline{W}^k||^2 + c\Delta t|| W^k||^2 \\ &\leq (1 + c^2 \Delta t)|| \underline{B}\hat{\underline{R}}^{k-1}\underline{W}^k||^2 \\ &\leq (1 + c^2 \Delta t)^{k+1-i} || \underline{B}\hat{\underline{R}}^{i-1}\underline{W}^i||^2 \\ &\leq e^{c^2(k+1-i)\Delta t} || \underline{B}\hat{\underline{R}}^{i-1}\underline{W}^i||^2 \\ &\leq e^{c^2(k+1-i)\Delta t} c|| \underline{W}^i||^2 . \end{aligned}$$

Therefrore

$$|| \underline{W}^{k+1}||^2 \leq c^2 e^{c^2(k+1-i)\Delta t} || \underline{W}^i||^2 .$$

Noticing

$$\underline{W}^{k+1} = \prod_{j=i}^{k} (\hat{\underline{R}}^j)^{-1} \hat{\underline{S}}^j \underline{W}^i,$$

we have

$$|| \prod_{j=i}^{k} (\hat{\underline{R}}^j)^{-1} \hat{\underline{S}}^j ||^2 \leq c^2 e^{c^2(k+1-i)\Delta t} .$$

Moreoever, we can obtain (5.9) from (5.14) if $\underline{B}$ is invertible. Therefore, the proof of stability can be reduced to finding an invertible B such that (5.13) and (5.14) hold.

In Section 4, we point out that (4.13) can be rewritten in the form (4.16). Similarly, every equation in (5.12) which corresponds to some difference equation, after multiplying by a $b_{n,\ell,m}$ defined in (4.15), can be written in the following form

$$\begin{aligned} &\sum_h (\gamma^{k+1}_{h,n,\ell,m} \bar{v}^{k+3/2}_{n,\ell,m+h} + O(\Delta t)\bar{V}^{k+3/2}_{\ell,m+h} + O(\Delta t)\bar{V}^{k+1}_{\ell,m+h}) + O^*(\Delta t)Y^{k+1} \\ &= \sum_h (s^{k+1}_{h,n,\ell,m} \bar{v}^{k+\frac{1}{2}}_{n,\ell,m+h} + O(\Delta t) \bar{V}^{k+\frac{1}{2}}_{\ell,m+h}) + O^*(\Delta t)Y^{k+\frac{1}{2}}, \end{aligned} \tag{5.15}$$

which is in the form (5.6) or

$$\begin{aligned} &\sum_h (\gamma^{k+\frac{1}{2}}_{h,n,\ell,m} \bar{v}^{k+1}_{n,\ell,m+h} + O(\Delta t) \bar{V}^{k+1}_{\ell,m+h}) \\ &= \sum_h (s^{k+\frac{1}{2}}_{h,n,\ell,m} \bar{v}^{k}_{n,\ell,m+h} + O(\Delta t)\bar{V}^{k}_{\ell,m+h} + O(\Delta t) \bar{V}^{k+\frac{1}{2}}_{\ell,m+h}) \\ &\quad + O^*(\Delta t) Y^k + O^*(\Delta t) Y^{k+\frac{1}{2}}, \end{aligned} \tag{5.16}$$

which is in the form (5.3).

Here $\bar{v}_{n,\ell,m} = b_{n,\ell,m} v_{n,\ell,m}$, $v_{n,\ell,m}$ being the n-th component of $V_{\ell,m} = \tilde{G}_{\ell,m}(U_{\ell,m} - \tilde{U}_{\ell,m})$, and $\bar{V}_{\ell,m} = (\bar{v}_{1,\ell,m}, \bar{v}_{2,\ell,m}, \ldots, \bar{v}_{N,\ell,m})^T$. Therefore, if B is a diagonal matrix whose diagonal element is $b_{n,\ell,m}$ when the corresponding equation is a difference equation related to λ_n at $\xi = \ell + m\Delta\xi$ or is c_3 when the corresponding equation is a boundary condition, then in

$$\underline{B}\underline{\hat{R}}^k \underline{W}^{k+1} = \underline{B}\underline{\hat{S}}^k \underline{W}^k$$

every difference equation has the form (5.15) or (5.16).

Through straightforward derivation, we have

$$\bar{T}^{k+3/2}_{n,\ell} \equiv \sum_{m\in\underline{g}(\lambda_{n,\ell})} \left(\sum_h \left(\gamma^{k+1}_{h,n,\ell,m} \bar{v}^{k+3/2}_{n,\ell,m+h} + 0(\Delta t)\bar{v}^{k+3/2}_{\ell,m+h} + 0(\Delta t)\bar{v}^{k+1}_{\ell,m+h}\right) + 0*(\Delta t)Y^{k+1}\right)^2 \Delta\xi$$

$$\geq \sum_{m\in\underline{g}(\lambda_{n,\ell})} \left(\sum_h \gamma^{k+1}_{h,n,\ell,m} \bar{v}^{k+3/2}_{n,\ell,m+h}\right)^2 \Delta\xi - c\Delta t \|\underline{\bar{v}}^{k+3/2}_{\ell}\|^2 - c\Delta t \|\underline{\bar{v}}^{k+1}_{\ell}\|^2 - c\Delta\xi |Y^{k+1}|^2 \tag{5.17}$$

and

$$T^{k+1}_{n,\ell} = \sum_{m\in\underline{g}(\lambda_{n,\ell})} \left(\sum_h \left(\gamma^{k+\frac{1}{2}}_{h,n,\ell,m} \bar{v}^{k+1}_{n,\ell,m+h} + 0(\Delta t)\bar{v}^{k+1}_{\ell,m+h}\right)\right)^2 \Delta\xi$$

$$\geq \sum_{m\in\underline{g}(\lambda_{n,\ell})} \left(\sum_h \gamma^{k+\frac{1}{2}}_{h,n,\ell,m} \bar{v}^{k+1}_{n,\ell,m+h}\right)^2 \Delta\xi - c\Delta t \|\underline{\bar{v}}^{k+1}_{\ell}\|^2, \tag{5.18}$$

where $\|\underline{\bar{V}}_\ell\|^2 = \sum_{m=0}^{M} |\bar{V}_{\ell,m}|^2 \Delta\xi$, $|Y|^2 = \sum_{\ell=0}^{L} |Y_\ell|^2$ and c is a constant.

Therefore, when (3.9) holds,we have

$$\bar{T}^{k+3/2}_{n,\ell} \geq -c|\bar{v}^{k+3/2}_{n,\ell,0}|^2 \Delta\xi\delta_0(\lambda_{n,\ell,0}) - c|\bar{v}^{k+3/2}_{n,\ell,M}|^2 \Delta\xi\delta_1(\lambda_{n,\ell,M}) - c\Delta t \|\underline{\bar{v}}^{k+3/2}_{\ell}\|^2 - c\Delta t \|\underline{\bar{v}}^{k+1}_{\ell}\|^2 - c\Delta\xi |Y^{k+1}|^2 + c_2 \sum_{m\in\underline{g}(\lambda_{n,\ell})} |\bar{v}^{k+3/2}_{n,\ell,m}|^2 \Delta\xi$$

$$T^{k+1}_{n,\ell} \geq -c|\bar{v}^{k+1}_{n,\ell,0}|^2 \Delta\xi\delta_0(\lambda_{n,\ell,0}) - c|\bar{v}^{k+1}_{n,\ell,M}|^2 \Delta\xi\delta_1(\lambda_{n,\ell,M}) - c\Delta t \|\underline{\bar{v}}^{k+1}_{\ell}\|^2 + c_2 \sum_{m\in\underline{g}(\lambda_{n,\ell})} |\bar{v}^{k+1}_{n,\ell,m}|^2 \Delta\xi.$$

From the definition of $\underline{\hat{R}}^k$, we know

$$\|\underline{B}\underline{\hat{R}}^k \underline{W}^{k+1}\|^2 = \sum_{n=1}^{N} \sum_{\ell=1}^{L} \left(\bar{T}^{k+3/2}_{n,\ell} + T^{k+1}_{n,\ell}\right) + c_3^2\left(\left|B^{k+3/2}_g \begin{pmatrix} V^{k+3/2}_b \\ Y^{k+3/2} \end{pmatrix}\right|^2 + \left|B^{k+1}_g \begin{pmatrix} V^{k+1}_b \\ Y^{k+1} \end{pmatrix}\right|^2\right)\Delta\xi$$

$$\geq \sum_{j=k+1}^{k+3/2} \left\{-c\left(\sum_{i\in I_1} |\bar{v}^j_{bi}|^2 + |Y^j|^2\right)\Delta\xi + c_2 \sum_{n=1}^{N} \sum_{\ell=1}^{L} \sum_{m\in\underline{g}(\lambda_{n,\ell})} |\bar{v}^j_{n,\ell,m}|^2 \Delta\xi\right.$$

$$- c\Delta t\|\underline{\bar{V}}^j\|^2 + c_3^2 \left|B_g^j \begin{pmatrix} V_b^j \\ Y^j \end{pmatrix}\right|^2 \Delta\xi\}$$

$$\geq \sum_{j=k+1}^{k+3/2} \{ \frac{c_2}{2} (\|\underline{\bar{V}}^j\|^2 + |Y^j|^2 \Delta\xi)$$

$$-(c + \frac{c_2}{2}) (\sum_{i \in I_1} |v_{bi}^j|^2 + |Y^j|^2) \Delta\xi$$

$$+ \frac{c_2 c_3^2}{2} \sum_{i \in I_0} |v_{bi}^j|^2 \Delta\xi + c_3^2 \left|B_g^j \begin{pmatrix} V_b^j \\ Y^j \end{pmatrix}\right|^2 \Delta\xi - c\Delta t \|\underline{\bar{V}}^j\|^2 \} ,$$

where $\|\underline{\bar{V}}\|^2 = \sum_{\ell=1}^{L} \|\underline{\bar{V}}_\ell\|^2$. Here some relations similar to (4.17) are used.

Because (3.22) holds and c_2 is positive, we can choose such a c_3 that

$$-(c + \frac{c_2}{2}) (\sum_{i \in I_1} |v_{bi}^j|^2 + |Y^j|^2)$$

$$+ \frac{c_2 c_3^2}{2} \sum_{i \in I_0} |v_{bi}^j|^2 + c_3^2 \left|B_g^j \begin{pmatrix} V_b^j \\ Y^j \end{pmatrix}\right|^2 \geq 0,$$

$$j = k + 1, \quad k + 3/2.$$

Consequently, if $c\Delta t < \frac{c_2}{4}$, then noting $c_3 > 1$, we have

$$\|\underline{\hat{BR}}^k \underline{W}^{k+1}\|^2 \geq \frac{c_2}{4} \sum_{j=k+1}^{k+3/2} (\|\underline{\bar{V}}^j\|^2 + |Y^j|^2 \Delta\xi)$$

$$\geq \frac{c_2}{4} \|\underline{W}^{k+1}\|^2 ,$$

that is

$$\|(\underline{\hat{BR}}^k)^{-1}\|^2 \leq \frac{4}{c_2} .$$

The boundedness of $\|\underline{\hat{BR}}^k\|^2$ is obvious. Therefore we have proved (5.14) holds.

Clearly, noticing (5.15) and (5.16), we can obtain the following inequalities on $\bar{T}_{n,\ell}^{k+3/2}$ and $T_{n,\ell}^{k+1}$:

$$\bar{T}_{n,\ell}^{k+3/2} = \sum_{m \in \underline{g}(\lambda_{n,\ell})} (\sum_h (\gamma_{h,n,\ell,m}^{k+1} \bar{v}_{n,\ell,m+h}^{k+3/2} + 0(\Delta t) \bar{V}_{\ell,m+h}^{k+3/2} + 0(\Delta t) \bar{V}_{\ell,m+h}^{k+1}) +$$

$$+ 0^*(\Delta t) Y^{k+1})^2 \Delta\xi$$

$$= \sum_{m \in \underline{g}(\lambda_{n,\ell})} (\sum_h (s_{h,n,\ell,m}^{k+1} \bar{v}_{n,\ell,m+h}^{k+\frac{1}{2}} + 0(\Delta t) \bar{V}_{\ell,m+h}^{k+\frac{1}{2}}) + 0^*(\Delta t) Y^{k+\frac{1}{2}})^2 \Delta\xi \tag{5.19}$$

$$\leq \sum_{m \in \underline{g}(\lambda_{n,\ell})} (\sum_h s_{h,n,\ell,m}^{k+1} \bar{v}_{n,\ell,m+h}^{k+\frac{1}{2}})^2 \Delta\xi + c\Delta t \|\underline{\bar{V}}_\ell^{k+\frac{1}{2}}\|^2 + c\Delta\xi |Y^{k+\frac{1}{2}}|^2$$

and

$$T_{n,\ell}^{k+1} = \sum_{m\in \underline{g}(\lambda_{n,\ell})} \left(\sum_h (\gamma_{h,n,\ell,m}^{k+\frac{1}{2}} \bar{v}_{n,\ell,m+h}^{k+1} + 0(\Delta t)\bar{v}_{\ell,m+h}^{k+1})\right)^2 \Delta\xi$$

$$= \sum_{m\in \underline{g}(\lambda_{n,\ell})} \left(\sum_h (s_{h,n,\ell,m}^{k+\frac{1}{2}} \bar{v}_{n,\ell,m+h}^{k} + 0(\Delta t)\bar{v}_{\ell,m+h}^{k} + 0(\Delta t)\bar{v}_{\ell,m+h}^{k+\frac{1}{2}})\right.$$

$$\left. + 0^*(\Delta t)Y^{k} + 0^*(\Delta t)Y^{k+\frac{1}{2}}\right)^2 \Delta\xi$$

$$\leq \sum_{m\in g(\lambda_{n,\ell})} \left(\sum_h (s_{h,n,\ell,m}^{k+\frac{1}{2}} \bar{v}_{n,\ell,m+h}^{k})\right)^2 \Delta\xi + c\Delta t \| \underline{\bar{V}}_{\ell}^{k} \|^2 +$$

$$+ c\Delta t \| \underline{\bar{V}}_{\ell}^{k+\frac{1}{2}} \|^2 + c\Delta\xi (|Y^{k}|^2 + |Y^{k+\frac{1}{2}}|^2). \tag{5.20}$$

Therefore when (3.10) holds, from (5.17)-(5.20) we have

$$\bar{T}_{n,\ell}^{k+3/2} + T_{n,\ell}^{k+1} - \bar{T}_{n,\ell}^{k+\frac{1}{2}} - T_{n,\ell}^{k} \leq$$

$$\leq \sum_{j=k}^{k+\frac{1}{2}} \Big(\sum_{m\in \underline{g}(\lambda_{n,\ell})} \big(\sum_h s_{h,n,\ell,m}^{j+\frac{1}{2}} \bar{v}_{n,\ell,m+h}^{j}\big)^2 \Delta\xi$$

$$- \sum_{m\in \underline{g}(\lambda_{n,\ell})} \big(\sum_h \gamma_{h,n,\ell,m}^{j+\frac{1}{2}} \bar{v}_{n,\ell,m+h}^{j} \big)^2 \Delta\xi$$

$$+ c\Delta t \| \underline{\bar{V}}_{\ell}^{j} \|^2 + c\Delta\xi |Y^{j}|^2 \Big)$$

$$\leq \sum_{j=k}^{k+\frac{1}{2}} [(c\delta_0(\lambda_{n,\ell,0}) - c_2 |\sigma_{n,\ell,0}^{j}| (1-\delta_0(\lambda_{n,\ell,0}))) |\bar{v}_{n,\ell,0}^{j}|^2 \Delta\xi$$

$$+ (c\delta_1(\lambda_{n,\ell,M}) - c_2 |\sigma_{n,\ell,M}^{j}| (1-\delta_1(\lambda_{n,\ell,M}))) |\bar{v}_{n,\ell,M}^{j}|^2 \Delta\xi$$

$$+ c\Delta t \| \underline{\bar{V}}_{\ell}^{j} \|^2 + c\Delta\xi |Y^{j}|^2],$$

$$n = 1,2,\cdots,N, \qquad \ell = 1,2,\cdots,L.$$

Furthermore, noticing $\underline{B}_g^{k+1}\begin{pmatrix} \underline{V}^{k+1} \\ Y^{k+1} \end{pmatrix} = B_g^{k+1}\begin{pmatrix} V_b^{k+1} \\ Y^{k+1} \end{pmatrix} = 0$, we have

$$\| \underline{\hat{BR}}^{k} \underline{W}^{k+1} \|^2 - \| \underline{\hat{BR}}^{k} \underline{W}^{k} \|^2$$

$$= \sum_{n=1}^{N} \sum_{\ell=1}^{L} (\bar{T}_{n,\ell}^{k+3/2} + T_{n,\ell}^{k+1} - \bar{T}_{n,\ell}^{k+\frac{1}{2}} - T_{n,\ell}^{k})$$

$$+ c_3^2 \Big(\Big|B_g^{k+3/2}\begin{pmatrix} V_b^{k+3/2} \\ Y^{k+3/2} \end{pmatrix}\Big|^2 + \Big|B_g^{k+1}\begin{pmatrix} V_b^{k+1} \\ Y^{k+1} \end{pmatrix}\Big|^2$$

$$- \Big|B_g^{k+\frac{1}{2}}\begin{pmatrix} V_b^{k+\frac{1}{2}} \\ Y^{k+\frac{1}{2}} \end{pmatrix}\Big|^2 - \Big|B_g^{k}\begin{pmatrix} V_b^{k} \\ Y^{k} \end{pmatrix}\Big|^2 \Big) \Delta\xi$$

$$\leq \sum_{j=k}^{k+\frac{1}{2}} \Big(\Big(c \sum_{i\in I_1} |\bar{v}_{bi}^{j}|^2 - c_2 \frac{\Delta t}{\Delta\xi} \sum_{i\in I_0} |\lambda_{bi}^{j}| |\bar{v}_{bi}^{j}|^2 + c|Y^{j}|^2 \Big) \Delta\xi$$

$$+ c\Delta t \| \underline{\bar{V}}^j \|^2 \;)$$

$$= \sum_{j=k}^{k+\frac{1}{2}} ((c \sum_{i\in I_1} |v_{bi}^j|^2 - c_2 c_3^2 \frac{\Delta t}{\Delta\xi} \sum_{i\in I_0} |\lambda_{bi}^j| \, |v_{bi}^j|^2 + c|\underline{Y}^j|^2)\Delta\xi$$

$$+ c\Delta t \| \underline{\bar{V}}^j \|^2).$$

When (3.20) holds and E = 0 we can choose such a constant c_3 that

$$c|\underline{Y}^j|^2 + c \sum_{i\in I_1} |v_{bi}^j|^2 - c_2 c_3^2 \frac{\Delta t}{\Delta\xi} \sum_{i\in I_0} |\lambda_{bi}^j| \, |v_{bi}^j|^2 \le 0.$$

Consequently, we have derived

$$\| \underline{B}\underline{\hat{R}}^k \underline{W}^{k+1} \|^2 - \| \underline{B}\underline{\hat{R}}^k \underline{W}^k \|^2$$

$$\le c\Delta t (\| \underline{\bar{V}}^k \|^2 + \| \underline{\bar{V}}^{k+\frac{1}{2}} \|^2)$$

$$\le c c_3^2 \, \Delta t \; \| \underline{W}^k \|^2 \quad ,$$

that is, we have proved that (5.13) holds. (5.13) and (5.14) hold, so (5.12) is stable. Consequently, (5.11) holds, which means $\begin{pmatrix} \underline{U} \\ Z \end{pmatrix}$ converges to $\begin{pmatrix} \tilde{\underline{U}} \\ \tilde{Z} \end{pmatrix}$ with a convergence rate of order Δt^2.

Finally, we would like to point out that for scheme (2.16), the same result can be obtained. That is, if the scheme (2.16) has Property A, the boundary condition has Property B, and the solution in every subregion has certain smoothness, then there exists a solution $\begin{pmatrix} \underline{U}^{k+1} \\ Z^{k+1} \end{pmatrix}$ of (2.16) which satisfies

$$\left\| \begin{pmatrix} \underline{U}^{k+1} - \tilde{\underline{U}}^{k+1} \\ Z^{k+1} - \tilde{Z}^{k+1} \end{pmatrix} \right\| \le c\Delta t^2, \quad \text{for} \quad (k+1)\Delta t \le T.$$

6. SOME DISCUSSION ON PROPERTY A

In Section 3, we point out that schemes (2.5) and (3.14) possess Property A for the case of constant coefficients. Indeed, this fact is still true for the case of variable coefficients and many schemes are of Property A. In this section, we shall prove schemes (2.5) and (2.6) possess Property A for the case of variable coefficients.

First we give two Lemmas

Lemma 1 If $\sum_{\ell} (-1)^{H_3(\ell)} d_{\ell}^*(\theta) d_{\ell}(\theta) \equiv 0$, then the matrix $Q = \sum_{\ell} (-1)^{H_3(\ell)} D_{\ell}^* D_{\ell}$ is a pseudo-null matrix, i.e., the sums of the elements on every diagonal line of the matrix are all equal to zero, where $H_3(\ell)$ is equal to either 0 or 1,

$$d_\ell(\theta) = \sum_{h=H_1}^{H_2} d_{\ell,h} e^{ih\theta}, \quad D_\ell = (d_{\ell,H_1}, d_{\ell,H_1+1}, \ldots, d_{\ell,H_2})$$

and the symbol "*" represents conjugate transposition for vectors and conjugation for scalar quantities.

Proof. Because of

$$\sum_\ell (-1)^{H_3(\ell)} \left(\sum_{h=H_1}^{H_2} d_{\ell,h} e^{ih\theta}\right)^* \left(\sum_{h=H_1}^{H_2} d_{\ell,h} e^{ih\theta}\right)$$

$$= \sum_\ell (-1)^{H_3(\ell)} \sum_{h=-H}^{H} \left(\sum_j d^*_{\ell,j} d_{\ell,j+h}\right) e^{ih\theta}$$

$$= \sum_{h=-H}^{H} \left(\sum_\ell (-1)^{H_3(\ell)} \sum_j d^*_{\ell,j} d_{\ell,j+h}\right) e^{ih\theta} \equiv 0,$$

for any h, we have

$$\sum_\ell (-1)^{H_3(\ell)} \sum_j d^*_{\ell,j} d_{\ell,j+h} = 0,$$

where $H = H_2 - H_1$, and j in summation formula runs over all the values satisfying both $H_1 \leq j+h \leq H_2$ and $H_1 \leq j \leq H_2$.

On the other hand

$$Q = \sum_\ell (-1)^{H_3(\ell)} D^*_\ell D_\ell$$

$$\doteq \sum_\ell (-1)^{H_3(\ell)} \begin{bmatrix} d^*_{\ell,H_1} d_{\ell,H_1}, & d^*_{\ell,H_1} d_{\ell,H_1+1}, & \ldots, & d^*_{\ell,H_1} d_{\ell,H_2} \\ d^*_{\ell,H_1+1} d_{\ell,H_1}, & d^*_{\ell,H_1+1} d_{\ell,H_1+1}, & \ldots, & d^*_{\ell,H_1+1} d_{\ell,H_2} \\ \cdots & \cdots & \cdots & \cdots \\ d^*_{\ell,H_2} d_{\ell,H_1}, & d^*_{\ell,H_2} d_{\ell,H_1+1}, & \ldots, & d^*_{\ell,H_2} d_{\ell,H_2} \end{bmatrix}$$

and the sum of elements on the h-th diagonal line just is $\sum_\ell (-1)^{H_3(\ell)} \sum_j d^*_{\ell,j} d_{\ell,j+h}$, where the h-th diagonal line denotes the main diagonal line if h=0, it denotes the h-th upper-diagonal line if $h > 0$, and it represents the $|h|$-th lower-diagonal line if $h < 0$. Therefore, the conclusion is proved.

Lemma 2 If $d^*_1(\theta)\, d_1(\theta) - d^*_2(\theta)\, d_2(\theta) \geq 0$, then the matrix $D^*_1 D_1 - D^*_2 D_2$ can be represented by a sum of one nonnegative definite matrix Z and one pseudo-null matrix Q.

Proof. Because of $d^*_1(\theta) d_1(\theta) - d^*_2(\theta) d_2(\theta) \geq 0$, one can find a $c(\theta) = \sum_{h=H_1}^{H_2} c_h e^{ih\theta}$ such that $d^*_1(\theta) d_1(\theta) - d^*_2(\theta) d_2(\theta) \equiv c^*(\theta) c(\theta)$ (The Fejér-Riesz Theorem).

Hence, we know from Lemma 1 that $Q = D_1^*D_1 - D_2^*D_2 - C^*C$ is a pseudo-null matrix, where $C = (c_{H_1}, c_{H_1+1}, \cdots, c_{H_2})$. Obviously C^*C is a nonnegative definite matrix. Therefore, the conclusion of this lemma is true.

From Lemma 2, it can be easily seen that the conditions (3.3) and (3.4) are closely related to (3.10) and (3.9) respectively.

In fact, according to the definition of T^k and noting (3.2), we have

$$\left\{\begin{aligned}
T^{k+1} &= \sum_{m\in \underline{g}(\lambda)} \Big(\sum_{h=H_1}^{H_2} \gamma_{h,m}^{k+\frac{1}{2}} u_{m+h}^{k+1}\Big)^2 \Delta\xi \\
&= \sum_{m\in \underline{g}(\lambda)} \Big(\sum_{h=H_1}^{H_2} s_{h,m}^{k+\frac{1}{2}} u_{m+h}^{k}\Big)^2 \Delta\xi \\
&= \sum_{m\in \underline{g}(\lambda)} (S^*SU, U)_m^k \Delta\xi + 0(\Delta\xi)\, \|u^k\|^2 \\
T^k &= \sum_{m\in \underline{g}(\lambda)} \Big(\sum_h \gamma_{h,m}^{k-\frac{1}{2}} u_{m+h}^k\Big)^2 \Delta\xi \\
&= \sum_{m\in \underline{g}(\lambda)} (R^*RU, U)_m^k \Delta\xi + 0(\Delta\xi)\, \|u^k\|^2.
\end{aligned}\right. \tag{6.1}$$

When deducing (6.1), we use the conditions (3.5) and (3.6), assume $\Delta t/\Delta\xi$ to be bounded and adopt the following symbols

$$\left\{\begin{aligned}
&R_m^k = (\gamma_{H_1,m}^k, \gamma_{H_1+1,m}^k, \cdots, \gamma_{H_2,m}^k)\big|_{\Delta\xi=\Delta t=0}, \\
&S_m^k = (s_{H_1,m}^k, s_{H_1+1,m}^k, \cdots, s_{H_2,m}^k)\big|_{\Delta\xi=\Delta t=0}, \\
&(AU,U)_m^k = \sum_{i=1}^{H+1} \sum_{j=1}^{H+1} (a_{i,j} u_{m+H_1-1+i}^k u_{m+H_1-1+j}^k) \\
&a_{i,j} \text{ being the element located on the i-th row and the} \\
&\text{j-th column of A.}
\end{aligned}\right.$$

From Lemma 2, it follows that if the conditions (3.4) and (3.3) are satisfied, then there exist certain nonnegative definite matrices $\bar{Z}$, Z and certain pseudo - null matrices $\bar{Q}$, Q such that

$$(R^*RU, U)_m - c_1 |u_m|^2 = (\bar{Z}\, U, U)_m + (\bar{Q}\, U, U)_m,$$

$$((R^*R - S^*S)U,U)_m = (Z\, U,U)_m + (Q\, U, U)_m .$$

Therefore, we have

$$\begin{aligned}
T^k \geq\; &[c_1 \sum_{m\in \underline{g}(\lambda)} |u_m^k|^2 + \sum_{m\in \underline{g}(\lambda)} (\bar{Q}\, U,U)_m^k]\Delta\xi \\
&- 0(\Delta\xi)\, \|u^k\|^2
\end{aligned} \tag{6.2}$$

and

$$T^{k+1} - T^{k} \leq - \sum_{m \in \underline{g}(\lambda)} (Q\, U, U)_m^k \Delta\xi + 0(\Delta\xi) \, \| u^k \|^2 . \qquad (6.3)$$

From the property of pseudo-null matrices, we know that if every element of $\bar{Q}$ and Q satisfies the Lipschitz condition as ξ varies, then in the above inequality

$$\sum_{m \in \underline{g}(\lambda)} (\bar{Q}\, U, U)_m^k \quad \text{and} \quad - \sum_{m \in \underline{g}(\Sigma)} (Q\, U, U)_m^k$$

may be replaced by sums of $0(\Delta\xi) \sum_{m=0}^{M} |u_m^k|^2$ and certain quadratic forms of u_m^k on points near the boundaries. Therefore, (6.2) and (6.3) are respectively similar to (3.9) and (3.10). Consequently, it is quite common for a scheme to possess Property A.

We now prove that scheme (2.5) possesses Property A for the problem

$$\begin{cases} \dfrac{\partial u}{\partial t} + \lambda(\xi,t) \dfrac{\partial u}{\partial \xi} = 0, \\ \lambda(\xi,t) \geq \varepsilon > 0, \; 0 \leq \xi \leq 1. \end{cases} \qquad (6.4)$$

For (6.4), scheme (2.5) is in the form

$$\gamma_{-1,m}^{k+\frac{1}{2}} u_{m-1}^{k+1} + \gamma_{0,m}^{k+\frac{1}{2}} u_m^{k+1} = s_{-1,m}^{k+\frac{1}{2}} u_{m-1}^{k} + s_{0,m}^{k+\frac{1}{2}} u_m^{k}, \qquad m = 1,2,\cdots,M, \qquad (6.5)$$

where

$$\gamma_{-1,m}^{k+\frac{1}{2}} = \frac{1}{2} - \frac{1}{2} \mu \sigma_{m-\frac{1}{2}}^{k+\frac{1}{2}},$$

$$\gamma_{0,m}^{k+\frac{1}{2}} = \frac{1}{2} + \frac{1}{2} \mu \sigma_{m-\frac{1}{2}}^{k+\frac{1}{2}},$$

$$s_{-1,m}^{k+\frac{1}{2}} = \frac{1}{2} + \frac{1}{2} \mu \sigma_{m-\frac{1}{2}}^{k+\frac{1}{2}},$$

$$s_{0,m}^{k+\frac{1}{2}} = \frac{1}{2} - \frac{1}{2} \mu \sigma_{m-\frac{1}{2}}^{k+\frac{1}{2}}.$$

For this type of scheme, the condition (3.4) is in the form

$$\gamma^*(\theta)\gamma(\theta) = 1 - 4\gamma_0(1 - \gamma_0) \sin^2 \frac{\theta}{2} \geq c_1 > 0, \qquad (6.6)$$

where γ_0 denotes $\gamma_0(0, 0)$. Moreover $\gamma^*(\theta)\gamma(\theta)$ can be further rewritten as

$$\gamma^*(\theta)\gamma(\theta) = c_1' + (a + be^{i\theta})^* (a + be^{i\theta}),$$

where

$$c_1' = \min\left\{ \frac{1}{2}, \frac{c_1}{2} \right\},$$

$$a = \frac{\sqrt{1-c_1'} - \sqrt{(2\gamma_0-1)^2 - c_1'}}{2},$$

$$b = \frac{\sqrt{1-c_1'} + \sqrt{(2\gamma_0-1)^2 - c_1'}}{2}.$$

Consequently, according to Lemma 1, the matrix

$$\bar{Q}_2 \equiv R^*R - \begin{pmatrix} 0 & 0 \\ 0 & c_1' \end{pmatrix} - Z_2$$

is a pseudo-null matrix, where $Z_2 = \binom{a}{b}(a \;\; b)$. Actually,

$$\bar{Q}_2 = \begin{bmatrix} (1-\gamma_0)^2 - a^2 & 0 \\ 0 & \gamma_0^2 - c_1' - b^2 \end{bmatrix} .$$

Moreover, when (6.6) holds, we have $(2\gamma_0-1)^2 - c_1' \geq \frac{c_1}{2} > 0$, which guarantees that every element of $\bar{Q}_2$ satisfies the Lipschitz condition with respect to ξ. Therefore, since Z_2 is nonnegative definite, we have the following inequality

$$\begin{aligned} T = \sum_{m=1}^{M} (R^*RU,U)_m \Delta\xi + 0(\Delta\xi)\|u\|^2 &\geq c_1' \sum_{m=1}^{M} |u_m|^2 \Delta\xi + \sum_{m=1}^{M} (\bar{Q}_2U,U)_m \Delta\xi - 0(\Delta\xi)\|u\|^2 \\ &\geq c_1' \sum_{m=1}^{M} |u_m|^2 \Delta\xi + ((1-\gamma_0)^2 - a^2)_1 \; |u_0|^2 \Delta\xi \\ &\quad + (\gamma_0^2 - c_1' - b^2)_M \; |u_M|^2 \Delta\xi - 0(\Delta\xi) \; \|u\|^2 \\ &\geq c_2 \sum_{m=1}^{M} |u_m|^2 \Delta\xi - c|u_0|^2 \Delta\xi - 0(\Delta\xi) \; \|u\|^2 , \end{aligned} \quad (6.7)$$

where $c_2 = \min \{ c_1', \; (\gamma_0^2 - b^2)_M \}$, $\quad c > |((1-\gamma_0)^2 - a^2)_1|$.

Noting

$$\frac{\partial b^2}{\partial c_1'} = 2b \frac{1}{4} \left[\frac{-1}{\sqrt{1 - c_1'}} + \frac{-1}{\sqrt{(2\gamma_0-1)^2 - c_1'}} \right]$$

$$= \frac{-b^2}{\sqrt{1 - c_1'} \; \sqrt{(2\gamma_0-1)^2 - c_1'}} < 0 ,$$

$$(\gamma_0^2 - b^2)\Big|_{c_1' = 0} = 0,$$

and $c_1' > 0$, we know $(\gamma_0^2 - b^2)_M > 0$. Therefore (6.7) can be written in form (3.9).

For scheme (6.5), the condition (3.3) is

$$\gamma^*(\theta)\gamma(\theta) - s^*(\theta)s(\theta) = 4(s_0(1 - s_0) - \gamma_0(1-\gamma_0))\sin^2 \frac{\theta}{2} \geq 0, \quad (6.8)$$

where s_0 denotes $s_0(0,0)$. Since $4 \sin^2 \frac{\theta}{2} = (1-e^{i\theta})^* (1-e^{i\theta})$, according to Lemma 1, the matrix

$$Q_2 \equiv \begin{bmatrix} (1-\gamma_0)^2 & \gamma_0(1-\gamma_0) \\ \gamma_0(1-\gamma_0)^2 & \gamma_0^2 \end{bmatrix} - \begin{bmatrix} (1-s_0)^2 & s_0(1-s_0) \\ s_0(1-s_0) & s_0^2 \end{bmatrix}$$

$$-(s_0(1-s_0)-\gamma_0(1-\gamma_0))\begin{pmatrix}1 & -1\\ -1 & 1\end{pmatrix}$$

is a pseudo-null matrix. Actually

$$Q_2 = \begin{bmatrix} s_0-\gamma_0 & 0 \\ 0 & \gamma_0-s_0 \end{bmatrix}$$

Clearly, if (6.8) is valid, $(s_0(1-s_0)-\gamma_0(1-\gamma_0))\begin{pmatrix}1 & -1\\ -1 & 1\end{pmatrix}$ is nonnegative definite. Therefore, we have

$$\sum_{m=1}^{M}((S^*S-R^*R)U,U)_m\Delta\xi \le -\sum_{m=1}^{M}(Q_2U,U)_m\Delta\xi$$

$$\le (\gamma_0-s_0)_0|u_0|^2\Delta\xi-(\gamma_0-s_0)_M|u_M|^2\Delta\xi+0(\Delta\xi)\,||\,u\,||^2,$$

from which and noticing $\gamma_0-s_0=\mu\sigma$, we can obtain the following inequality in form (3.10):

$$T^{k+1}-T^k\le\sigma_0|u_0^k|^2\Delta\xi-\sigma_M|u_M^k|^2\Delta\xi+0(\Delta\xi)\,||\,u^k\,||^2.$$

In what follows, we prove that scheme (2.6) possesses Property A when it is applied to

$$\begin{cases}\dfrac{\partial u}{\partial t}+\lambda(\xi,t)\dfrac{\partial u}{\partial\xi}=0,\\ \lambda(0,t)\ge 0,\quad \lambda(1,t)\le 0, \qquad 0\le\xi\le 1.\end{cases}\tag{6.9}$$

In this case, the difference equations are

$$u_m^{k+1}=s_{-1,m}^{k+\frac12},\,u_{m-1}^k+s_{0,m}^{k+\frac12},\,u_m^k+s_{1,m}^{k+\frac12},\,u_{m+1}^k\,,\tag{6.10}$$

$$m=1,2,\cdots,M-1\,,$$

where

$$s_{-1}=\frac12(1+\sigma)\sigma,\quad s_0=(1+\sigma)(1-\sigma),\quad s_1=-\frac12(1-\sigma)\sigma.$$

For explicit schemes, (3.9) always holds. Therefore we only need to prove (3.10) is satisfied. Now condition (3.3) is

$$\gamma^*(\theta)\gamma(\theta)-s^*(\theta)s(\theta)=4(1-\sigma^2)\sigma^2\sin^4\frac{\theta}{2}\ge 0.\tag{6.11}$$

Since $\sin^4\frac{\theta}{2}=\frac{1}{16}(e^{-i\theta}-2+e^{i\theta})^*(e^{-i\theta}-2+e^{i\theta})$, according to Lemma 1, the matrix

$$Q_3\equiv\begin{bmatrix}0&0&0\\0&1&0\\0&0&0\end{bmatrix}-\begin{bmatrix}s_{-1}^2 & s_{-1}s_0 & s_{-1}s_1\\ s_0s_{-1} & s_0^2 & s_0s_1\\ s_1s_{-1} & s_1s_0 & s_1^2\end{bmatrix}-Z_3$$

$$=\begin{bmatrix}-s_{-1}^2-a & -s_{-1}s_0+2a & 0\\ -s_0s_1+2a & 1-s_0^2-4a & -s_0s_1+2a\\ 0 & -s_1s_0+2a & -s_1^2-a\end{bmatrix}$$

is a pseudo-null matrix, where

$$Z_3 = a \begin{bmatrix} 1 \\ -2 \\ 1 \end{bmatrix} (1 \quad -2 \quad 1),$$

$$a = \frac{1}{4}(1-\sigma^2)\sigma^2 .$$

Because Z_3 is a nonnegative definite matrix when (6.11) holds and because Q_3 is Lipschitz continuous with respect to ξ, we have

$$\sum_{m=1}^{M-1} ((S^*S - R^*R)U, U)_m \Delta\xi \leq - \sum_{m=1}^{M-1} (Q_3U, U)_m \Delta\xi$$

$$\leq (Q_uU,U)_0 \Delta\xi + (Q_\ell U,U)_M \Delta\xi + 0(\Delta\xi) \|u\|^2 , \tag{6.12}$$

where

$$(Q_uU, U)_0 = (u_0, u_1) \begin{bmatrix} s_{-1}^2 + a & s_{-1}s_0 - 2a \\ s_0s_{-1} - 2a & s_{-1}^2 + s_0^2 - 1 + 5a \end{bmatrix}_0 \begin{pmatrix} u_0 \\ u_1 \end{pmatrix},$$

$$(Q_\ell U,U)_M = (u_{M-1}, u_M) \begin{bmatrix} s_1^2 + s_0^2 - 1 + 5a & s_{-1}s_0 - 2a \\ s_{-1}s_0 - 2a & s_1^2 + a \end{bmatrix}_M \begin{pmatrix} u_{M-1} \\ u_M \end{pmatrix}.$$

Since

$$\begin{bmatrix} s_{-1}^2+a & s_{-1}s_0-2a \\ s_0s_{-1}-2a & s_{-1}^2+s_0^2-1+5a \end{bmatrix}_0 = \frac{1}{2} \begin{bmatrix} (1+\sigma)\sigma^2 & (1-\sigma^2)\sigma \\ (1-\sigma^2)\sigma & \sigma^2(\sigma-1) \end{bmatrix}_0$$

$$= \frac{1}{2} \begin{bmatrix} (1+\sigma)^2(\sigma-1) & (1-\sigma^2)\sigma \\ (1-\sigma^2)\sigma & \sigma^2(\sigma-1) \end{bmatrix}_0 + \begin{bmatrix} \frac{1}{2}(1+\sigma) & 0 \\ 0 & 0 \end{bmatrix}_0$$

and the first matrix on the right hand side of the last sign of equality is nonpositive definite, it follows that

$$(Q_u U, U)_0 \leq \frac{1}{2}(1 + \sigma_0)|u_0|^2.$$

Similarly, we can obtain

$$(Q_\ell U, U)_M \leq \frac{1}{2}(1 + \sigma_M)|u_M|^2 .$$

Therefore from (6.12) we can get (3.10), which means that scheme (2.6) possesses Property A when it is applied to (6.9).

Indeed, many schemes have Property A. For more results, the reader is referred to book [3].

References

[1]. Yong, W.-a. and Zhu, Y.-l., Convergence of difference methods for nonlinear problems with moving boundaries (to appear in English).

[2]. Zhu, Y.-l., Difference schemes for initial-boundary-value problems of the first order hyperbolic systems and their stability, Mathematicae Numericae Sinica, 1979, No.1, pp 1-30(in Chinese).

[3]. Zhu, Y.-l., Zhong, X.-c., Chen, B.-m. and Zhang, Z.-m., Difference methods for initial-boundary-value problems and flows around bodies, Science Press, Beijing, China, 1980 (in Chinese), the English edition is in the press.

[4]. Zhu, Y.-l., Stability and convergence of difference schemes for linear initial-boundary-value problems, Mathematicae Numericae Sinica, 1982, No.1, pp 98-108 (in English).

LECTURE NOTES IN MATHEMATICS
Edited by A. Dold and B. Eckmann

Some general remarks on the publication of proceedings of congresses and symposia

Lecture Notes aim to report new developments - quickly, informally and at a high level. The following describes criteria and procedures which apply to proceedings volumes.

1. One (or more) expert participant(s) of the meeting should act as the responsible editor(s) of the proceedings. They select the papers which are suitable (cf. points 2, 3) for inclusion in the proceedings, and have them individually refereed (as for a journal). It should not be assumed that the published proceedings must reflect conference events faithfully and in their entirety. Contributions to the meeting which are not included in the proceedings can be listed by title. The series editors will normally not interfere with the editing of a particular proceedings volume - except in fairly obvious cases, or on technical matters, such as described in points 2, 3. The names of the responsible editors appear on the title page of the volume.

2. The proceedings should be reasonably homogeneous (concerned with a limited area). For instance, the proceedings of a congress on "Analysis" or "Mathematics in Wonderland" would normally not be sufficiently homogeneous.

   One or two longer survey articles on recent developments in the field are often very useful additions to such proceedings - even if they do not correspond to actual lectures at the congress. An extensive introduction on the subject of the congress would be desirable.

3. The contributions should be of a high mathematical standard and of current interest. Research articles should present new material and not duplicate other papers already published or due to be published. They should contain sufficient information and motivation and they should present proofs, or at least outlines of such, in sufficient detail to enable an expert to complete them. Thus resumes and mere announcements of papers appearing elsewhere cannot be included, although more detailed versions of a contribution may well be published in other places later.

   Surveys, if included, should cover a sufficiently broad topic, and should in general not simply review the author's own recent research. In the case of surveys, exceptionally, proofs of results may not be necessary.

   The editors of a volume are strongly advised to inform contributors about these points at an early stage.

.../...

4. Proceedings should appear soon after the meeeting. The publisher should, therefore, receive the complete manuscript within nine months of the date of the meeting at the latest.

5. Plans or proposals for proceedings volumes should be sent to one of the editors of the series or to Springer-Verlag Heidelberg. They should give sufficient information on the conference or symposium, and on the proposed proceedings. In particular, they should contain a list of the expected contributions with their prospective length. Abstracts or early versions (drafts) of some of the contributions are very helpful.

6. Lecture Notes are printed by photo-offset from camera-ready typed copy provided by the editors. For this purpose Springer-Verlag provides editors with technical instructions for the preparation of manuscripts and these should be distributed to all contributing authors. Springer-Verlag can also, on request, supply stationery on which the prescribed typing area is outlined. Some homogeneity in the presentation of the contributions is desirable.

   Careful preparation of manuscripts will help keep production time short and ensure a satisfactory appearance of the finished book. The actual production of a Lecture Notes volume normally takes 6 -8 weeks.

   Manuscripts should be at least 100 pages long. The final version should include a table of contents.

7. Editors receive a total of 50 free copies of their volume for distribution to the contributing authors, but no royalties. (Unfortunately, no reprints of individual contributions can be supplied.) They are entitled to purchase further copies of their book for their personal use at a discount of 33 1/3%, other Springer mathematics books at a discount of 20% directly from Springer-Verlag.

   Commitment to publish is made by letter of intent rather than by signing a formal contract. Springer-Verlag secures the copyright for each volume.

Vol. 1201: Curvature and Topology of Riemannian Manifolds. Proceedings, 1985. Edited by K. Shiohama, T. Sakai and T. Sunada. VII, 336 pages. 1986.

Vol. 1202: A. Dür, Möbius Functions, Incidence Algebras and Power Series Representations. XI, 134 pages. 1986.

Vol. 1203: Stochastic Processes and Their Applications. Proceedings, 1985. Edited by K. Itô and T. Hida. VI, 222 pages. 1986.

Vol. 1204: Séminaire de Probabilités XX, 1984/85. Proceedings. Edité par J. Azéma et M. Yor. V, 639 pages. 1986.

Vol. 1205: B.Z. Moroz, Analytic Arithmetic in Algebraic Number Fields. VII, 177 pages. 1986.

Vol. 1206: Probability and Analysis, Varenna (Como) 1985. Seminar. Edited by G. Letta and M. Pratelli. VIII, 280 pages. 1986.

Vol. 1207: P.H. Bérard, Spectral Geometry: Direct and Inverse Problems. With an Appendix by G. Besson. XIII, 272 pages. 1986.

Vol. 1208: S. Kaijser, J.W. Pelletier, Interpolation Functors and Duality. IV, 167 pages. 1986.

Vol. 1209: Differential Geometry, Peñíscola 1985. Proceedings. Edited by A.M. Naveira, A. Ferrández and F. Mascaró. VIII, 306 pages. 1986.

Vol. 1210: Probability Measures on Groups VIII. Proceedings, 1985. Edited by H. Heyer. X, 386 pages. 1986.

Vol. 1211: M.B. Sevryuk, Reversible Systems. V, 319 pages. 1986.

Vol. 1212: Stochastic Spatial Processes. Proceedings, 1984. Edited by P. Tautu. VIII, 311 pages. 1986.

Vol. 1213: L.G. Lewis, Jr., J.P. May, M. Steinberger, Equivariant Stable Homotopy Theory. IX, 538 pages. 1986.

Vol. 1214: Global Analysis – Studies and Applications II. Edited by Yu.G. Borisovich and Yu.E. Gliklikh. V, 275 pages. 1986.

Vol. 1215: Lectures in Probability and Statistics. Edited by G. del Pino and R. Rebolledo. V, 491 pages. 1986.

Vol. 1216: J. Kogan, Bifurcation of Extremals in Optimal Control. VIII, 106 pages. 1986.

Vol. 1217: Transformation Groups. Proceedings, 1985. Edited by S. Jackowski and K. Pawalowski. X, 396 pages. 1986.

Vol. 1218: Schrödinger Operators, Aarhus 1985. Seminar. Edited by E. Balslev. V, 222 pages. 1986.

Vol. 1219: R. Weissauer, Stabile Modulformen und Eisensteinreihen. III, 147 Seiten. 1986.

Vol. 1220: Séminaire d'Algèbre Paul Dubreil et Marie-Paule Malliavin. Proceedings, 1985. Edité par M.-P. Malliavin. IV, 200 pages. 1986.

Vol. 1221: Probability and Banach Spaces. Proceedings, 1985. Edited by J. Bastero and M. San Miguel. XI, 222 pages. 1986.

Vol. 1222: A. Katok, J.-M. Strelcyn, with the collaboration of F. Ledrappier and F. Przytycki, Invariant Manifolds, Entropy and Billiards; Smooth Maps with Singularities. VIII, 283 pages. 1986.

Vol. 1223: Differential Equations in Banach Spaces. Proceedings, 1985. Edited by A. Favini and E. Obrecht. VIII, 299 pages. 1986.

Vol. 1224: Nonlinear Diffusion Problems, Montecatini Terme 1985. Seminar. Edited by A. Fasano and M. Primicerio. VIII, 188 pages. 1986.

Vol. 1225: Inverse Problems, Montecatini Terme 1986. Seminar. Edited by G. Talenti. VIII, 204 pages. 1986.

Vol. 1226: A. Buium, Differential Function Fields and Moduli of Algebraic Varieties. IX, 146 pages. 1986.

Vol. 1227: H. Helson, The Spectral Theorem. VI, 104 pages. 1986.

Vol. 1228: Multigrid Methods II. Proceedings, 1985. Edited by W. Hackbusch and U. Trottenberg. VI, 336 pages. 1986.

Vol. 1229: O. Bratteli, Derivations, Dissipations and Group Actions on C*-algebras. IV, 277 pages. 1986.

Vol. 1230: Numerical Analysis. Proceedings, 1984. Edited by J.-P. Hennart. X, 234 pages. 1986.

Vol. 1231: E.-U. Gekeler, Drinfeld Modular Curves. XIV, 107 pages. 1986.

Vol. 1232: P.C. Schuur, Asymptotic Analysis of Soliton Problems. VI, 180 pages. 1986.

Vol. 1233: Stability Problems for Stochastic Models. Proceedings 1985. Edited by V.V. Kalashnikov, B. Penkov and V.M. Zolotarev. VI, 223 pages. 1986.

Vol. 1234: Combinatoire énumérative. Proceedings, 1985. Edité par G. Labelle et P. Leroux. XIV, 387 pages. 1986.

Vol. 1235: Séminaire de Théorie du Potentiel, Paris, No. 8. Directeurs: M. Brelot, G. Choquet et J. Deny. Rédacteurs: F. Hirsch et G. Mokobodzki. III, 209 pages. 1987.

Vol. 1236: Stochastic Partial Differential Equations and Applications Proceedings, 1985. Edited by G. Da Prato and L. Tubaro. V, 25 pages. 1987.

Vol. 1237: Rational Approximation and its Applications in Mathematic and Physics. Proceedings, 1985. Edited by J. Gilewicz, M. Pindor an W. Siemaszko. XII, 350 pages. 1987.

Vol. 1238: M. Holz, K.-P. Podewski and K. Steffens, Injective Choice Functions. VI, 183 pages. 1987.

Vol. 1239: P. Vojta, Diophantine Approximations and Value Distribution Theory. X, 132 pages. 1987.

Vol. 1240: Number Theory, New York 1984–85. Seminar. Edited by D.V. Chudnovsky, G.V. Chudnovsky, H. Cohn and M.B. Nathanson. V, 324 pages. 1987.

Vol. 1241: L. Gårding, Singularities in Linear Wave Propagation. III 125 pages. 1987.

Vol. 1242: Functional Analysis II, with Contributions by J. Hoffmann Jørgensen et al. Edited by S. Kurepa, H. Kraljević and D. Butković. VII 432 pages. 1987.

Vol. 1243: Non Commutative Harmonic Analysis and Lie Groups Proceedings, 1985. Edited by J. Carmona, P. Delorme and M. Vergne V, 309 pages. 1987.

Vol. 1244: W. Müller, Manifolds with Cusps of Rank One. XI, 15 pages. 1987.

Vol. 1245: S. Rallis, L-Functions and the Oscillator Representation XVI, 239 pages. 1987.

Vol. 1246: Hodge Theory. Proceedings, 1985. Edited by E. Cattani, F Guillén, A. Kaplan and F. Puerta. VII, 175 pages. 1987.

Vol. 1247: Séminaire de Probabilités XXI. Proceedings. Edité par Azéma, P.A. Meyer et M. Yor. IV, 579 pages. 1987.

Vol. 1248: Nonlinear Semigroups, Partial Differential Equations an Attractors. Proceedings, 1985. Edited by T.L. Gill and W.W. Zachar IX, 185 pages. 1987.

Vol. 1249: I. van den Berg, Nonstandard Asymptotic Analysis. IX, 187 pages. 1987.

Vol. 1250: Stochastic Processes – Mathematics and Physics I Proceedings 1985. Edited by S. Albeverio, Ph. Blanchard and L. Stre VI, 359 pages. 1987.

Vol. 1251: Differential Geometric Methods in Mathematical Physic Proceedings, 1985. Edited by P.L. García and A. Pérez-Rendón. V 300 pages. 1987.

Vol. 1252: T. Kaise, Représentations de Weil et GL_2 Algèbres d division et GL_n. VII, 203 pages. 1987.

Vol. 1253: J. Fischer, An Approach to the Selberg Trace Formula vi the Selberg Zeta-Function. III, 184 pages. 1987.

Vol. 1254: S. Gelbart, I. Piatetski-Shapiro, S. Rallis. Explicit Constructions of Automorphic L-Functions. VI, 152 pages. 1987.

Vol. 1255: Differential Geometry and Differential Equations. Proceedings, 1985. Edited by C. Gu, M. Berger and R.L. Bryant. XII, 24 pages. 1987.

Vol. 1256: Pseudo-Differential Operators. Proceedings, 1986. Edited by H.O. Cordes, B. Gramsch and H. Widom. X, 479 pages. 1987.

Vol. 1257: X. Wang, On the C*-Algebras of Foliations in the Plane. V, 165 pages. 1987.

Vol. 1258: J. Weidmann, Spectral Theory of Ordinary Different Operators. VI, 303 pages. 1987.

Vol. 1145: G. Winkler, Choquet Order and Simplices. VI, 143 pages. 1985.

Vol. 1146: Séminaire d'Algèbre Paul Dubreil et Marie-Paule Malliavin. Proceedings, 1983–1984. Edité par M.-P. Malliavin. IV, 420 pages. 1985.

Vol. 1147: M. Wschebor, Surfaces Aléatoires. VII, 111 pages. 1985.

Vol. 1148: Mark A. Kon, Probability Distributions in Quantum Statistical Mechanics. V, 121 pages. 1985.

Vol. 1149: Universal Algebra and Lattice Theory. Proceedings, 1984. Edited by S. D. Comer. VI, 282 pages. 1985.

Vol. 1150: B. Kawohl, Rearrangements and Convexity of Level Sets in PDE. V, 136 pages. 1985.

Vol 1151: Ordinary and Partial Differential Equations. Proceedings, 1984. Edited by B.D. Sleeman and R.J. Jarvis. XIV, 357 pages. 1985.

Vol. 1152: H. Widom, Asymptotic Expansions for Pseudodifferential Operators on Bounded Domains. V, 150 pages. 1985.

Vol. 1153: Probability in Banach Spaces V. Proceedings, 1984. Edited by A. Beck, R. Dudley, M. Hahn, J. Kuelbs and M. Marcus. VI, 457 pages. 1985.

Vol. 1154: D.S. Naidu, A.K. Rao, Singular Pertubation Analysis of Discrete Control Systems. IX, 195 pages. 1985.

Vol. 1155: Stability Problems for Stochastic Models. Proceedings, 1984. Edited by V.V. Kalashnikov and V.M. Zolotarev. VI, 447 pages. 1985.

Vol. 1156: Global Differential Geometry and Global Analysis 1984. Proceedings, 1984. Edited by D. Ferus, R.B. Gardner, S. Helgason and U. Simon. V, 339 pages. 1985.

Vol. 1157: H. Levine, Classifying Immersions into $\mathbb{R}^4$ over Stable Maps of 3-Manifolds into $\mathbb{R}^2$. V, 163 pages. 1985.

Vol. 1158: Stochastic Processes – Mathematics and Physics. Proceedings, 1984. Edited by S. Albeverio, Ph. Blanchard and L. Streit. VI, 230 pages. 1986.

Vol. 1159: Schrödinger Operators, Como 1984. Seminar. Edited by S. Graffi. VIII, 272 pages. 1986.

Vol. 1160: J.-C. van der Meer, The Hamiltonian Hopf Bifurcation. VI, 115 pages. 1985.

Vol. 1161: Harmonic Mappings and Minimal Immersions, Montecatini 1984. Seminar. Edited by E. Giusti. VII, 285 pages. 1985.

Vol. 1162: S.J.L. van Eijndhoven, J. de Graaf, Trajectory Spaces, Generalized Functions and Unbounded Operators. IV, 272 pages. 1985.

Vol. 1163: Iteration Theory and its Functional Equations. Proceedings, 1984. Edited by R. Liedl, L. Reich and Gy. Targonski. VIII, 231 pages. 1985.

Vol. 1164: M. Meschiari, J.H. Rawnsley, S. Salamon, Geometry Seminar "Luigi Bianchi" II – 1984. Edited by E. Vesentini. VI, 224 pages. 1985.

Vol. 1165: Seminar on Deformations. Proceedings, 1982/84. Edited by J. Ławrynowicz. IX, 331 pages. 1985.

Vol. 1166: Banach Spaces. Proceedings, 1984. Edited by N. Kalton and E. Saab. VI, 199 pages. 1985.

Vol. 1167: Geometry and Topology. Proceedings, 1983–84. Edited by J. Alexander and J. Harer. VI, 292 pages. 1985.

Vol. 1168: S.S. Agaian, Hadamard Matrices and their Applications. III, 227 pages. 1985.

Vol. 1169: W.A. Light, E.W. Cheney, Approximation Theory in Tensor Product Spaces. VII, 157 pages. 1985.

Vol. 1170: B.S. Thomson, Real Functions. VII, 229 pages. 1985.

Vol. 1171: Polynômes Orthogonaux et Applications. Proceedings, 1984. Edité par C. Brezinski, A. Draux, A.P. Magnus, P. Maroni et A. Ronveaux. XXXVII, 584 pages. 1985.

Vol. 1172: Algebraic Topology, Göttingen 1984. Proceedings. Edited by L. Smith. VI, 209 pages. 1985.

Vol. 1173: H. Delfs, M. Knebusch, Locally Semialgebraic Spaces. XVI, 329 pages. 1985.

Vol. 1174: Categories in Continuum Physics, Buffalo 1982. Seminar. Edited by F.W. Lawvere and S.H. Schanuel. V, 126 pages. 1986.

Vol. 1175: K. Mathiak, Valuations of Skew Fields and Projective Hjelmslev Spaces. VII, 116 pages. 1986.

Vol. 1176: R.R. Bruner, J.P. May, J.E. McClure, M. Steinberger, H_∞ Ring Spectra and their Applications. VII, 388 pages. 1986.

Vol. 1177: Representation Theory I. Finite Dimensional Algebras. Proceedings, 1984. Edited by V. Dlab, P. Gabriel and G. Michler. XV, 340 pages. 1986.

Vol. 1178: Representation Theory II. Groups and Orders. Proceedings, 1984. Edited by V. Dlab, P. Gabriel and G. Michler. XV, 370 pages. 1986.

Vol. 1179: Shi J.-Y. The Kazhdan-Lusztig Cells in Certain Affine Weyl Groups. X, 307 pages. 1986.

Vol. 1180: R. Carmona, H. Kesten, J.B. Walsh, École d'Été de Probabilités de Saint-Flour XIV – 1984. Édité par P.L. Hennequin. X, 438 pages. 1986.

Vol. 1181: Buildings and the Geometry of Diagrams, Como 1984. Seminar. Edited by L. Rosati. VII, 277 pages. 1986.

Vol. 1182: S. Shelah, Around Classification Theory of Models. VII, 279 pages. 1986.

Vol. 1183: Algebra, Algebraic Topology and their Interactions. Proceedings, 1983. Edited by J.-E. Roos. XI, 396 pages. 1986.

Vol. 1184: W. Arendt, A. Grabosch, G. Greiner, U. Groh, H.P. Lotz, U. Moustakas, R. Nagel, F. Neubrander, U. Schlotterbeck, One-parameter Semigroups of Positive Operators. Edited by R. Nagel. X, 460 pages. 1986.

Vol. 1185: Group Theory, Beijing 1984. Proceedings. Edited by Tuan H.F. V, 403 pages. 1986.

Vol. 1186: Lyapunov Exponents. Proceedings, 1984. Edited by L. Arnold and V. Wihstutz. VI, 374 pages. 1986.

Vol. 1187: Y. Diers, Categories of Boolean Sheaves of Simple Algebras. VI, 168 pages. 1986.

Vol. 1188: Fonctions de Plusieurs Variables Complexes V. Séminaire, 1979–85. Edité par François Norguet. VI, 306 pages. 1986.

Vol. 1189: J. Lukeš, J. Malý, L. Zajíček, Fine Topology Methods in Real Analysis and Potential Theory. X, 472 pages. 1986.

Vol. 1190: Optimization and Related Fields. Proceedings, 1984. Edited by R. Conti, E. De Giorgi and F. Giannessi. VIII, 419 pages. 1986.

Vol. 1191: A.R. Its, V.Yu. Novokshenov, The Isomonodromic Deformation Method in the Theory of Painlevé Equations. IV, 313 pages. 1986.

Vol. 1192: Equadiff 6. Proceedings, 1985. Edited by J. Vosmansky and M. Zlámal. XXIII, 404 pages. 1986.

Vol. 1193: Geometrical and Statistical Aspects of Probability in Banach Spaces. Proceedings, 1985. Edited by X. Femique, B. Heinkel, M.B. Marcus and P.A. Meyer. IV, 128 pages. 1986.

Vol. 1194: Complex Analysis and Algebraic Geometry. Proceedings, 1985. Edited by H. Grauert. VI, 235 pages. 1986.

Vol.1195: J.M. Barbosa, A.G. Colares, Minimal Surfaces in $\mathbb{R}^3$. X, 124 pages. 1986.

Vol. 1196: E. Casas-Alvero, S. Xambó-Descamps, The Enumerative Theory of Conics after Halphen. IX, 130 pages. 1986.

Vol. 1197: Ring Theory. Proceedings, 1985. Edited by F.M.J. van Oystaeyen. V, 231 pages. 1986.

Vol. 1198: Séminaire d'Analyse, P. Lelong – P. Dolbeault – H. Skoda. Seminar 1983/84. X, 260 pages. 1986.

Vol. 1199: Analytic Theory of Continued Fractions II. Proceedings, 1985. Edited by W.J. Thron. VI, 299 pages. 1986.

Vol. 1200: V.D. Milman, G. Schechtman, Asymptotic Theory of Finite Dimensional Normed Spaces. With an Appendix by M. Gromov. VIII, 156 pages. 1986.